P9-DMR-539

FIASCO

FIASCO

Jack Anderson
with James Boyd

Published by TIMES BOOKS, a division of
The New York Times Book Co., Inc.
Three Park Avenue, New York, N.Y. 10016

Published simultaneously in Canada by
Fitzhenry & Whiteside, Ltd., Toronto

Library of Congress Cataloging in Publication Data

Anderson, Jack, 1922–
Fiasco

1. Petroleum industry and trade—Government policy—United States—History—20th century. 2. United States—Economic conditions—1971– . I. Boyd, James, 1929– . II. Title.
HD9566.A65 1983 333.8'232'0973 83-45037
ISBN 0-8129-0943-7

Designed by Doris Borowsky

Manufactured in the United States of America

83 84 85 86 87 5 4 3 2 1

Contents

Introduction

Two oil shocks in the 1970's essentially destroyed the prosperity of the industrial democracies.

—Lester C. Thurow, 1982
Professor of economics,
Massachusetts Institute of Technology

The current estimates of potential supply of conventional oil in the world seem to converge at about two trillion barrels . . . a seventy-five year supply at current rates of production.

—James W. McKie, 1978
Professor of economics,
University of Texas

The world "energy crisis" or "energy shortage" is a fiction. But belief in the fiction is a fact. It makes people accept higher oil prices as imposed by nature, when they are really fixed by collusion.

—M. A. Adelman, 1972
Professor of economics,
Massachusetts Institute of Technology

The economic tragedy of the 1980's calls out for an inquest into the oil politics of the 1970's.

That a catastrophe has occurred, spreading beyond economic devastation to threaten the basic cohesion of the West, is hardly subject to dispute. That the catastrophe in substantial measure is the result of a thirtyfold leap in oil prices during the 1970's has the ring of common sense to it but requires demonstration, which we assay here at the outset. That the oil price explosion itself, the cause of incalculable misery and mischief, was the work of failed, self-serving politics, mainly American politics, will be the burden of all the chapters that follow.

For several years the Western economies have been unable to maintain the 4 percent growth rate that is necessary if existing job and prosperity levels are to be preserved and new jobs created for each year's influx of school graduates, a growth rate consistently achieved before the onset of hyperinflation in oil prices. In 1983 the industrial democracies counted 33 million unemployed workers and projected 35 million for 1984, almost four times the level of the early 1970's. Of these, 12 million were in the United States, where steel mills were operating at 31 percent of capacity, lower than the 33.1 percent of 1933. Automobile sales for 1982 were the lowest in twenty years. Overall, business profits had fallen for three successive years. Bankruptcies were at the highest level since the 1930's, and the economy's 1982 growth rate for goods and services had fallen below zero.

The nations of Western Europe had been struggling for three years with an inflation rate of 13 percent and a growth rate that hovered around zero. "Negative growth" had been forecast for 1983. So there were 21 million unemployed European workers, and the estimate was for 23 million within two years.

What is the place of world oil prices in all this? There are other factors, of course—inflationary spending by governments, for instance, and the dislocating shifts from traditional manufacturing toward "services" and electronics. But the role of oil inflation, followed by the cutbacks it forced in other areas, has been critical.

Between 1969 and 1979 the open market price of crude oil rose from $1.20 a barrel to $41, a peak from which it has lately receded by about 25 percent. The effect of this stupendous leap in the cost of the oil the West runs on can be simply put: As the price of energy has skyrocketed, consumers have had less and less disposable income to buy other goods, which cost more and more to produce; hence there has been a progressive shrinkage of economic activity that must go on until the oil price bulge is digested.

Economists address the cause and effect relationship more rigorously. "The rapid rise in oil prices has driven inflation in several ways," writes oil analyst Daniel Yergin, summarizing the views of a dozen economists participating in a four-year study sponsored by the Atlantic Institute for International Affairs.* ". . . So pervasive is the place of oil in the economy that its price has also had a significant impact on the cost of much partly finished and finished production. The prices of substitutes or alternatives have also been bid up, whether they be coal or housing insulation or Hondas. Finally, increases in oil prices can, through compensating wage hikes and shifts in expectations, become embedded in the 'home-grown' or underlying inflation rate."

Artificially high prices and wages lowered real production. ". . . the sudden increases in energy prices have had a particularly unfortunate effect on such high-employment industries as steel and autos," says Yergin. Faced with "oil-driven" inflation, governments reacted by cutting spending and raising interest rates, which had the effect of further

*From this study came the book *Global Insecurity* (Houghton Mifflin, 1982), edited by Daniel Yergin and Martin Hillenbrand.

crimping economic activity. As the price of imported oil quadrupled to $9 a barrel in 1974, crept to $13 in the mid-seventies, and then tripled in the late 1970's, hundreds of billions of dollars of Western money that would otherwise have been spent on other things were drained out of the economy by the "OPEC tax." As purchasing power was thus reduced, economic activity fell and unemployment rose.

It is significant that 1973, the last year before the "first oil shock" took hold, was also the last year of 7 percent growth in goods and services in the industrial democracies. For several years thereafter the average was stunted to about half that rate. Then came the "second oil shock" of 1979, followed by three years of close to zero growth.

Indeed, 1973 marked a statistical watershed. In the high-oil-cost years after 1973 the average inflation rate has been almost three times as high, the unemployment rate twice as high, the economic growth rate one-half as high as in the preceding low-oil-cost years—ratios that have sharply worsened in the past three years. An especially ominous result of this post-oil-shock decline is the turning away from free trade, while embattled politicians try to bolster foundering domestic industries today by blocking out competing imports, though the price will be paid tomorrow, with interest, in losses suffered by their own *export* industries—as everyone loses. In 1982, for the first time since World War II, the volume of world trade actually contracted. Robert Strauss, former U.S. special trade representative, sees in this an unraveling of the whole structure of Western security: "A NATO alliance among nations waging economic war instead of competing fairly will be virtually worthless."

The impact of oil inflation is seen even more clearly in the looming crises of international debt default and third world disintegration. Peter Drucker reminds us that under the classic rules of cartels, whenever the price of a product is forced upward to monopoly heights, the price of the other products in its general class must fall—for one reason because the money that would have purchased them is less available. Oil is in the class of primary goods, and the rule has been visited with crushing force on exports associated, strangely enough, with the United States and the third world—foodstuffs, cotton, timber, metals, minerals—the prices of which in the main have been dropping since 1974. This downward trend has been intensified since 1979 by the falloff in demand for primary goods as a result of the continuing recession in the industrial nations.

Developing nations, while paying thirty times the 1969 price for their imported oil, are getting the lowest prices in thirty years for their own exports. The non-oil-exporting nations in the third world are suffering through the lowest growth rates in *several decades.* In Latin America, for instance, the average inflation rate is 80 percent, and in some South American countries unemployment has reached 45 percent (with underemployment factored in). So they have turned to debt, which now hangs like the sword of Damocles over the West's financial structure.

In ten years the debt owed to the banking system of the West by the third world has multiplied by more than five times to more than $500 billion. Oil prices are at the center of this debt, reports an in-depth study by *Time:* "Oil-rich developing nations, such as Mexico, Nigeria, Venezuela and Indonesia, wanted to borrow for their development plans, in effect cashing in on as yet unpumped crude reserves. Developing countries without oil wealth, the majority, needed the money to offset higher energy prices that were squeezing their fledgling industries and threatening them with recession."

By 1981 thirty-two countries were in arrears on their debts. By 1982 even the bigger "developing" economies like Mexico, Brazil, Venezuela, and Argentina were flying distress signals. "The Third World owes the First World $500 billion dollars," said Morton Kondracke of *The New Republic.* "It's not coming back."

Thus the full destructive impact of the oil price explosion, partially concealed for years, is yet to come. Only when the "debt bomb" explodes or is defused by a bailout from the taxpayers of the West, with all the further bleeding away of disposable income that this implies, can we begin to measure the ultimate cost of the oil politics of the 1970's.

If it may be said, then, that much of our current hardship and future danger is traceable to the oil price explosion, the authorized explanations of that event merit renewed scrutiny. It was maintained that the "oil crisis" resulted from declining oil resources overtaken by surging demand, combined with the emergent strength of the Arab-led oil-producing nations united in OPEC; thus it was rooted in forces of nature, economics, and history that simply could not be resisted by the West's statesmen and had to be submitted to. As Henry Kissinger told us: "World conditions of supply and demand shifted inexorably against the consumer." And again: "It must be stressed that the price explosion was not [traceable to] a personal decision; one way or another, market conditions would have produced a price explosion, though perhaps over a longer period of time."

How successful this rationale has been can be seen in the peculiarity that though Richard Nixon and Henry Kissinger are regularly assailed with a wide variety of accusations, they have almost entirely escaped censure for the oil catastrophe—the most disastrous legacy of their reign. But it is a defense that has always postured uneasily alongside contradictions, and lately these contradictions have grown to gigantic size.

Consider:

Instead of "inexorable" oil shortage, we see the greatest oil surplus in history—a minimum 20 million barrels a day in unused production capacity being sat on by the oil states so as to maintain the price at 100 times the cost of production. This prompts a look at the figures that were available all along, and they show that between 1950 and 1973 the world's oil reserves expanded eightfold while demand was expanding only fivefold.

Instead of strong Arab leadership, after a decade's surfeit of wealth and arms purchases, we see an Arab world still inherently fearful, weak, and helpless, afraid of Iran's Ayatollah Khomeini, powerless before Israel, and totally dependent on the West for defense against the Soviets. As an American diplomat told *The New York Times* in November 1982: "The lack of any Arab governmental response to what took place in Lebanon this summer [the invasion of an Arab land and the rout of the PLO by Israel] has exposed the inadequacy of practically every Arab regime in a way that is wider in its revolutionary potential than the war of 1948, which led to the downfall of almost every important Arab leader."

Eqbal Ahmad, the Pakistani writer on Middle Eastern affairs, has commented on this phenomenon: "They have acquired wealth without working and made enormous profits without producing. Their countries are littered with expensive machines but they have no technology. Their economies are run by foreigners. Their investments have linked them, symbiotically, with America. They own billions of dollars but control no capital. They lack the will and capacity to translate their wealth into power."

Instead of a united and irresistible OPEC, there stands revealed in early 1983 a teetering OPEC frequently unable to enforce or even to *make* oil decisions because some of its members are cheating on the production ceiling and others are cheating on the price floor. A look back shows it has consistently been so with OPEC, except when temporary circumstance or Western nonresistance lent it the appearance of unity and decisiveness.*

There was no respectable reason for the oil calamity that took full shape undeterred during Richard Nixon's first term and burst upon the world unopposed in his second. There was no causal chain unwinding out of nature's limits or the imperatives of economics or the attritions of history or even the whorl of conspiracy which could account for the thirtyfold jump in oil prices after 1969 that was to numb productive activity across the West and wrench beyond the reach of many of today's generation the common achievements and plausible hopes of yesterday's. There was no pattern of necessity at all but, rather, a jumble of politicians' blunders, negligences, and timidities the connecting thread of which was the personal opportunism of the moment and the postponement of the inconvenient consequences to the next fellow's term.

It is a theme that, of course, invites stout resistance—and not just from those whose official or intellectual fingerprints are all over the debacle. For since we are already victims, we find it less demeaning to be victims of predestination than of folly, less rankling if our diminishment was dictated by geology or history rather than self-inflicted by our own cheered-on leaders, less demanding of us all along if we could regard ourselves as optionless casualties of an iron fate rather than as yielders to economic aggression by Lilliputians.

It is thus an argument to be pressed with the stonemason's caution, and in the laying of its foundation we ask the reader to go back a very short time to the late 1960's, to the "golden age of oil," when nature, economics, and geopolitics had scheduled for the 1970's and 1980's an era of continued material progress that would exceed even the unexampled gains of the previous twenty years, progress to be fueled by oil supplies ample to every need and available at near $1 a barrel.

*True to form, OPEC was enabled to scrape through its latest crisis of disunion, in early 1983, only by the forbearance of Western governments, which supported OPEC's attempts to restore its authority and refused to fish in troubled waters. It is widely agreed that the British could have easily torpedoed the OPEC price-and-production-quota accord of March 14, 1983, and thus sent the price of oil plunging, by a timely cut in the price of its North Sea oil; instead, the British cooperated with OPEC.

FIASCO

Chapter One

Defeat for the Desert Hawks

The Koran says on fertile land, a tithe of one-tenth, on unfertile land, one-half as much. The Americans are offering about one-fourth. Are you unsatisfied with one-fourth when God is satisfied with one-twentieth?

—King Ibn Saud to his Royal Council in 1932

There remains no other course for the national and progressive forces except that of struggle in all its forms, even if this were to lead to the cutting of oil supply lines . . . and the closing down of wells in order to deprive the monopolists, the embezzler, the despot, of this oil.

—*Al-Ba'ath,* Damascus, 1966

If there is no oil worth developing, the producer's money has been wasted. If there will be large development producing profits, the landowner wishes he had held out for much more. . . .

This is the great divide of the petroleum industry: a rich discovery means a dissatisfied landlord.

—Professor M. A. Adelman, 1972

Upon a dozen capitals of Islam, along the sunlit curve of the crescent that spans two continents, the central reality about oil descended in late 1967 like an enshrouding miasma. For those expansive souls who dreamed of monopoly and power, it produced a fog of frustration, closing off fancied horizons, clouding dreams of gain and glory, cooling acquisitive blood with the chill of caution: *There was just too much oil.* Too much oil in nature and too much oil-producing capacity installed by people. Too many billions of barrels available from too many threadbare countries hungry for the royalties. Too much crude overhanging the market to allow the exorbitant price that only scarcity can extort. Too much in too many hands for any one country's share to be important at the bargaining table. Too much to permit even an alliance of nations to control a large enough percentage to intimidate the masters of manipulation who sat in oil company boardrooms.

Because the Persian Gulf and North Africa were providing 13 million barrels of oil a day to fuel the Western economies, the more romantic movers of the region had presumed themselves to have attained at least a degree of indispensability to the great oil companies and their parent nations. But the events of 1967 had exposed how fatuous was that conceit. A contentious year had brought to decision a series of challenges by individual states to their resident oil consortiums and a collective confrontation of all the Arab states against all the great oil companies. In every case the end result had been the same:

flexible but successful defense by the oil companies of their key positions; eventual backdown by the oil states from theirs.

Did not Syria force a shutdown of the pipeline carrying Iraqi oil across its territory, demanding a "nonnegotiable" 21-cent-a-barrel increase in its transit fee (a tripling), while proclaiming a war against foreign cartels "that will stop at no limit" and vowing that "not one drop of oil will cross Syrian territory until the people's rights are satisfied in full"? The oil consortium merely yawned, signaled it was willing to pay maybe 5 cents, replaced the shut-off oil by having its members lift more from Iran and Saudi Arabia, and waited for the Syrian government to eat its holy vows—as it did after ten weeks, in February 1967, accepting the 5-cent increase.

Did not a faction of the Kuwaiti National Assembly obstruct, during the mid-1960's, the terms offered by the Kuwait Oil Company (British Petroleum and Gulf Oil) and threaten nationalization? The consortium cut Kuwait's revenue growth rate to one-fifth of what it had been in 1964, pointed to the demise of past oil-rich lands which had thought they could set up shop on their own, and in 1967 the obstreperous faction was driven from control, the terms of the company were accepted, and amity and full production growth were restored.

Did not the Nasserite oil and dock workers and street agitators rise in Libya against the oil policies of the pro-Western monarchy and force it to halt oil production in the summer of '67? The oil companies quietly stood by the monarchy while it regrouped, and by autumn the agitators were in prison and oil production was headed for another record year.

Did not virtually all the Arab nations, impelled to a rare unity by the shock of the stunning Israeli victory in the June '67 war, join together in an oil cutback-and-embargo directed against the "allies" of Israel, during which half of all the oil flowing from the Arabs to the West was shut off? The oil companies, seeing this as a threat to the system that assured their dominance, mobilized the excess capacities of their non-Arab oil provinces (the United States, Canada, Venezuela, Iran, Indonesia), rerouted their fleets of tankers on the high seas, drew down their stockpiles, tempted the more vulnerable Arabs with the profits that could be had from secret defection, and waited for the boycotting challengers to grow fearful at the sight of markets being lost and to abandon their brothers one by one without notice and crawl back to their old places at the company table—as they did within weeks. For the reigning oil companies (Exxon, British Petroleum, Texaco, Shell, Standard of California, Gulf, and Mobil) knew the endurance quotient of their hosts, having been careful over the years to ladle out oil revenues at a measured pace that kept appetites growing and pantries next to bare.

Such costly brushes with Oil's reality widened the divisions among the landlords of the desert, reducing them the more to its thrall. The "conservatives" (the Saudis, the sheikhdoms of the Persian Gulf, the Kuwaiti establishment, the Libyan monarchy) reacted to the defeat of militant schemes with an "I told you so" acceptance of the oil companies' schedule for their gradual enrichment, under which host government revenues had doubled in the past decade and would likely triple in the next. They accepted with an increased tolerance the rationale that Too Much Oil meant that the price per barrel could not rise and that revenue gains must come mainly from increased production as demand rose—production which in turn must be disciplined to what the world market could bear without collapsing the price.

The "radicals" (the revolutionary regimes of Egypt, Syria, Iraq, and Algeria, the obstreperous parliamentarians of Kuwait, the Nasserite movement in every Arab land, which had borne the brunt of the recent defeats) blamed conservatives made docile by their wealth for betraying all their dreams—dreams of a hundredfold rise in revenue, of burgeoning oil wealth shared in by the whole Arab people, of control seized from the foreigners, of Oil transformed from the puppeteer manipulating the lands of Islam into the means of Islam's renaissance and redemption from Israeli invasion and Western exploitation.

It was at Khartoum in September 1967, at one of their grandiloquently styled "Arab Summits," that the Arab leaders formally and collectively ended the oil cutback-and-boycott against the industrial nations, out of need of oil company money and fear of losing their oil company quotas.

From Khartoum they returned to their capitals twice humiliated—by the Israelis on the battlefield and by the oilmen in the counting-houses—to cope with the disintegration fissuring across Arab Islam and to reexamine the separate strategies they had pursued against the Oil Colossus, all of which had led to defeat.

In the summing up of a tumultuous year, indeed of what might seem to other than oilmen a contentious decade, two general conclusions were difficult to resist, so consistent were the particulars. First, in the skirmishing going on in each Arab land to better its posture vis-à-vis the resident oil combine, gains were achievable in many things: revenue growth; profit split; patronage; the proliferation of projects for civic uplift. But on questions of basic control, over price, production level, and pace of development, the oil companies had proved unyielding and impregnable. Secondly, in the conflict within the Arab world itself, between radicals who burned to seize control of oil from the foreigner and conservatives who feared what the radicals would do with wealth and power if they got it and who deduced that their own interests lay in partnership with the omnicompetent, wealth-bearing oil companies, the radicals were everywhere in defeat by the end of 1967 and the conservatives in clear ascendancy.

Across all these strivings and jockeyings had stretched an inhibiting shadow from the preceding decade—the shadow of Mohammed Mossadegh. That in his seventy-first year Mossadegh materialized as a fantastic player on the world stage was due to the slow-footed stuffiness of the British Petroleum Company in Iran. In 1950 the American oil consortium in next-door Saudi Arabia (Aramco) offered the Saudi government a 50-50 profit split, a marked boost in terms of both revenue and self-esteem that was already in effect in Venezuela and was spreading throughout the American oil empire. Instead of jumping aboard the inevitable, British Petroleum, known in Iran as Anglo-Iranian, pettifogged around for months with an obscure proposal of its own, which it claimed, when all the plusses and minuses were added and subtracted, was just as generous, if not more so, but which lacked the magical appeal of "50-50."

Was Iran, the oldest and largest oil supplier in the region, to accept a lower place than Arab nomads? Mossadegh, the leader of a splinter faction in the Iranian National Assembly, or Majlis, seized the moment. He was a peculiar protagonist by Western lights. Frail, hairless, wrinkled, banana-nosed, given to wearing pajamas at official functions and to weeping in public, perpetually on the verge of either feigned or actual collapse, this exotic old bird mounted daily attacks on the British that seemed to Occidentals preposterously emotional but that struck a deep chord in the Iranian psyche. Soon, as thousands massed in the streets to cheer his passage to the Majlis, he was demanding that the Iranian nation reassert itself, throw out the imperialists, nationalize Anglo-Iranian *without compensation,* and claim for itself the riches that were flowing away.

The young shah, insecure on the Peacock Throne given him by the British, bowed to this eccentric embodiment of the people's will and named him prime minister. "How could anyone be against Mossadegh?" he would later explain. "He would enrich everyone, he would fight the Foreigner, he would secure our rights. No wonder students, intellectuals, people from all walks of life, flocked to his banner."

So the Anglo-Iranian Oil Company was nationalized, on Mossadegh's terms, in abrogation of the concession agreement and in violation of the commonly accepted norms of international law.

At this point Iran ran into the conundrum that awaited any state that laid expropriating hands on an international oil company. It was one thing to surround a refinery or a tank farm with troops and declare nationalization; the smallest principality possessed the armed force to occupy the largest foreign corporation. But if its goal was something larger than the exhibition of a trophy, like a head on a pike, if the state wanted the *fruits* of its oil, then seizure was only the first step down a street controlled by the evictee.

The oil companies, prescient as they were, had a defense against host regimes that would welcome them in with fifty-year pacts when oil was all risk, expense, and uncertainty, but that, after the oil had been found, the world market been built, and the profits rolled in, would become discontented with their share, would begin to feed on the notion that it was the possession of raw crude under the ground that was the be-all, not the many-sided process of exploration, extraction, refining, and transport, the painstaking buildup of marketing outlets, the gradual creation of a world geared to run on oil.

That defense was pacific but effective: to see to it that no other corporation in the world would touch oil that was contaminated by unlawful seizure. Oil traders who were not restrained by their sense of a common stake in enforcing the sanctity of contracts, or by their desire to continue doing business with the oil giants over the long future, were kept in line by Anglo-Iranian's threat to bring costly suit against any trafficker in hot Iranian oil. Thus it was slowly borne in upon the celebrants in the Iranian Majlis that "control of our own resources" was but an empty slogan unless they planned only to sit on the oil fields, that though they could commandeer an oil company's existing machinery and operate it until it ran down, it was access to the world's markets, use of the other fellow's distribution system, that gave "our own resources" their value.

There was a precedent for all this, an object lesson that had proved sufficient for a dozen years. Mexico had once been the world's second leading oil exporter. Then a procession of revolutionary governments began to harass the "foreign imperialists," and in 1938 Mexico nationalized the oil companies. It was a famous victory, but the customers for Mexico's oil turned to other suppliers, the expertise was evacuated, and investment pulled out. Before long Mexico, for all its prodigious oil resources waiting to be tapped, had dropped off the charts as a significant oil producer, not to reappear for forty years.

The emotional Mossadegh, in the course of his crowd-enrapturing orations, had painted himself into a corner from which he could not return to the table and renew negotiations with the British. But with his gift for histrionics and his apt audience for theater, he was able for a considerable time to capitalize politically on the commercial isolation and unfolding bankruptcy of his country. In a typical performance, he would be carried to the rostrum on his bed (apparently a motif of Iranian statecraft), and from the pillows he would sound laments that raised the assembled masses to hysteria. As his climax drew near, he would dissolve into spasms of weeping and collapse in a heap, then to be carried about through admiring hordes that would put up a great wailing at the sight of his inert body.

This was a spectacular but losing game. Two and a half years passed while Mossadegh tried to penetrate the oil company freeze-out and dickered with the American government for a bailout. Oil production had fallen from an annual rate of 665 million barrels in 1950 to 21 million in mid-1953. The treasury was empty, credit unobtainable, and the country sinking into destitution. When theater eventually failed to control the streets, Mossadegh became dependent on the Tudeh Communists. The shah, threatened by Tudeh mobs, fled into exile.

But the solid front of the oil companies was not to be penetrated, nor did their parent governments undercut them. British Petroleum by itself made up for most of the lost Iranian oil through increased liftings from its wells in Kuwait and Iraq. As for the American companies that were cooperating in the boycott, they would just as soon have seen Iran's oil permanently gone from the market to ease the annual strain of holding down production. When asked by their government if they would consider taking over part of the Iranian oil industry to save that country from mounting bedlam and Communist takeover, they at first resisted on the grounds that having to dispose of Iranian oil would require them to cut back their expansion plans for Saudi Arabia, which they regarded as much the greater prize. Iran, which had at the start brandished its bargaining chips as the hemisphere's premier oil nation, had by 1953 sunk without leaving a ripple on the oil waters, to find after three years that its disappearance was considered a general convenience.

Only when Mossadegh was overthrown by a street uprising with the not-so-covert aid of CIA agents, and the shah safely reinstalled, and the oil industry invited back on a basis that gave the companies the controls over volume, price, development pace, and management did Iranian oil start to move again to the markets of the world, in late 1954.

The long shadow of Mossadegh had served to mark the limit of what was attemptable in the innate striving of host governments to gain an ever larger payoff for their oil and an ever greater measure of control over it. It was a limit that allowed considerable elbowroom for the former endeavor, but little for the latter.

In terms of payoff, the regimes of North Africa and the Persian Gulf had progressed from 20 cents a barrel for 315 million barrels in 1947 to 80 cents for almost 5 *billion* in 1967, or, to put it another way, from a 25 percent share of the oil profits to 50-50 and now to 70-30. But profit, argued the vanguard Islamics, was just a percentage of a too low price and was circumscribed by volume; it was in the control over price and volume that the real wealth lay, along with the pride and the power; and it was the uniform failure of efforts at gaining control, culminating in the setbacks of 1967, that darkened the oil vista with the aura of defeat.

Defeat brings reassessment. A realignment was under way in the Islamic oil world, a sorting out of the winners from the losers. The brief history of relations between oil companies and host states by now revealed three loose but distinct patterns: Host regimes that had coexisted tolerably with the oil companies, keeping the inevitable skirmishing within limits, such as in Nigeria, Libya, and post-Mossadegh Iran, had made steady gains in oil production and revenues. It was a continuous upward progress unique in the distressful fortunes of providers of primary goods, as the histories of copper and tin, of cotton and wool attested. Hosts that had marred a reasonably cooperative record with a fractious period of mischief—Saudi Arabia and Kuwait come to mind—had suffered commensurate losses in revenue growth; when they returned to good standing, they were restored to the growth ladder, but to a place several rungs lower than they otherwise would have occupied.

And hosts that had frontally challenged the oil order—Mexico, the old Iran, the new Iraq—had paid a grave and in some respects terminal price in the coin of withered oil development, economic disintegration, political instability, ruined careers, and broken lives.

What made these outcomes seem the more fated and inescapable, the more rooted in the natural order of things, was that they resulted not so much from a calculated design of reward and punishment as from the businessman's natural hesitancy to continue investing money in an environment turned insecure and the allocator's routine tendency, when faced with more offers of supply than the market can absorb, to accommodate as best he can his most reliable suppliers.

The Iran of the restored shah was a conspicuous example of the gains to be had from bowing to the prerogatives of the oil companies. The abyss facing the shah on his return to power in August 1953 was succinctly described in a message from his new prime minister, Fazlollah Zahedi, to President Dwight Eisenhower: "The treasury is empty; foreign exchange resources are exhausted; the national economy is deteriorated. Iran needs immediate financial aid to enable it to emerge from a state of economic and financial chaos."

The principal vehicle for rescue was the revival of Iran's oil industry. Under Mossadegh (who was to serve a prison term and then, as if to spoof his frailty motif, live on under house arrest until 1967, when he died at the age of eighty-seven) 650 million barrels a day of Iranian oil had disappeared from the world market and been promptly replaced by oil from other countries. If Iran were to be rehabilitated, not only must the old level of oil exports be resumed, but even higher levels had to be reached—at the expense of those oil-exporting countries which had filled the Iranian void. Moreover, it was a political impossibility for the shah to allow the British Petroleum Company, for years the butt of popular hatred, to return in blatant triumph to its former status.

The shah created a propitious atmosphere by avowing a pro-Western alignment in place of Mossadegh's nonalignment that at the end tilted toward Moscow and by adopting a posture of as much hospitality toward the oil companies as was consistent with Iran's interests and his own need to establish a visible independence of foreign control.

Under the leadership of the U.S. government, an arrangement was finally hammered together: The five American oil giants bought 40 percent of the old Anglo-Iranian from British Petroleum; BP retained 40 percent; the British-based Shell held 14 percent, with the remaining 6 percent picked up by the French Compagnie Française des

Pétroles.* Iran was to hold legal title to its oil, but the new consortium explicitly and firmly retained those operating controls which oil companies had traditionally deemed essential.

The first tanker carrying Iranian oil for export under the new arrangement left harbor in November 1954; by 1955 oil production had returned to half the pre-Mossadegh peak; by the end of 1956 it had drawn even with its former high; and in 1957 it handsomely surpassed it, averaging 716 million barrels a day. Thus far the larger designs of Western diplomacy had influenced this remarkable turnabout; henceforth whether Iran would be favored above its peers would much depend on the character of the shah's continuing relationship with the oil companies.

In shaping that relationship, the shah was torn by ambivalences and tugged at by built-in conflicts. He had at age twenty-two been given a crown by the British, but it was a crown they had taken from his father (for siding with Hitler in World War II). He had known from youth both gratitude for oil revenues from the foreigner's hand and the resentment that dependence engenders in a monarch. "We were an independent country," he reminisced somewhat disingenuously in 1975, "and then all of a sudden the Russians invaded our country and the British took my father into exile. Then we were hearing that the oil company was creating puppets—people just clicking their heels to the orders of the oil company—almost a kind of government within the Iranian government."

He knew, and had once been swept aside by, the power of anti-Oil emotion, the appeal of demagogues who promised that Oil was a fat cow to be milked. But he knew also the outcome—that Oil was not a cow but an octopus, slippery and unreachable in its murky environment, with tentacles able to reach out thousands of miles beyond the apparent point of engagement and pull levers that shut off its adversary's oxygen.

He would wait, then, for the unfolding of his imperial schemes to strengthen him and for time and events somehow to weaken the mighty oil companies. In the meantime, he would battle with the companies for Iran's interests, vigorously but within the accepted limits, while showing himself their reliable partner and useful ally. So Iranian oil had flowed unstinted to help the companies undo the Arab cutback in 1957

*The American companies were Exxon, Texaco, Gulf, Socal, and Mobil; later, under pressure from the Justice Department, a number of American independents were permitted to participate on a minor basis.

and again in 1967, and the shah had been a prominent foe of radical, anti-Jewish, anti-Western causes in the Middle East.

In consequence, Iran continued to expand its oil production and reserves at a faster pace than its Persian Gulf rivals. Over the long haul Iran's production grew at 15 percent a year, twice as fast as world demand for oil. In the years from 1954 to 1967 Iran's growth rate exceeded Saudi Arabia's ten times and equaled it once. The shah's annual oil revenues grew from about zero in 1953 to $90 million in 1955 to $482 million in 1964 to $750 million in 1967, providing a solid basis for him to consolidate his rule and finance an economic buildup.

Iran's steep, unbroken climb in oil production after 1954 was made possible in part by the slowdowns imposed by the oil companies on the other major oil states of the Middle East, two of which—Kuwait and Saudi Arabia—sinned against Oil but reformed, while the third, Iraq, waned unrepentant.

Kuwait had for a while sulked obstructively in its small tent by the Persian Gulf, muttering mutinously of expelling its consortium and bringing in a new one, but a few years of slowed revenue growth and of wondering where, if the leap were taken, the newcomer who could run the oil company blockade could be found had dampened the stirrings of revolt. When the instigators were ousted, the revenue spigot was promptly turned up.

The difficulties with Saudi Arabia had been more serious and prolonged. Throughout the late 1950's and early 1960's, a time of fratricidal struggle between two sons of old King Ibn Saud (the dissolute Saud and the abstemious Faisal) for the ultimate succession to the throne, the Saudi oil ministry was in the hands of two ministers vehemently hostile to the Arabian-American Oil Company (Aramco), a consortium made up of Exxon, Texaco, Standard of California, and Mobil, which controlled all the oil in Saudi Arabia. The first was Abdullah Sulaiman, near retirement, who pressed unsuccessfully but persistently within Saudi councils for the expropriation of Aramco. The second was Abdullah Tariki, Sulaiman's rising young protégé, educated at the University of Texas, who seconded Sulaiman's aspirations at home and used his position to campaign throughout the Arab world for oil state seizure of the oil properties. Tariki was named Saudi oil minister in 1960, giving him the more prestigious platform from which to propagate his slogan: "Arab Oil for the Arabs."

Aramco responded to the steady harassment of the Sulaiman-Tariki oil ministries in a manner that is by now familiar. From 1955 through

1964, with the modest exception of two years, Saudi Arabia's growth in oil production was severely restricted, astonishingly so considering its incomparable resources. By 1964 the combination of declining revenue growth and runaway profligacy by the royal family had brought the kingdom to the verge of bankruptcy.

By this time Faisal had completed his gradual supplanting of his brother Saud and had at last become the unchallenged ruler of Saudi Arabia. He moved vigorously to clean up the royal finances and to restore good relations with Aramco. Tariki had been forced into exile, and his international oil policy of fomenting collective action against the oil companies, which had led to the founding in 1960 of the Organization of Petroleum Exporting Countries (OPEC), was changed to a pro-oil-company policy which frustrated collective action and soon diminished OPEC to the point that its most fervent supporters lost hope in it.

The results of Faisal's rapprochement with Aramco were immediate and stunning. In the first year after Faisal had solidified his rule the annual oil production growth rate jumped from 5 percent to 18 percent, the highest increase in fifteen years. Saudi Arabia had begun the most extraordinary period of exploration finds and production growth in the history of oil.

But what of obstinate Iraq?

By 1967 Iraq was in the tenth year of the foremost Middle Eastern experiment since Mossadegh's to break loose from an oil pact with the great companies and go into the oil business on its own. In August the Baghdad junta took its furthest step by turning over to its own state company, the Iraq National Oil Company, 99½ percent of the vast acreage claimed by the Western oil companies (British Petroleum, Shell, Exxon, and Mobil, which together constituted the Iraq Petroleum Company). But behind this façade of revolutionary vigor lay a sad history of impotence that smelled more of the oil demise of Mexico and the paralysis of Iran than of a bold march into a new day.

Iraq had been one of the first Middle Eastern countries to attract the oil companies, beginning in the World War I period, and was one of the more spectacularly endowed. In 1958 its reactionary monarchy fell to a revolutionary takeover by the Arab Socialist Ba'ath party, which included, among its milder elements, Communists. The existing concession agreement of the Iraq Petroleum Company gave it control of all the land in Iraq that had the remotest possibility for oil development, and the new government, under General Abdul Karim Kassem, demanded that it relinquish 60 percent of that land. The consortium refused, and in 1961 Kassem declared government ownership of virtually all concession lands except those which the oil companies were operating. On these, the government demanded that oil production be doubled.

The Iranian precedent was obviously on the minds of the Baghdad regime. By leaving untouched all holdings where oil was actually under development, they had stopped short of the kind of provocation that might have prompted a freeze-out of Iraqi oil by the world oil community. They hoped to enjoy the best of both worlds: While the revenues rolled in from the crude the oil companies had already developed, the government would be escaping production controls and opening up the rest of the country by inviting in other oil companies to explore and develop the seized lands on Ba'athist terms.

The consortium responded with becoming flexibility. As for the going operations left in its hands, it would conduct them at its convenience, reaping the profit of past investment but ceasing further development and minimizing production gains. The oil growth of Iraq was henceforth to be stunted to make way for faster growth by host countries which were reliable. As for the areas the Ba'athists claimed, the consortium would exert its powers of persuasion to scare away any developer tempted by Iraqi offers.

And so it was. Every significant scheme of the Iraqi government to lure in foreign oil developers was broken up. Over the years several independent American oil companies expressed interest in going into Iraq, but in each case pressure from Exxon or Mobil or the State Department caused them to back off. The 1967 Iraqi decision to turn these lands over to its own "oil company" was thus a rather transparent confession of defeat.

The price Iraq paid for its policy can be seen in this: In 1960 Saudi Arabia, Iran, and Iraq all hovered close to 1 million barrels a day in oil production, with Kuwait in the lead at 1.6 million. By 1967 Saudi Arabia's allowed production had risen to 2.6 million barrels a day, Iran's to 2.5 million, and Kuwait's to 2.3 million. But Iraq's production stood at only 1.2 million.

The loss of revenue implicit in these figures—to half of what it might have been—goes a long way toward explaining the economic stagnation and political instability that blighted Iraq through the decade of its confrontation with Oil (three revolutions in eleven years—one chief of state murdered outright, one killed in a mysterious accident, a third charged with treason and exiled) and left it not only low man on the oil totem pole but a poor relation among the Arab states, an object lesson whose example was shunned. For the oil companies, the convenient submersion of Iraq had enabled them to raise production in places of higher priority.

Nor did the future offer any foreseeable reprieve from the rigors of Too Much Oil. The Middle East was the prime illustration in support of this foreboding. As of the end of the 1960's that region was the source of 66 percent of the free world's oil reserves, 35 percent of its oil production, and 80 percent of all the oil moving in international trade. Its price, therefore, was the "world price"; if control of Oil was achievable at all, it was achievable here.

Two implacable facts, at once wonderful and terrible, overhung the decades ahead: First, as against the 4.5 billion barrels of oil pumped in the Middle East in 1969, the oil companies were sitting on 334 billion barrels there—seventy-four years' supply at current consumption even if not another field was ever found, though in truth, three giant fields, each containing billions of barrels, were being discovered every year. Secondly, the average Middle Eastern production cost of a barrel of oil —which contained 42 gallons and weighed about one-seventh of a ton —was 10 cents. Only 10 cents a barrel financed the finding, drilling, bringing up, shipment to embarkation ports, and the return on investment.

Taken together, these two conditions heralded the fulfillment of the age-old dream that has recurred alike in adults' schemings and children's fairy tales, the dream of some prosperity-bearing, happiness-bringing substance, flowing from the earth in unquenchable profusion, possessable for pennies.

But the implications of these two facts, bright as they were for the world at large and for Middle Easterners of less than Brobdingnagian appetite, were so many shafts of gloom to those who, but for Oil's diffuse and unmanageable plenty, could see themselves within striking distance of controlling enough of it to render it scarce and make the world pay dear. Because of undentable proved reserves and minimal production costs, Islamics who nursed the vision of control of Oil, whether in minareted palaces or in junta command posts or in Nasserite cabals lately driven underground, could see it fading progressively.

Proved reserves of 334 billion barrels, alas, were not theoretical barrels of remote significance, but barrels clearly located by drilling, credibly measured, and as quickly hooked up to the production apparatus as demand permitted. This meant not only a seventy-four-year reserve but also a promptly activatable production capacity far greater than current demand. And since this excess capacity was spread among several Middle Eastern states, some of them visceral enemies of others, as, for instance, the Iraqis and the Iranians, any attempt by some to reduce production so as to raise prices or exert political pressure could be readily countered by revving up just a portion of the excess capacity of the nonparticipants.

Similarly, the production cost of 10 cents a barrel became the *natural* floor price of oil in world trade, toward which all the competitive forces around the globe were pushing, against the gradually yielding defenses of oil companies and oil states. Infinitesimal production costs were an irresistible lure to increased exploration—by the major companies, for which it was slightly more profitable to open new fields than to beef up old ones, and by invading entrepreneurs, increasingly ingenious at finding ways to break the lock of the majors on the region's oil lands. Thus the surplus had risen steadily, and the price of oil loaded at Persian Gulf ports had grudgingly receded, from $2.08 per barrel in 1957 to $1.30 in 1967, and seemed certain to inch down to around $1.20 by 1969. Oil company profits per barrel had fallen off from 55 cents in the 1950's to 34 cents in the 1960's. Only the tax take of the host governments—80 cents a barrel, eight times the cost of production—was still propping up the selling price above $1. But for how much longer?

Rising consumer demand for Middle Eastern oil was the hope, the countervailing force. Indeed, there was no doubt that demand would rise sharply. In the previous fifteen years it had multiplied four times. In the coming fifteen years (1968 through 1982) it could rise, to give demand the best of it, by almost six times, say, 11 percent a year. If so, by 1982 the region would be exporting not 3.5 billion barrels, as in 1967, but 18 billion!

An 11 percent growth sustained for fifteen years was within the realm of responsible prophecy but was at its upper limit. It assumed a strong, steady, long-term expansion of Western economies that historically had been dogged by stalls and dips. It had to contend with the political interventions of the industrial nations to protect their own domestic energy industries threatened by imported oil and to allay overdependence on a Middle East that had used the embargo weapon twice in eleven years, however ineffectually. And it was threatened by the unfolding plans of the great oil companies to reduce overconcentration in one region, plans already taking shape in a worldwide effort to expand oil activity in North Africa, West Africa, the North Sea, the Alaskan Arctic, South America, offshore North America, and the Far East.

The unrelenting thing about proved reserves was that they were already there—the past ruling the future. The figures on oil already found dictated that even if the maximum projections of demand were realized, there would still be more proved reserves in the 1970's than in the 1960's and more in the 1980's than in the 1970's. This would remain true even if an invisible hand were to stop all further exploration and not a single new field were to be found. For it was inherent in existing reality, according to the statistics of the past, that routine development of fields already operating—the tapping of new wells in old fields, the application of improved extraction processes—would enhance present proved reserves by about 60 percent over fifteen years, adding about 200 billion barrels to the 350 billion already available during this period. This enhancement would by itself far outrun the total demand projected in the 11 percent scenario—140 billion barrels over the fifteen years. Thus, after an ocean of oil had flowed out, the next generation of Islamic leaders would be confronted with proved reserves not in the 300 billions of barrels, but in the *400* billions.*

So long as world oil affairs continued to operate on any basis resembling, however loosely, the laws of supply and demand, of competition and incentives, of diverse participation and the sanctity of contracts, of great nations defending their lawful interests against weak ones, then so long would the presence of gigantic surpluses and dirt-cheap production costs continue to mock the hope of desert dreamers that Oil could be somehow captured and the world made to behave as though it had suddenly become scarce and costly. If the immense weight of the al-

*With the disadvantage of hindsight, we know how it turned out. For the first five years of the projection period, Middle East oil output *did* rise at about 11 percent a year, as a result of politically imposed curbs on production in North Africa and the breakdown of import discipline in the United States. Then, after a peak demand growth year in 1973, during which Middle East oil production reached 7.6 billion barrels, output leveled off and then receded. Meanwhile, the accumulation of proved reserves was dwarfing the growth in demand. Even during the years when demand was booming, between 1968 and 1973, the figures on proved reserves in Saudi Arabia alone rose by over 100 billion barrels as against a cumulative output *in the entire Middle East* of only 34 billion. By 1982 Middle East output, far from reaching the 18 billion barrels of the scenario, had dropped back to the 1967 level of 3.5 billion barrels, and Middle East reserves were sufficient to supply current demand for far longer than the 100-year span of 1967.

ready-in-place surplus were to be lifted so that the price of oil could rise and the autonomy of oil-endowed states could expand, then nature and economics themselves would have to be somehow overthrown.

Only politics can repeal reality. But the great oil-consuming industrial nations whose corporations held concessionary title to the oil surpluses also dominated the non-Communist world's politics. Even in the land of dervishes, it was difficult to conjure up the circumstances in which an American President and his NATO counterparts would let slip away the great cornerstone of their economic progress and political cohesion, which had been set in place over the years with such painstaking care and could be kept in place with such ease.

Above this shifting panorama of rising and falling national fortunes, beyond the reach of wounded pride, repressed avarice, thwarted idealism, and cultural antipathy, loomed the reigning oil companies. The writer Anthony Sampson has captured, in a memorable passage, the image they projected:

> They gave every appearance of permanence and stability, with their self-perpetuating boards and bureaucracies and their evident ability to survive two world wars and countless revolutions across the continents. Their engineering achievements commanded the awe of governments and publics. Great refinery complexes rose up along the coastlines, with their grotesque skylines of strange shapes—spheres, towers, and cylinders interwoven with twisting pipes and surrounded by white tanks—which looked like giants' kitchens which had outgrown any human fallibilities. They seemed to mark the triumph of technology over man. The whole style of the corporations, grown smoother and more confident over the decades, suggested a lofty superiority to all governments.

For a generation seven gigantic corporations—the legendary Seven Sisters—had controlled about 90 percent of all the oil that moved in international trade. Five of them were American (Exxon, Texaco, Mobil, Standard of California, and Gulf), and two were British (British Petroleum and Shell).* This exclusive club of seven was of fortunate

*Shell, or more properly Royal Dutch/Shell, was British-based and British-led, though its ownership was divided between the British and the Dutch.

size—large enough to blanket the globe, small enough for its members to cooperate discreetly in carving up that globe.

The undertakings of each of the Seven Sisters were a mixture of the heroic, the chancy, the managerial, and the conspiratorial: to find oil and wrest it from ice-locked tundra, tropical jungle, scorched desert, or tumultuous ocean; to gamble millions on the mysteries of a geological formation and tens of millions on the stability of a regime; to develop marketing outlets with their globe-arching spawns of pipelines, terminals, refineries, tank farms and filling stations, fleets of ships and trucks; and to maintain through all the vicissitudes of economic cycles and political upheavals a rather precise balance between supply and demand.

From skyscraper citadels in New York, Pittsburgh, Houston, San Francisco, and London, the Seven Sisters directed international operations of a scope not before seen in the world. Prodigies from childhood, they soon had gushers rising from five continents. By adolescence, during World War I, they were already equal to Marshal Foch's plea "We must have oil or we shall lose the war" and rose to it so fully that Lord Curzon would later attribute to them the triumph: "The Allies floated to victory on a wave of oil." One corporation alone, Exxon, furnished one-fourth of all the oil used by the Allies. At early maturity, during World War II, the Sisters repeated that phenomenon on an incomparably vaster scale. Reaching their prime in the 1960's, they produced, refined, and sold the overwhelming bulk of all the oil traded among the nations.

In their office towers the oilmen determined the annual revenues of a dozen oil-producing countries by deciding how much oil they would lift from each, occasionally toppling or preserving a government thereby, withdrawing income from one and adding to another with a Jovian hand that was beyond the effective questioning of parliaments and kings. They would estimate the industrial potential of each oil-consuming nation in a future year and call into being by decrees of their boards the new harbors, depots, pipelines, refineries, and shipping that would make those growth rates attainable. It excited wonderment, not to speak of resentment, that the oil Sisters had larger revenues than most of the concessionary states they operated in, more tonnage on the high seas than a superpower's navy, that on barren desert reaches they built and maintained cities with green-lawned suburbias, that in the Arctic dark at the edge of the world their crews ate steak every day and watched the latest movies during off hours.

The oil majors had even risen beyond the restraints imposed by the normal workings of business and the natural laws governing commerce. Generating huge revenues and cash surpluses, they could often plan their expansions without the leave and supervision of the great banks or recourse to the stock and bond markets. In a position to assure steady dividends, they need not fear the restraints of stockholders. Controlling oil from its source until it reached its ultimate user, they had forced a livable truce with the dread forces of supply and demand.

The harmonizing of supply and demand was the key to all. To provide *too little* oil, though it would run up prices nicely and bring Jay Gouldish profits temporarily, would provoke political interference from angry Western governments, which would then invade their control, inject competitors, and destroy their prized equilibrium. For though the Seven Sisters were mighty in their realms and though they could usually have their way with nations in calm weather, let it storm and they were but corporations after all, functioning in a world ultimately ruled by governments and ideologies. To provide *too much* oil, on the other hand, would be even worse, driving down prices, destroying profits, turning an ordered oil universe into a free-for-all. In the 1960's, as in the 1980's, the specter ever glowering over Oil was overproduction. Then as now, more than enough oil had been discovered and was being sat on to flood the world with several times the demand for it—if producers operated at full throttle.* A surplus of a mere million or two barrels a day, in a market that consumed 30 million, could turn profit into loss, order into panic.

The trick, as Oil's statesmen saw it, was to provide *just enough* oil, just enough each year to satisfy the growing demand fully, and to do so at a stable price, one high enough to be profitable but not so high as to be a magnet for invading companies or a target for populists, a price low enough to discourage production by non-Sisters in the high-cost areas they could gain access to, thus to keep down glut and rivalry, but not so low that it reduced Oil to the profit margin common to other industries. With ample supply at a steady price, the consuming publics of the world had nothing visible to kick about, would-be rivals and scholarly antimonopolists had small, listless audiences for their complaints, the governments of the industrial states were quies-

*In 1938 world oil production was 5 to 9 percent of reported world reserves; in 1974 production was only 2.9 percent of reserves; in 1982, about 2.5 percent. Saudi Arabia, with reserves that would support an output of 20 million barrels a day, would be producing at a rate of 3.5 million in 1988.

cent and cooperative, and the gradual industrialization of an expanding global population, largely fueled by oil that was cheap but not too cheap, assured that the market for the Seven Sisters would grow with almost boring regularity, at 7 percent a year, and that their profits would remain steady in the aggregate, even as they declined per barrel.

To hold back the ever-threatening tide of Too Much Oil, the multinational oil giants had to frustrate continually the designs of two distinct but symbiotic groups: first, the oil-endowed states, which, though they be as diverse as Saudi Arabia, Iran, Nigeria, Libya, and Iraq, all demanded higher and higher oil production quotas because they would mean higher and higher revenues in the short run; secondly, the smaller oil companies, the independents—Marathon, Continental, Sinclair, Union Oil, Occidental, and two dozen more—who wanted "in" on the international scene where the cheap-to-produce oil was.

The Seven Sisters had conspired in every way this side of prison gates to keep out the independents for three basic reasons: (1) They did not want to share their profits with them or the control over world trade that made those profits perpetual and automatic; (2) they comprehended that the independents, if they got in, would wildly overproduce because, on intimate terms with greed, they knew that the greed of the small tends to be short-range, whereas the large and already sated are in a position to take the longer view of how to squeeze the lemon totally dry; and (3) an independent company was likely to have its chief source of oil in one host country, as opposed to a Sister with resources spread over several host countries. Thus the independent would soon become the captive of that one host, instead of the other way around, a captive that would be used to compete against the Sisterhood and break its grip—the grip that kept the concessionary states from seizing control of the world oil market.

By and large the multinationals had succeeded in fending off the groups that threatened their power and the profitable stability they saw as the fruit of that power. The Sisters had kept out newcomers by sewing up almost all the key oil lands and consumer markets. (It was not unusual for two or four of the majors to join together in a consortium that owned exclusive oil rights to *all* the acreage in an entire country for a half century or longer, as with Gulf and British Petroleum in Kuwait; Exxon, Texaco, Socal, and Mobil in Saudi Arabia; and Exxon, Shell, British Petroleum, and Mobil in Iraq.) As for the more difficult task of resisting the determined will of host governments, they had succeeded by being accommodating in the lesser things but unyielding in the greater. When a ruler became too abrasive, as we have seen, his country found its production share reduced, its development stopped, its revenues slumping, and if a ruler went so far, in the hubris of sovereignty, as to lay expropriating hands on a Sister for refusing to heed him, his country would henceforth be a pariah among trading nations, unable to sell its oil, float loans, or engage unhindered in commerce beyond its shores, until it relented.

In such extremities the oil giants had looked to the American and British governments for undergirding and had always received it. But 99 percent of the time they themselves were able to run the oil show throughout the world, supplying oil products to all who wished to buy, seeing to it that the future oil needs of 100 nations would be precisely met at a gradually declining price, and furnishing a score of countries with the larger part of their hard cash.

The maintenance of the status quo in oil seemed assured for as long as the United States maintained its existing oil posture abroad and at home.

Abroad the American government had been vigilant over the years to come to the support of its oil companies when an equalizer was needed in a confrontation with a foreign government. From the days of World War I onward, Washington had used its growing punch in international affairs to establish American oil companies, or to smooth their paths (indeed, by waiving antitrust laws) in the oil-rich regions abroad—Latin America, Iraq, Kuwait, Bahrain, Saudi Arabia, Iran, Libya. Its initial motive—beyond the traditional urge to promote American commercial interests—was concern for the future, concern that the U.S. industrial machine would one day run out of oil at home and have to develop secure access elsewhere.

John J. McCloy, lawyer to all the American majors and the industry's ambassador to seven U.S. presidents, would recall this era at a Senate inquiry in 1974:

> MR. MCCLOY: I see somebody has criticized the government for not having an oil policy. I don't agree. . . . We had a very good and aggressive oil policy . . . where we really did move into a point where the U.S. influence on the energy situation was very important.

SENATOR FRANK CHURCH: When did we have such a policy?

MR. MCCLOY: I think it goes back to President Wilson's days. It certainly was in Franklin Roosevelt's days when Harold Ickes and Jim Forrestal were so concerned about the fact that we were drawing down our oil reserves.

After World War II, when America became the reviver and protector of the Western family of industrial nations, the rationale broadened and the success expanded. A Senate Foreign Relations subcommittee chaired by Senator Frank Church described the challenge and assessed the performance: "U.S. foreign policy objectives were threefold. First, the U.S. desired to provide a steady supply of oil to Europe and Japan at reasonable prices for post-World War II recovery and sustained economic growth. Second, the U.S. desired to maintain stable governments in the non-Communist, pro-Western oil *exporting* countries. Third, the U.S. desired that American-based firms be a dominant force in world oil trade. These three U.S. foreign policy goals were generally attained during the 1950's and 1960's."

In pursuit of these goals, every postwar President had made difficult interventions at critical junctures. Harry Truman quashed a mammoth antitrust prosecution of the international oil companies in order to preserve those companies as effective instruments of U.S. policy in a disintegrating Iran and a mercurial Middle East. To replenish the Saudi treasury, among others, Truman enabled the oil companies to increase painlessly their payments to host governments by allowing tax breaks that, in effect, financed them out of the U.S. Treasury. Dwight Eisenhower refused to save Mossadegh with U.S. aid expressly because of Iran's failure to honor its legal obligations to British Petroleum, and when the Ba'athists, full of the rhetoric of expropriation, toppled the Iraqi government in 1958, Eisenhower promptly ordered American troops into next-door Lebanon, stationed the Sixth Fleet nearby, and kept these forces poised to strike until the sobered Iraqis recanted their more extravagant designs on Oil and their neighbors. John Kennedy orchestrated irresistible pressures on the NATO allies in order to break up a proposed Soviet pipeline project that would carry Russian oil into West Germany and Austria, and he frustrated other efforts to bring Communist oil into the Western sphere supplied by the Seven Sisters. Lyndon Johnson cut off foreign aid to Peru in 1964 to influence a dispute between the Peruvian government and Exxon, tipping the scales toward temporary compromise, and he bolstered the oil companies in repeated disputes with the lately inflated shah of Iran. Thus a firm, consistent, bipartisan foreign policy of long standing, aimed at preserving the oil status quo, was in place for the guidance of those who came after.

Chapter Two

The Faustian Bargain

The 1968 Presidential race between the late Hubert Humphrey and Richard Nixon was most certainly decided by money.

—Mark Shields, *The Washington Post* political analyst

The trouble with this country is that you can't win an election without the oil sector, and you can't govern with it.

—Franklin D. Roosevelt

The United States had been able to pursue an effective oil policy abroad in part because it maintained a strong oil posture at home. The world's greatest producer of oil, America not only was free of any real dependence on oil from beyond its hemisphere but kept up an unused reservoir of shut-in oil, a spare capacity of almost 4 million barrels a day, which it could mobilize on short notice to flood out any artificial shortage the concessionary states might contrive.

So the oil status quo of plentiful, cheap oil—rooted in the abundance of nature and the effective organization of that abundance by the oil companies and protected by a U.S. foreign policy that was itself buttressed by a strong domestic oil posture—seemed manifestly secure as 1967, its many challenges so easily met, gave way to the new year.

Successful policies, of course, are but the creatures of changeable politics, and dependence on one country as the backstop of a worldwide system had its inherent shortcomings—hazards soon to be tested again, for 1968 was a presidential election year in the United States. But the foreign and domestic policies that stood behind the oil status quo had been the handiwork of both the major parties, and their prospective presidential candidates—President Johnson and Richard Nixon—had throughout their careers been identified with Oil and its politics. So unthreatened was U.S. oil policy, in fact, that in the early presidential primary contests the subject was not even being mentioned.

Among the many constellations shaken by the March 1968 announcement of President Johnson that he was withdrawing from the presidential race was the galaxy of Oil. To its more disinterested occupants, its statesmen and scholars, Johnson's withdrawal meant that come January 1969 someone new and untried would become the protector of the oil security of the Western world and of the continued prosperity and stability which that term embraced. To the nuts-and-bolts oil barons of narrower perspective, who had been dealing with Lyndon Johnson for more than thirty years, his impending exit was a signal that Oil's money and electoral power had to be mobilized to see to it that the "someone new" be someone who was bound to preserve the protections and subsidies that gave American oil much of its safety and profit—quota barriers against the importation of cheap foreign oil, which kept domestic oil prices 40 percent higher than the world market, and tax write-offs that shrank Oil's income tax load to one-sixth the stated rate for corporations.

So within the outer campaign that was to be fought out via posturings over Vietnam War policy, rising inflation, racial inequality and unrest, permissiveness toward crime and disorder, and matters of personality, there was to be a race within a race, about the fundamentals of oil profits; if the outer race narrowed down to a standoff, the inner race would be decisive.

"As a Congressman twenty-one years ago, as a Senator, as Vice President of the United States, as a candidate for President in 1960, I opposed the reduction of the depletion allowance because I want these great resources developed in Texas and across the nation."

It was September 6, 1968, the opening week of Richard Nixon's presidential campaign against Democrat Hubert Humphrey, and Nixon was already in Houston, driving home to oildom's contributors and voters the difference between him and Humphrey on the issue that meant most to them. During the two decades that he had been toiling to keep Oil's tax rate down, Nixon would tell oil country audiences, Humphrey had been toiling to push it up. The argument had the added virtue of being tolerably true. Humphrey had been a charter member of the band of Senate liberals who since the late 1940's had been warring against the oil depletion allowance, sometimes seeking to cut it down to the depletion rate allowed for other minerals, sometimes trying to make it less of a pocketable giveaway and more of an incentive to risk the costly search for oil.

By the campaign's second week, on September 13, in Cleveland,

Nixon was ennobling Oil's tax forgiveness with the mantle of national survival: "I continue to believe that America's security requires the maintenance of the current oil depletion allowance."

To oilmen, Nixon's choice of the word "current" exuded a special insider's balm. He was promising that as President he would make no change in either the domestic or foreign depletion tax arrangements under which the fifteen largest oil companies paid an effective corporate income tax rate of only 8 percent. In a Nixon presidency "oil depletion" would continue to save companies and investors about $1.5 billion each year that would otherwise have to be paid in federal taxes. Because of this provision, there were a number of millionaires who year after year did not pay a dollar of federal income tax and hundreds of big dividend collectors who paid taxes that would make a bus driver blush.

To the end of the campaign Nixon would hammer away on this issue to selective audiences, spending two days of the last precious week in the heart of oil country. But this was only the public phase, the blander phase, of an operation that moved quickly and surefootedly to corner the oil industry's financial support. Early and late, whenever the Humphrey men approached Oil's financial citadels, they found themselves on the defensive. The Nixon men had preceded them, with their projections, based on the indelible past, of what a Humphrey presidency would cost them—and of how much a Nixon presidency would save them.

As a reporter covering that campaign and trying to penetrate the Oil connection, I found that Nixon himself took part in the private proselytizing, among other things promising oil millionaire John Shaheen that "oil depletion" would always be sacrosanct to him, a pledge promptly conveyed to the Sisterhood.

If there was a premier lesson that Richard Nixon's campaign experience had taught him, it was that in close elections money was decisive. And Nixon did not let his big Labor Day lead over Humphrey divert him from the judgment that by November, when the Democrats had pulled themselves together, this one would be neck and neck. He liked to hearken back, during the planning sessions, to his 1960 race against John F. Kennedy. He had begun that race with a big lead, too, but Kennedy, helped by plenty of advertising money and the organization that money can buy, caught up with him in mid-campaign and then went ahead of him in the polls. Nixon almost regained the lead in the final days—this is what haunted him—but was unable to close the hair-thin gap because, as Nixon analyzed it, *he* ran out of television money and Kennedy didn't. Whenever he talked campaign strategy to his staff in 1968, he was sure to caution them that he must never again be outspent or "outtelevised."

The 1968 Oil contribution would be a prize of cardinal importance, in the millions of dollars. As important as getting that money was, it was more important to keep the hard-up Humphrey from getting it or even from getting the minor share, the hedge money that Oil might give to the candidate it didn't want just in case he won. For the money factor was even more critical to Humphrey than to Nixon. Humphrey was not only far behind Nixon in the opening polls, but trying to pick himself out of the shambles of a Democratic party that was split down the middle over the Johnson Vietnam War policy and demoralized by the spectacle of riot and bloodletting outside the police barricades that guarded the Chicago convention. And Humphrey was starting the campaign flat broke, worse than flat broke. Both the Humphrey campaign organization and the Democratic National Committee had big deficits.

If Humphrey were to break free of the Chicago miasma of division and ineffectuality, he must dispel it at the outset—by spreading around some unity money and by changing the scene through a splurge of television advertising that would unleash the "real Humphrey." He made an impressive start on the first task by having the Democratic National Committee assume the multimillion-dollar debts run up by his rivals for the nomination. That was relatively easy for one with Humphrey's ingrained nonchalance about deficits, involving only the rolling over of dubious paper into the indefinite future.* But the television splurge required hard cash—in advance—and Humphrey cast about for a place to get it in a hurry.

There was no up-front Texas money, usually a factor in Democratic campaigns, and the Humphrey people decided to go after some, for they did not concede the oil money to Nixon, not all of it anyway. Even the hind-teat money could be Humphrey's salvation. Just a loan would do. A case could be made, after all, that Oil had done very well under the Democrats. Its first great governmental bailout had come under Franklin Roosevelt, and most of its privileges had been engineered and safeguarded by that long line of Texans and Oklahomans and Louisianans whom Democratic Congresses had elevated to leadership in both the House and the Senate. Was not Humphrey now the leader of

*It would be twelve years before the Democratic National Committee paid off these debts.

that Democratic establishment? Was he not being supported by the *Texas* Democratic establishment, led by men who had long supped at Oil's table—President Lyndon B. Johnson and Governor John B. Connally? Why, even the maverick Eugene McCarthy, Humphrey's old protégé, had just been warmly received at the Houston Petroleum Club and, notwithstanding his "kids' crusade" and all it represented, had picked up $40,000 in donations. Braced by such arguments, Humphrey's fund raisers, headed by millionaire businessman Jeno Paulucci of Duluth, Minnesota, flew to Houston, to the Petroleum Club atop the Exxon U.S.A. headquarters building, to beard the oil moneymen for a quick loan of $1 million.

The Humphrey people recognized, of course—it was implicit in the nature of their request—that Hubert Humphrey could not be Oil's preferred candidate, though in office he would be chaperoned by Russell Long and Carl Albert and Hale Boggs and whispered to from beyond the pale by Sam Rayburn and Tom Connally and Bob Kerr. A Humphrey presidency must seem to oilmen a worrisome prospect, but it need not be an abhorrent one, and since it was one that might well come to pass, it warranted some insurance, a repayable investment, if you will, in tolerable relations.

The Paulucci mission was therefore shocked at the Petroleum Clubmen's disdain for complexity and subtlety where Humphrey was concerned. Acting as if Oil did not have many problems and interests at home and around the globe, the Houston oilmen reduced their future with Humphrey to one blunt question: What was Humphrey prepared to do about the oil depletion allowance? Only if he recanted his past, in the crassest and most humiliating of circumstances, would they deal with him, even in consolation money.

The extremity in which Humphrey found himself—of having either to sell out on the dotted line or to be totally estranged from one of the great blocs in the old Democratic coalition—is not satisfactorily explained by recalling that long before, as a freshman senator, he had mounted a fight against "oil privilege" that never seemed to get anywhere and had finally petered out. A national politician, especially one of Hubert Humphrey's all-embracing breadth and longevity in presidential politics, could be expected in the years since to have woven enough ambiguity into his postures so that he could not be feared as irrevocably hostile by *any* major constituency group. If circumstance prevented him from being seen as an undoubted friend, he should appear, at the worst, to offer the kind of sympathetic neutrality asked of the Lord in Alben Barkley's story about the country boy who was running along a railroad track inside a long, narrow tunnel while a train bore down on him from behind and who, as hope of escape withered, cried out, "Lord, if you can't help *me,* for God's sake don't help the train!"

But such was Humphrey's posture that he could not offer even neutrality and seemed to the oilmen committed to helping the train. Humphrey was not unaware of or unconcerned about this vulnerability. Over the years his sensitivity to it had grown apace, as necessity rubbed his nose in the realities of campaign finance and the arithmetic of electoral votes. The oil industry was the heftiest financier in politics, and though his reformist soulmates sniffed at fat cat contributors, they had as yet provided no substitute.

Beyond money, Oil delivered its own bloc of votes. There were fifteen states* in which the oil industry—with its array of entrepreneurs, shareholders, landowners large and small who had oil or hoped they did, jobbers, production and refinery and distribution employees, and satellite dependencies—could field a single-issue vote respectable enough to be a commanding consideration to politicians who lived or died by one or two percentage points. In a few of these states the oil phalanx commanded a vote formidable by any standard—in particular Texas, where oil-owning landowners alone numbered over 300,000, each of whom presumably had relatives, and Oklahoma, of which it was said by its senior senator, A. S. ("Mike") Monroney, only a bit facetiously, that he was the one Oklahoman he knew who didn't own some piece of oil. Perhaps the most telling evidence of Oil's political clout in these fifteen states is this: Of their thirty U.S. senators, twenty-nine faithfully voted the oil industry line, year in and year out; the exception that proved the rule had been Senator Paul H. Douglas of Illinois, a seminal figure in the revolt against Oil's dominance of the U.S. Congress and a key factor in the shaping of Humphrey's Houston dilemma.

The fifteen states that made up oil country—three in the South, two in the Midwest, ten in the West or Southwest—were by no means hostile territory for Democrats. Six of the states consistently elected Democrats to both of their Senate seats. Eight tended to have one

*The fifteen states, in the order of their oil production: Texas, Louisiana, California, Oklahoma, Wyoming, New Mexico, Kansas, Alaska, Mississippi, Illinois, Montana, Colorado, North Dakota, Utah, and Arkansas.

senator from each party; only one, Kansas, seemed immune to the pull of Democratic ties and blandishments. The three great presidential prizes—Illinois, Texas, and California—were winnable by either party but had lately done better by Democrats: Illinois and Texas had given their electoral votes to Kennedy in 1960 and Johnson in 1964, while California had divided—a Nixon squeaker in 1960, a Johnson landslide in 1964.

For more than ten years, as a Humphrey presidency bid had emerged, he had been edging toward a softening of the hostile image he had in oil country. First, he had ceased being the star performer in the periodic Senate debate that challenged the oil industry's tax privileges—a challenge that, with or without him, was always voted down by margins of two or three to one. Then he had stopped taking even a token part in these Senate teach-ins. But there his image change mysteriously stalled. Should his old confederates bring the issue to the Senate floor and force it to a vote—as they were wont to do every year or two—Humphrey would gamely, if glumly, honor paternity and fraternity by voting with the thirty.

This quasi demarche was too subtle to be pacifying to Oil. If Humphrey was no longer a point man against them, he was still in the reserve platoon. In his almost four years as Vice President he did not have to vote, and so offered no fresh offense, but neither had he recanted.

By the norms of his profession, Humphrey was shedding this particular skin with excessive delicacy. President Nixon would state the rule for such defenestrations when he opened a high strategy council on political-economic policy in 1971 by telling his advisers, "In this discussion, nobody is bound by past positions."

The oil depletion allowance issue, as it stood in 1968, was in some respects a textbook justification for finking—a cause that was viable only as a spoiler, too often routed to be more than a protest symbol, too arcane to rally general interest, yet pivotal to campaign finance and critical to the carrying of some battleground states.

What was it, then, that had thus far impeded Humphrey from following through on his instinct and left him frozen in the posture, unsatisfying to all, of half-completed retreat? The answer, significant not only to the outcome of the Nixon-Humphrey campaign but also to the growing polarization between Oil's critics and its defenders that by the mid-1970's would paralyze U.S. energy policy, was tangled up in two phenomena: the rugged character of one man, Paul Douglas, who over twenty years elevated oil tax reform from an obscure issue to a moral litmus test among liberals, and the propensity of Oil to spill political blood in rigid defense of its subsidies to the end that for Humphrey—and for liberals who came after him—to make practical compromise with Oil was to invite the charge of betrayal.

Humphrey had himself been the instigator of his present dilemma. Almost twenty years before, as a brand-new senator from Minnesota, chafing under the tradition of silent deference imposed upon freshmen and tempted beyond endurance by the approach of a major Senate overhaul of the tax code, Humphrey had determined to crash the debate, which would otherwise be dominated by elders he regarded as too close to the plaints of business interests and the discomforts of wealthy taxpayers. Looking around for allies, his eye lighted upon another irreverent freshman, tall, craggy, white-haired Paul Douglas of Illinois, long a renowned professor of economics at the University of Chicago, who had more than once used his analytical prowess to penetrate the collusions of subsidized business and corrupt politics. Douglas, age fifty-seven, joined up with Humphrey enthusiastically, the more so because of his admiration for the Minnesotan, who was more than twenty years his junior.

"With more physical and nervous energy than any man I have ever known in public life," Douglas wrote of the early Humphrey, "he could simultaneously push scores of measures, tend to the personal and political chores of his office, do his committee work and speak wisely and incessantly both in and out of the Senate. . . . During these years he was also the best rough-and-tumble debater in our whole political arena."

Before the tax bill reached the Senate floor, it first had to undergo the lengthy process of clearing the House of Representatives and then the Senate Finance Committee. Humphrey used the time to prepare, by staging seminars so that he and his small band of neophytes could learn enough about the tax code to stay in the ring for a few rounds against the Senate's tax heavyweights, such as the feared and caustic Eugene Milliken, the Colorado Republican, and the legendary Tom Connally, the Texas Democrat who had arranged for the federal government to enforce, under his "Hot Oil Act," the month-by-month collusion in limiting oil production by Oil-dominated bodies in the separate states so as to prevent a competitive market.

Some of Humphrey's seminars were held in secret, in order to conceal from the Senate elders the identity of the lecturers, moonlighting Treasury Department tax technicians who could point out the widest and least justifiable loopholes and how to close them. Douglas assumed the toughest task—the tax provisions for the oil industry—and in due course he and Humphrey singled out for attack the 27½ percent oil depletion allowance.

The percentage depletion allowance was Oil's creative leap beyond the hackneyed, confining principle that business, in arriving at its taxable income, may deduct its capital expenses over a fixed time period. "Depletion" opened up the exciting, unconfined universe of deductions that were not connected to expenses, that soared infinitely beyond expenses, and that, while the oil flowed on, knew no limitation of time.* Under its provisions the owner of an oil well could subtract from his income, before figuring his tax bill, 27½ percent of the value of the oil it pumped out. Put another way, percentage depletion exempted from federal taxation 27½ percent of the gross income from oil and gas each year, no matter how long the well kept pumping or how many times the investment in that well had already been recompensed by the deduction. The only limitation on it was that in any given year the owner could not write off more than half of his taxable income from the well. The way economies of scale worked out in the oil business, this limitation was so well fitted to the operations of the big companies that there was only minor overlapping. (The smaller companies found it difficult to realize more than a 16 or 18 percent depletion benefit.)

In East Texas there were wells that would be pumping 27½ percent deductible oil forty years and more before they were pumped out. For the country as a whole "oil depletion" forgave taxes that amounted to between thirteen times and nineteen times the total drilling and developmental costs. It was a prime factor behind the figure cited earlier—that the fifteen largest oil companies paid only 8 percent of their net incomes in federal taxes, in contrast with an average tax rate of 42 percent for all corporations.

When Humphrey and Douglas first took up the challenge, at midcentury, the American oil industry suffered from an excess of capital investment that had led to a good deal of uneconomic activity. Oil production from the most abundant wells was being held down to one-half and sometimes one-third of normal capacity by state-federal prorationing regulation. The oil security of the country was not in question. What was in question, to Humphrey and Douglas, was simple equity: Why—in the absence of some compelling reason of national interest—should Oil have tax privileges that other industries, even

*Oil's operating expenses were also deducted in a uniquely advantageous way.

other mineral extraction industries, did not have and be forgiven billions of dollars in taxes that had to be made up by ordinary taxpayers?

Oil had built walls around depletion that would not crumble before the trumpets of would-be reformers. The industry's search-and-destroy defense of this 1926 law was so suffocatingly effective that in a quarter century no challenge to it had ever reached the floor of the House of Representatives for a vote, and none would for another two decades. When in 1950 Humphrey and Douglas looked toward the House, whose agreement was needed for anything they were able to achieve in the Senate, they saw the squat, implacable figure of Sam Rayburn, his gavel raised against them. During his long domination over such matters *no* congressman who was not a certifiable supporter of the oil depletion allowance could hope to gain appointment to the twenty-five-member Ways and Means Committee, and until the day came when there was not just one but thirteen antidepletion votes on Ways and Means, no such bill could ever come properly before the full House.

In the Senate a comparable defense had been constructed, if one makes allowances for the limited anarchy which tradition indulges in that body. There was no way to keep a Douglas from at least raising the issue on the floor, but it could be seen to that he could not do so successfully. Many a prospective senator was signed up by the oil lobby while still a candidate in return for oil contributions to his campaign —often under the auspices of the Senate leadership and its creature the Senate Campaign Committee. (In the nature of things, only the refusals came to public attention. In the 1958 Senate campaign in Utah, for example, Democratic candidate Frank Moss was offered a $10,000 contribution, then a sizable amount for a Utah campaign and one which the hard-pressed Moss needed desperately. The catch was that he had to promise in advance to support 27½ percent depletion against any attack. Moss refused the money and voted alongside Douglas in his 1960 challenge. But many accepted.) For all the years of the Douglas challenge, Oil was consistently the foremost industrial contributor to congressional campaigns and could always produce not only its twenty-nine senators from oil states but thirty to fifty more from states where Oil was not a constituency, only a distant financier.

Rayburn's counterpart in the Senate for much of this period, Majority Leader Lyndon Johnson, maintained a cordon sanitaire around the Senate's tax-writing committee, the Finance Committee, which, though more flexible than Rayburn's, was no less effective.

For many years there was not a single opponent of "depletion" on the Senate Finance Committee, and Douglas's efforts to be the first were repeatedly rebuffed. I should here acknowledge a personal interest in the matter. At that time I was working for the late Drew Pearson, and one of my beats was Capitol Hill. Drew, from his first sighting of presidential aspirations in Lyndon Johnson, determined to make use of them to pressure Johnson—in his national radio broadcast on Sunday evenings and in the 600 papers that carried his daily column—in behalf of Drew's pet causes, one of which was tax reform. Sometimes his pressure took the carrotlike form of gushing support of Johnson, but periodically it was manifested in the stick of attack on Johnson's links to Oil. I was under standing orders to bring in items about Johnson and Oil, and the saga of Douglas's exclusion from the Finance Committee was a recurring source. One day, at a luncheon for former President Harry Truman, Johnson approached Drew and lit into him. "You forgive Hubert Humphrey when he champions the dairy people in Minnesota," said Lyndon. "You forgive Tom Hennings his faults. You forgive Kefauver his problems. But you don't forgive me mine."

"He made quite a little argument," Drew recalled, "that both he and I had the same things in common, but that he had one disadvantage in that he had to represent the oil and gas people."

Fair or not, Drew kept it up, as did columnist Doris Fleeson, who made a personal crusade of the Douglas affair. Eventually Johnson relented and named Douglas to Finance, though who can tell, out of the fathomless mix of motives, trade-offs, and pressures that characteristically moved Johnson the bargainer, which was uppermost?

Yet for years thereafter Douglas was an isolated pariah within the committee, and up to the end he had to prosecute his battle from the Senate floor, like a backbencher.

Whenever rump challenges to the Finance Committee monolith and the Senate leadership were made on the floor on this issue, if they were accorded the courtesy of a roll-call vote at all, they were annihilated. On the first antidepletion roll call achieved by Humphrey and Douglas in 1951, their amendment was routed by 80 to 10. Seven years of patient "education" and agitation passed before the next recorded vote, which won a third of the Senate—a level Douglas generally sustained but could not improve on in several attempts thereafter.

But the issue would not quite give up the ghost. After each beating it would, like a bludgeoned rattlesnake left on the ground for dead, at length begin to stir and twitch and then slink off unseen, to return another day, knitted back together and resuscitated. This durability was due to two circumstances.

The first was the slow but inexorable penetrating power of inconvenient facts: the yearly publicity from Douglas about the latest batch of millionaires who paid no income tax; the fact that the depletion allowance was granted on oil ventures abroad, which competed with domestic oil ventures for the capital supposedly loosed by depletion; the contradiction between Oil's insistence on prorationing controls, to hold down production, and on tax incentives, which were supposed to build up production, leading to the suspicion that the oilmen's real interest was in government-protected profit that was actually enervating to the industry; and the Treasury Department and academic findings that began to appear in the 1960's to the effect that 27½ percent depletion, though it was producing ever-increasing tax rebates to the oil companies—more than $1 billion a year—had ceased to have any significant effect on domestic oil exploration.

The second factor was the durable appeal of Douglas in keeping the echo of those facts and arguments reverberating in the public dialogue for almost twenty years until they became a part of liberal orthodoxy and populist demonology. Few who watched him over the years from the Senate press gallery will doubt the crucial influence of personality upon history. His visage, looming cheerfully above the wreckage of a hundred battles, was a testament to Lincoln's adage "Every man over the age of forty is responsible for his own face." The Douglas nose was bent, broken in childhood baseball. The ears were jugged, a predicament which the close cropping of his white thatch seemed designed to emphasize. The lips were often curled in a gargoyle's inward grin as he savored some intended jest. Yet the disconcerting parts yielded up a whole of appealing fineness and nobility.

The same effect was at work in his voice. Closer in its basic timbre to the high honk of his native Maine woods than to the mellow sonority that was the usual ticket to the Senate chamber, it was an instrument made pleasing to the ear by the traits, affections, enthusiasms, and sorrows that played through it: a strength, quiet and sturdy, that bespoke tested conviction; a warming strain of kindliness; a precision and refinement—of claim, word, and inflection—as if the scholar's fairness and the pilgrim's self-discipline had at length gained mastery over the revolutionary's outrage and impulse; and a buoyancy which, ever rising out of contention toward humor, seemed to affirm that the grounds for hope slightly outweighed the proof of man's greed and folly. Given time to tune up, Douglas's spare tenor would begin to swell—with the adventurer's enthusiasm for new nostrums, like the fish flour he consumed to bear witness that it could rescue the world's malnourished and didn't taste all that bad, either, or with the sharer's delight in declaiming old poems or recalling old places, where Lincoln had walked with Ann Rutledge, where Jane Addams rested. Sometimes, when it seemed that an interest-group Senate was forever deaf to argument and indifferent to injustice, anguish would well up in Douglas, reedlike, and choke off his words, but after an embarrassed pause there would return the note of equable dignity that characterized Douglas the advocate.

The tenacity he showed in his oil battles was foreshadowed in an anecdote. In World War I Douglas had been rejected by the army for poor eyesight. He had always regretted this and frequently reproached himself for not having somehow gotten past this rejection so as to serve in the defense of democracy. When the United States entered World War II, Douglas was fifty years old, with the same eyesight and a body softened by the sedentary life of a professor. He determined to fight in the war and to fight as an infantryman, a private—the most meaningful and unpretentious form of service he could imagine.

He used what influence he had to get himself accepted by the U.S. Marine Corps—a branch small enough, he hoped, to indulge his eccentricities. The corps wanted to make him an officer, but he persuaded it to take him as a private and put him through the rigorous boot training at Parris Island, South Carolina, which the white-haired recruit survived with vast pride. When at length he was made a staff officer, as befitted his abilities, he tried to spend a couple of hours each day in the front line of combat, where he would take off his insignia and fight as an infantryman. In the last year of the war the fifty-three-year-old Douglas was wounded on Okinawa. As he looked down at the wound that would hospitalize him for two years and cripple his left arm for life, "a wave of exaltation surged through me that, at my advanced age, I had shed blood in defense of my country."

During my 1965 investigation into the campaign finance abuses of Senator Thomas Dodd, the Connecticut Democrat, I got a peculiar insight into the silent spread of the Douglas influence and its portent for the future. In the midst of Dodd's 1958 run for the Senate he told his aides of an offer by oil interests that would dwarf anything else in his campaign coffers, but an offer that assumed his future support of

27½ percent depletion. The immediate problem was that under the Connecticut election laws the contribution had to be reported and would be published. After a few days of anguished teetering Dodd turned the offer down. "I can't take it," he told his men. "The press would have a picnic with it. The liberals would crucify me." Considering that Dodd's general indelicacy about political contributions led to his formal condemnation by the Senate in 1967 and to the ruin of his political career, his refusal to touch oil money, for fear of public reaction, indicated that in certain areas at least Douglas and his band were winning the battle of public opinion.

But where it immediately counted, in the Senate and in selected election campaigns, they were losing, and badly.

After more than a decade of bruising Senate defeats of his oil depletion bill, Douglas remained undaunted. "Nothing is ever settled," he would say, "until it is settled right." But he had begun to feel guilty about exposing his comrades to the retributions of the oil lobby. So in the mid-1960's Douglas contented himself with raising the issue on the floor, making the argument, and allowing it to be decided by a voice vote. But it was too late to shield his allies.

By the end of the 1960's, almost without exception, every active ally of Douglas's had been defeated, was on the verge of defeat or retirement, or had recanted. The liberal Democrat John Carroll of Colorado was defeated in 1962 at the hands of Oil-backed Peter Dominick. "Most people knew of my stand in favor of developing energy sources," Dominick later euphemized, adding that Carroll's position on depletion "galvanized the independent oil people behind me." Democrat Joseph Clark of Pennsylvania, a key partner of Douglas's in the tax loophole effort, was beaten in 1968. John Williams of Delaware, the sardonic Republican foe of tax evasion, announced his retirement. Tennessee's Albert Gore, Douglas's right arm in the debate after Humphrey became mute, was on his last political legs and would be defeated in 1970 by the Oil-blessed Republican William Brock.

Moreover, Oil need not deal only in fear. It could exhibit for purposes of suasion a number of edifying specimens captured from liberal ranks —a reformed Douglasite, a liberal professor and counterculture hero, and an authentic populist: Frank Moss, the liberal Democrat from Utah who had rejected the tainted contribution from oilmen and who in 1960 had cast his first depletion vote on the side of Douglas, had thereafter seen the light and voted consistently with the oil bloc. Eugene

McCarthy of Minnesota, whom Douglas had welcomed to the Senate and to his cause as a "young Samson," turned up inexplicably absent on his first depletion roll call, and in the years thereafter he voted with Oil. Fred Harris, an authentic radical-at-large in most things, minced in tamed lockstep with his Oklahoma predecessors, Bob Kerr and Mike Monroney, when it came to the depletion allowance. "I rise to speak for my state and its interests," intoned Harris in 1967 in defense of oil depletion, "as does the Senator from Wisconsin [Proxmire] when he talks about the interests of the dairy industry." It was only after he gave up hope of running for reelection that Harris switched to the antidepletion side.

The most instructive trophy of all for Oil was the 1966 defeat of Douglas himself, at age seventy-four, in his bid for a fourth term. There were other factors, but the oil lobby played its part and loomed large in Douglas's recollection:

> This issue [depletion] was especially important in the southeastern part of Illinois, where there was an entire Congressional district largely dependent on the oil industry. . . . Local Democratic candidates for Congress were terrified by my argument and tried to make it clear that they did not agree. They wanted my help but hoped I would be silent on oil. The big companies circularized their stockholders and dealers over the state, while large amounts of money poured in at election time. . . . As I look back on it, it seems a miracle that I lasted as long as I did.

Among the Senate agitators for the oil tax reform, only William Proxmire—hard for Oil to get at in his Wisconsin redoubt—survived. But Proxmire, for all his intelligence and perseverance, was then known as a gadfly and a loner in the Senate, not one to build and sustain a coalition that the oil barons need fear.

When Douglas left the Senate in January 1967, oil tax reform seemed a hopeless cause, even to him: "We were always beaten. . . . It seemed impossible for flesh and blood to break the hold of the massed billions of economic power." Not only was the issue dead, but its drawn-out death throes seem not to have attracted the slightest notice. Rowland Evans and Robert Novak could write that, as of the end of 1968, "there had been no political interest whatever in tax reform for five years."

Such was the history that Vice President Humphrey had to repudiate if he were to make truce with an oil lobby that, having so painstakingly exterminated the antidepletion cause in Congress, was not about to risk its resurrection in the White House.

The Humphrey finance mission had come to Houston in the hope that Oil, with its worldwide interests and concerns, had a political perspective broader than the single issue of the tax depletion allowance.

But, as has been seen, the oilmen coldly responded that before they considered any aid to Humphrey, they wanted to know his position on the 27½ percent oil depletion allowance, which Nixon was promising to preserve intact. Humphrey had been willing to soft-pedal his past, but he would not disown it. He instructed his people to make no guarantee on oil depletion. Paulucci told the oilmen, "The Vice President won't promise you a thing." The Petroleum Clubmen were not bluffing. "Oil won't give us a dime," Paulucci reported back to Humphrey.

Whether or not that early burst of television exposure to the "real Humphrey" and the other preliminary campaign essentials for which the Oil loan was sought could have rescued Humphrey's campaign from what he called the "shipwreck" of the Chicago convention and given it the 1 percent boost that would have turned defeat to victory cannot be definitively known, of course. What *is* measurable is the degree of paralysis inflicted on the Humphrey campaign during most of its ten-week life for the want of an early million dollars, and the 15-point half time lead built up by a Nixon campaign that outspent Humphrey's 10 to 1 in the first half and would outspend it 4 to 1 overall, and the way Humphrey began to gain dramatically as soon as he *did* get some money.

"It's not the amount of money you get," Humphrey later told reporters; "it's when you get it." On the day of his nomination, August 28, the Humphrey camp commissioned elaborate polls in all regions of the country to identify where and why Humphrey was weak and strong, the absolute prerequisite to any coherent master plan of issue making, advertising, and personal scheduling. But when the polls were done, Humphrey did not yet have the money to pay the pollsters—mere thousands—and some of them withheld their results. Ads were prepared, which the professionals regarded as highly effective, but they could not be printed or aired for lack of money.

Not until September 30, when the Gallup Poll showed Nixon leading

Humphrey 43 to 28, with Wallace at 21, did the Humphrey campaign get together the $100,000 needed for its first national telecast, which began to turn the campaign around. Not until October 10 was the first million raised. Not until October 24, when the campaign was two months old with less than two weeks to run, was the Humphrey campaign able to schedule a single advertising spot on national television or radio. Meanwhile, the Nixon campaign had spent and scheduled to be spent $10 million worth of media advertising and an equal amount for other promotions.

As mid-October drew near, Humphrey's gifts as a campaigner, a growing trickle of contributions, and the gradual healing process within the Democratic party began to close the gap. By the home stretch Humphrey was coming up fast, closing on Nixon by 3 points a week in the polls. By the last week Nixon's Gallup lead had shrunk to 2 points, and the Harris Poll had it "a dead heat." Cartoonists, who had been depicting a woebegone Humphrey swamped in futility, now fashioned a heroic figure, a foot from the finish line, carrying all alone the dead weight of his bickering party.

It was for just this unhappy circumstance that Richard Nixon had periodically warned his campaign managers to keep in hand a large reserve of cash. He was not again to be "outtelevised" in the last week of the campaign, and he deluged the air waves in an effort to flood out and stall the Humphrey momentum. Humphrey, with a half million dollars suddenly in hand for television spots in crucial California, could not place them; all the available air time was already booked by others. To prevent any last-minute backsliding of oil money and oil votes to Humphrey, Nixon made a maximum personal effort. Though a half dozen critical regions were begging for his visit in the last five days of the campaign, Nixon spent two of them stumping the heart of oil country, warning Texas crowds and those reached by the media spillover into adjacent New Mexico and Oklahoma that their oil tax write-offs were at stake. "As he campaigned in Austin and El Paso today," reported a November 2 dispatch in *The New York Times,* "as in Fort Worth, Lubbock and San Antonio yesterday . . . he promised to retain the 27½ percent oil depletion allowance."

The success of Nixon's campaign for Oil's support was overwhelming. It revealed a striking difference between the general electorate and the oil voters and contributors. Among the electorate at large, Nixon squeaked past Humphrey by less than 1 percentage point—43.4 percent of the vote for Nixon to 42.7 percent for Humphrey. (Wallace got 13.5

percent.) But of the fifteen "oil states," Humphrey lost fourteen.* In 1960 Kennedy had divided the oil state electoral vote evenly with Nixon. In 1964 Lyndon Johnson had won 140 of the 157 electoral votes in oil country. In 1968 Humphrey got only 25.

The Nixon effort to corner Oil's campaign fund and freeze out Humphrey from obtaining any significant hedge money, early or late, was an even more spectacular success. In oil contributions of $10,000 or more, Nixon outdrew Humphrey by 15 to 1. In other contributions recorded by the American Petroleum Institute, Nixon led Humphrey by 14 to 1. In contributions reported by the Independent Petroleum Association, it was 30 to 1. While these ratios may be regarded as accurate, the total amount of money given by Oil is only partially known. In *reported* contributions Nixon received $1.3 million to Humphrey's $78,000.

The future exacts a price for such successes. The day was not far off when the problems of oil exploration and production would require objective, sympathetic national understanding but would face a public that equated Oil with dishonesty and a Congress in which the embittered forces long repressed by Oil were suddenly ascendant. Moreover, Oil had given its money and votes to Richard Nixon too cheaply. Instead of getting a President who after twenty-one years understood and empathized with its functions and problems as the premier industry, and the stake of the Western world in them, it had in seven campaigns trained a Nixon who thought of Oil only in terms of its subsidies and who, so long as he upheld his bargain to preserve its *subsidies,* would feel free to subordinate its *functions* to the claims of yet other lobbies and interests which were ever beckoning.

*The curious exception was Texas, where about 600,000 Wallace votes cut into Nixon's oil vote enough to allow Humphrey—backed by the Johnson-Connally Democratic establishment and a rising minority population—to edge Nixon by 39,000 votes out of 3 million cast, though 60 percent of Texas voted against Humphrey. Wallace won in two of the fifteen states, Arkansas and Louisiana.

Chapter Three

The Inheritance and the Prodigals

The federal government owns most of the fuels that are likely to be economically recoverable for the foreseeable future. The federal government's gross failure to exercise its responsibility as the proprietor of these fuels, even to know what it owns, and to devise speedy methods of bringing these fuels to market without ruining the earth is a major cause of our energy crisis.

—S. David Freeman, 1974
Director of the Energy Policy Staff
of the Office of Science and Technology
in the Nixon White House

In January 1969, when the incoming Nixon Republicans and the tenacious congressional Democrats assembled in Washington to take up their respective dominions, the pivotal calamity of our time lay almost five years into the future. Whatever may have been the hazards confronting the country at that hour, what we have since come to call the energy crisis was not one of them.

Taken in the aggregate, the combined recoverable energy resources of the United States were greater by several times than any conceivable demand that could be made on them for the next fifty years. Taken individually, each of the principal resources—crude oil, natural gas, and coal—was sufficient to meet the foreseeable demands on it far enough into the future to permit the nation's planners the luxury of ample time and flexibility in making the adjustments that would keep America the master of its energy destiny—whether the nation chose the path of quasi self-sufficiency or the path of a modest reliance on cheap foreign oil imports, modest enough to avoid either a dangerous dependence or a tightening of the world energy market, the chronic oversupply of which had for a generation meant price stability.

There were but two small conditions impinging on this vista of prolonged plenty. Industry had to "do its thing" in developing and producing the U.S. resources known to be recoverable, if not with éclat, then at least with run-of-the-mill regularity. And government occasionally had to exhibit some minimal level of awareness and cohesion, so that if one or another of its policies was obviously bollixing up energy production in any important way or encouraging overuse of one fuel or underuse of another, it would be somewhere taken note of and minor adjustments would be made to clear the blockage or right the balance.

So overwhelming were our natural resources, however, so many our alternatives if production of one or another source faltered temporarily, that U.S. energy was impervious even to frequent high-level mistakes; only an unbroken series of blunders, persisted in over a protracted period, could shrink this bulging overabundance into inadequacy.

Lest this assertion of big-shouldered plenty provoke disbelief, let us take a brief inventory of the three major energy resources as they stood at the beginning of 1969.

The United States was the world's greatest single producer of crude oil, greater by far than any other country, greater even than any continent, region, or bloc of nations save the Middle East, the oil resources of which were in substantial measure American-developed and American-owned. Though also the world's biggest oil *consumer,* it was capable of producing just about as much as it used. And its production was still rising, the peak years yet ahead. During the preceding decade daily output of crude oil had risen by 30 percent, though, for reasons revolving around a long-standing world oil glut, we were not really trying very hard to find oil in the United States and were, through various state regulatory bodies, blocking the production of about 3 million barrels a day that were pumpable. Even so, 60,000 American wells were producing 10 million barrels a day, up from the 7 million range of a decade before and headed toward an assured 11 million in 1970. For various reasons of convenience—a reluctance to incur the expenses of transporting southwestern oil to remote New England, or to permit the building of odious refineries, or to bother with refining the low-profit "residual oil" used in industrial boilers—the nation imported 19 percent of its oil consumption. Since oil provided about 40 percent of total energy use, we were thus depending on imported oil for less than 8 percent of our energy needs. And since our oil imports were at about the same volume as the "shut-in" production of our half-idle wells that would be available if really needed, we were, in fact, roughly self-sufficient.

In natural gas, too, the United States was the world's greatest producer, furnishing 50 percent more than all the rest of the world put together. Providing 36 percent of all U.S. energy, natural gas was riding the crest of history's most prodigious energy boom. In the previous twenty years it had accounted for half of all the energy growth in the

United States; in the decade before 1969 gas production had almost doubled, though here, too, the industry was only half trying—deliberately slowing development and holding huge quantities off the interstate market while it battled the Federal Power Commission over price control decisions that typically took eight or ten years to clear. Notwithstanding the red-tape constrictions, natural gas output kept jolting upward on its previous momentum; in the five years before 1969 annual production had climbed from 16 trillion cubic feet to 22 trillion, and it would go on reaching new annual highs throughout the Nixon term. Proved and probable reserves—the two surefire categories that left out a huge amount of natural gas that would certainly be discovered in the future—totaled 547 trillion cubic feet at the end of 1968, about a thirty-year supply at the 1968 consumption rate. (Nine years later the CIA, a respected energy prognosticator, would place the longevity of our natural gas reserves, at 1977 consumption, at between fifty and sixty years.)

As for coal, the United States was the world's greatest repository, containing about one-third of the earth's known supplies. Our actual production of coal was shrunken by the difficulty in *selling* it in a nation awash with oil and gas that were cheaper, or easier to use, or easier on the environment. Even so, America produced more coal than any other nation, or any group of nations, in the free world. Its greatest value now was as the guarantor of American energy security. With identified reserves of at least 200 *billion* tons that were recoverable by commercially feasible means and at the 1968 consumption rate of about 500 million tons, the United States could quadruple its coal use and still have a 100-year supply—and its *likely* recoverable reserves, considering the probable discoveries and mining techniques of the future, were several times as great as the *known* recoverables.

Oil imports, later so debilitating an addiction, were in January 1969 a manageable and wholesome affair. As we have seen, they amounted to less than one-fifth of our oil consumption and were more a convenience for special situations than a necessity. Most of these imports came from secure sources in our own hemisphere; only 4 percent of our crude oil came from the Arab world in 1968, and that figure would drop to 3 percent in 1970.

The energy abundance of the United States was reinforced by the resources of our neighbors in the Western Hemisphere, resources those neighbors were panting to develop and sell to us in far greater volume than we had use for. North America alone possessed half of all the world's recoverable energy reserves; Latin America contained more crude oil than North America.

Canada of late had been exploring for oil more aggressively than the United States, and if its upward push were to continue, it had to have access to an expanding foreign market. Canadians expected that market to be the United States. Thus far in the 1960's their oil exports to the Americans had quadrupled, from 123,000 barrels a day in 1960 to 500,000 in 1968, and they were prepared to triple that to 1.5 million barrels a day in the early 1970's. But since the American market for foreign crude did not rise appreciably in the 1960's, Canadian gains were coming at the expense of South America, mainly Venezuela, whose exports to the United States had been stagnating. Like Canada, Venezuela sought a long-term pact that would assure a steadily rising level of oil sales to the United States. As 1969 opened, Latin America as a whole was able and eager to sell America a million barrels a day more than it could absorb. And the gigantic Mexican oil boom was yet to come.

The excess oil capacities on which both North and South America floated were surpluses within a larger glut that enveloped the world and would grow throughout the presidential term, so that by mid-1972, according to the impeccable M. A. Adelman of MIT, "excess producing capacity . . . was almost universal."

A dozen oil-producing countries in Africa, the Middle East, and Asia were seeking to sell their unwanted oil to the United States. Anxieties were heightening which would before long impel the two giants among them—Saudi Arabia and Iran—to propose long-range "special relationships" with the United States designed to forge permanent linkages beyond the oil trade and, in return, guarantee to America all the crude oil it was willing to take for decades at stable prices.

That it was an energy utopia is further demonstrated—and with a hurtful pungency—by the prices then charged for fuel. On the world market the price of oil had been in a gradual decline for twenty-two years, so that by 1969 the average world price, in constant dollars, had dropped 65 percent below the 1947 level; crude could be had for as low as $1 a barrel in the Persian Gulf, one-fortieth of the price the same barrel would command in the same market ten years later.

In the United States the price of oil had been traditionally kept well above the world price, thanks to the arrangements between industry and government by which domestic production was held down and foreign imports were limited to a volume that would not undermine the domestic price. Nonetheless, the American price could not wholly resist the downward trend worldwide. In January 1969 the price of oil was 10 percent lower than it had been eleven years earlier, compared with the price of other products at wholesale. A barrel of crude oil that cost $3.07 in 1950 cost only $2.88 twenty years later. With a forty-two gallon barrel of crude hovering at around $3 and the cost of refining it only $1, a gallon of gasoline purchased at the refinery in 1969 cost 10 cents; purchased at the filling station gas pump, with taxes added, it cost 35 cents.

So it was, too, with the other basic fuels—natural gas and coal. At the end of the 1960's natural gas was cheaper than it had been in 1965 and only a penny more per thousand cubic feet than in 1960. A ton of coal sold for $4.65 in 1969 (at the mine), scarcely a nickel more than the price of twenty years before.

For the better part of a generation, then, the peoples of the world had been enabled to build on a foundation of low-cost energy, the supply of which expanded to meet any need and which could be counted on by planners and builders to be as cheap tomorrow as it had been yesterday or ten years before—and probably cheaper. Upon this base of easy, confident access to the stuff that turns the wheels of industry, convenience, and pleasure, there was created in the 1950's and 1960's the greatest and most widely spread improvement in living standards ever experienced.

The American economy of the sixties seemed to have broken free of the old cyclical curses and to have at last found the formula for that stable, ever-increasing prosperity glimpsed by Abraham Lincoln a century before, "whose course shall be onward and upward, and which, while the earth endures, shall not pass away." Western Europe and Japan, miraculously risen from the ashes, were hot on our trail, their masses aspiring to, and beginning to realize, that enlargement of material life and expansion of personal horizons heretofore identified with Americans. In the underdeveloped world, where it had seemed through long ages that nothing much could be done, the worldwide glut of low-cost fuels had made the beginnings of general progress a visible reality for numberless millions—Punjabi farmers putting aside their ancient incantations in favor of oil-driven pumps which brought up from deep wells the water that was banishing drought, Malay women braving their fears of the new to flock to paying jobs in electronic plants, Kenyans aspiring to be mechanics when the new GM truck factory opened in Nairobi. It was not unduly sentimental to say (though today it appears so) that in half a hundred emerging societies in Africa, Asia, and Latin America the millennial victims of history at last had their feet upon the first rungs of the magic ladder, their eyes on the stars of hope.

An illustration in the *Business Week* issue that appeared during the fortnight of the Nixon inauguration captured America's enviable oil position in a world of overproduction—a nation sitting on its own 25 percent excess capacity, preparing perplexedly for a looming flood of new oil which was welling up in Alaska, beseeched on the north by Canada and to the south by Venezuela to admit ever larger imports on a permanent basis, surrounded by siren-singing salesmen dressed in Arab burnooses who offered oil at well below the American price, while U.S. oilmen struggled feverishly to hold in place the barriers that kept home production down and imports out.

If America wanted to buy judiciously, it was a buyer's market, and we were everyone's favorite customer, in a position to sew up a future of oil security and price stability not only for ourselves but for the oil-consuming world.

This is not to say that there were no baby serpents wriggling about in the energy paradise of the late 1960's. Decay and change are rules of nature, and it follows that any optimum situation will be nibbled away at by erosions and challenges. But as erosions and challenges go, the energy problems of 1969 were of a rather agreeable sort, the downside of overabundance and oversolicitude for particular groups of consumers and stockholders.

Too much fuel was the problem, and naturally enough, it was slowing down the search for more, especially in higher-cost areas like North America; too cheap oil and gas were prompting wasteful use and causing a careless abandonment of coal; too many low-cost oil bonanzas around the globe were inviting the outmigration of American investment; too low wellhead prices for natural gas were saving consumer pennies at the cost of discouraging future production; a too sympathetic indulgence by government toward the prerogatives of the international oil companies was permitting, indeed encouraging, them to shape a world oil order that in some respects was more in their corporate interest than in the national interest; the too easy satisfaction of energy demands was creating a complacent public and officialdom which, assuming energy to come painlessly from gas pump or wall socket rather than disruptively from the torn bowels of the earth, were becoming intolerant of such ugly necessities as mines, refineries, pipeline construction, coal smog, and offshore drilling mishaps.

There was no reason, however, for these embryonic dangers to multiply unchecked, for they could not easily escape the attention of policy makers. The essential facts about energy were monitored too remorselessly for that. The many segments of the industry, and the many government agencies that watched over them, kept count of a vast array of minutiae, an accounting that constituted an immense early warning system for the emergence of untoward developments. To be sure, thirty-year projections of supply and demand were notorious for errors, for the far futures of dynamic economies and unexplored continents will not be divined; moreover, the industry was known for its purposeful obfuscations in narrow areas and for its concealments of profit-sensitive information. Even so, the data base for projections of half a decade was more than accurate enough for the ball park precisions that suffice for statesmen, and the astonishing variety and depth of reliable information served up every month by energy statisticians were ample for the spotting of incipient catastrophes. In the main, the barrel of oil produced or kept in the ground, the gas tract leased or withheld from leasing, the small coal mine or stripper well opened or shut down, the energy dollar invested in the United States or shifted abroad, the balance between the amount of oil or gas discovered in a given year and the amount consumed, between the amount imported and the amount produced at home, between our projected appetites and the world's capacity to satisfy them—all was kept track of, worked into tables, graphed in every conceivable juxtaposition, and formulated into exhaustive series of best- and worst-case scenarios.

As 1969 progressed, the negative fallout from plenty showed up in a number of small curiosities scattered among these data. Investment for developing oil and gas in the United States was slipping. At the end of 1968 it still stood at an all-time peak, but suddenly we crossed a divide, and the graphs showed that 1969 would be the first year not to show an increase over the previous year. Over at the Federal Power Commission, a study confirmed that American companies had been sharply expanding their explorations abroad while slacking off at home. At the Department of the Interior the final 1968 figures for domestic oil and natural gas production, exploration, and consumption were in: Production had set another record in 1968 (and would do so again in 1969 and 1970); still, it was the first year that growth in newly proved oil reserves was smaller than growth in consumption—for the first time we were using it up faster than we were discovering it. And a close look at consumption revealed some minor upspurts that were running ahead of past projections: In 1968 public utilities and industrial users had replaced 7 million tons of coal with foreign oil, and the rate of annual increase in gasoline consumption, which between 1960 and 1965 had been 2.8 percent, was lately accelerating at twice that level, mostly because cars were getting heavier, bigger-engined, and more accessory-laden.

Flickers of concern over a few sour straws found in so large and sweet a bale were tempered by this consideration: Their origin lay in man-made decisions, which were easily altered, not in any basic lack in nature or inferiority in geopolitics.

Oil production and exploration, for instance, could be raised substantially, at will. Various state governments—Texas, Louisiana, New Mexico, Oklahoma—were avowedly and with the most delicate calculation holding down national crude oil production by about 25 percent, through prorationing, which set an allowable limit on individual wells so as to keep surpluses down and prices up; in a typical year less than half the oil that was pumpable from the larger wells was being pumped. This had the side effect of retarding oil exploration in the United States by the big companies, for why go to the expense of finding and bringing in more wells when the ones you already had were turned off by law more than half the time? Another disincentive to U.S. production that could be readily changed was the tax provision that made it more profitable for U.S. firms to find and produce oil in the Middle East than at home by permitting them to deduct a dollar from their U.S. taxes for every dollar they paid to foreign treasuries for oil in the guise of taxes. Why produce oil here that raised taxes when you could produce oil abroad that lowered your taxes?

U.S. natural gas development could be easily made far more productive, for it was being arbitrarily held down by policy decisions of another kind. The pricing of natural gas—a crucial factor in the development of any resource—had fallen into the hands of the courts, the bureaucracies, and the Congress, a trio from whose amiable ministrations no productive enterprise emerges unmangled. Prices set low enough to please voters and public tribunes are not likely to be high enough to please investors and owners of land containing oil and gas, so more and more moneymen and landowners had been squatting on their assets in the 1960's, holding out for higher prices. But as old price decisions had turned off the valve, so new price decisions could turn it back on. The immutable factor, *the resource,* was there, in unimaginable profusion.

Even the upward trends in gasoline consumption by motorists and in oil use by utilities switching from coal were reversible, if they got out of hand, by public decisions about car size and oil imports.

There were a few puffs of cloud forming on the energy horizon that were beyond the ken of petroleum economists and energy statisticians. A growing crusade to protect the environment from the ravages of industrialization was beginning to grind out challenges that would soon involve several layers of government in deciding such questions as whether oil exploration, low-sulfur coal mining, or the emerging network of nuclear power plants would go forward on a schedule that would keep domestic energy production reasonably abreast of rising demand and hold imports at their current safe level. Here, too, there was no valid reason for protracted impasses. Substantial concessions to environmental concerns could be made by the producing interests without endangering their enterprises, and ordinary political leadership should be able to get these obstacles resolved in time to meet the energy needs of half a decade hence.

Fortunately these first faint wrinkles in the fabric of energy abundance were being identified years before they could pose real embarrassment to our economic aplomb and, so long as they were taken seriously, should be relatively simple to iron out, for no fundamental physical scarcity was involved. Whenever a dip in production or an unexpected surge in consumption or a falloff in exploration or a flight of investment capital or a patently unrealistic price appeared on the charts, it was usually caused by some past political intervention, intended originally to help some producing or consuming interest, but now in need of adjustment. This had the happy faculty of placing the cure within handy reach for, as compared to overcoming something really intractable, like a raw materials shortage, to amend a mere regulation or close an old loophole ought to be doable.

The energy inheritance of the newly elected President was enhanced by two spectacular additions, even as he campaigned for the office, additions so vast that they would seem to have solved in advance any critical oil problem that could arise in the Nixon era, abroad as well as at home, for it was axiomatic that there could be no oil crisis abroad as long as America remained in surplus.

The first of these happenings began to take shape in January 1968, up near the end of the world at an ice-clad inlet on Alaska's Arctic shore known to a few adventurers and map fanciers as Prudhoe Bay. For months a small band of oil drillers had clung to life and sanity there under the buffetings of the earth's bitterest of winters. Winter had whirled in on Alaska's North Slope in October, and at its approach the great inland herds of caribou had departed on their seasonal migration, "moving across the tundra in a living tide." The Barren Ground grizzlies, after a season of fattening up, disappeared into hibernation. Millions of waterfowl had filled the sky, winging south, diminishing with each flight the oilmen's sense of connection to the world of the living. Even the sun abandoned them to months of perpetual night. To the south, lakes in the Brooks Range were solid ice to the bottom and 1,000-foot waterfalls froze in midfall.

Through the dark months they had drilled, sometimes in temperatures of sixty-five degrees below zero, sometimes in blinding, snow-throwing winds called whiteouts, gouging down into ground so hard that often the bit had to be hauled up and changed after only a few feet of gain, down to depths of 4,000 feet, 6,000, 8,000.

They stayed with it because it was the last gamble for the Atlantic Richfield Company on the North Slope. Years of expensive exploration and drilling had produced nothing. In the previous year Arco had spent $4.5 million drilling a dry hole at a site far more promising than this one, near the Sagavanirktok River. But it is in the nature of big losers to throw good money after bad in a last chance to recoup. This was it.

At 8,500 feet the men at "Prudhoe Bay State Number One" struck a flare-out of natural gas; at 9,000 feet the core samples they brought up showed the sand to be saturated with oil! But how much? It would have to be an "elephant" to justify the costs of operating in the Arctic. So in March another well was drilled seven miles away to see if the pool was as large as its Triassic formation indicated it might be. In June the famed Dallas consulting firm of DeGolyer and MacNaughton made the announcement that would set off the Great Alaska Oil Rush: "This important discovery could develop into a field with recoverable reserves of some five to ten *billion* barrels of oil, which would rate as one of the largest petroleum accumulations known to the world today."

Only a few weeks after the oily sand came up at Prudhoe, a second vast option opened in California—as though the providential cycle, which brought forth new cornucopias for America just as they were needed, were not to end after all. On February 6, 1968, the Interior

Department opened sealed bids for exploration and development rights to seventy-one tracts in the Santa Barbara Channel and realized almost two-thirds of a billion dollars—"the highest return to the government for any lease sale in the history of the continental shelf."

The combined message of the Johnson administration's auction and the oil industry's response was that the long neglect of federal offshore oil had ended and that a major effort to develop the outer continental shelf was under way. Interior Secretary Stewart L. Udall hailed the auction as a national response to political turmoil in the Middle East and to "a realization that new domestic reserves must be developed to assure uninterrupted supplies for our burgeoning energy market."

The importance of either one of these new oil sources, had its development been pushed forward on schedule, defies exaggeration. There is widespread agreement among oil authorities that the addition to American oil production of 1.5 to 2 million barrels a day by 1973 would have prevented the great oil price explosion of that year. (And had the 1973 crisis been avoided, ignited as it was by a concatenation of one-of-a-kind circumstances, it is not likely to have recurred in anything resembling the same form, for there was no true basis for a quadrupling of prices either in oil statistics or in the power relationships among nations. And certainly the accidental and culminating price explosion of 1979, which we shall see to have been entirely a fluke, could not have occurred on anything like the same scale without its 1973 predecessor, which created all the key conditions that made 1979 possible.)

Behind this judgment are reasons of history, psychology, economics, and simple mathematics which we have already touched upon and shall return to frequently hereafter. The skeleton is this: The condition of the world oil market, pre-1974, was such that, though 50 million barrels a day might be consumed, the addition or subtraction of only 1 or 2 million from the market made the difference between the psychology of glut—timid, price-cutting sellers—and the psychology of shortage—arrogant, extortionate sellers. When the oil supply in question was that of the United States, the rule had especial force. With 2 million barrels a day more in 1973 the United States would have required *no* Middle East imports, causing a 1.5-million-barrel-a-day surplus in Middle East production and giving the United States the capacity to divert another half million to potential trouble spots. Moreover, it was the cardinal fact of that period of oil history that a United States with no dependency on Arab oil was an uninhibited superpower sure to move forcefully to break up any Middle Eastern cabal against oil stability.

After the OPEC nations' climactic victory of 1973, which promptly magnified their revenues manyfold, a whole new book of rules would apply. They could thenceforth afford to cut their own oil output to offset gains in non-OPEC production, for instance, from Alaska or the North Sea, as well as to offset declines in Western *consumption.* After 1973 the power of the West to apply the pressures of the market would be diluted enormously until such time as profligacy worked its magic and OPEC's spending caught up with its revenues and financial reserves. Before 1973, oil glut meant a price retreat toward $1 dollar a barrel while economies boomed throughout the West; after 1973, oil glut would mean a limited fallback from $12 a barrel, and then from $40, as OPEC held most of its price gains by halving its oil production against a background of 20 to 35 million unemployed in the Western industrial nations.

Thus, up until the events of December 1973 forever truncated the future from the past, any great infusion of new oil such as Prudhoe Bay or the American offshore offered, or any equivalent cut in oil consumption by conservation or substitution, would have prolonged the stability of the past indefinitely. A dozen such rescues would offer themselves between 1969 and 1973; Alaska and offshore were only the first.

There was a prerequisite to the orderly carrying through of both the Alaskan and the offshore plans for American oil sufficiency—cooperation from the U.S. government and, in particular, its Department of the Interior. The oil under Prudhoe Bay would never be produced unless it could be shipped to market; since Alaska's northern shore was frozen-in most of the year, that meant a pipeline had to be constructed across Alaska, and since the federal government owned 96 percent of Alaska, a pipeline to anywhere would need a permit from the Department of the Interior. Similarly, all oil development of the outer continental shelf was dependent on the department's leasing policy, its timetables, its regulations.

So it was that when a new President was elected in November 1968, his selection of a secretary of the interior was awaited with anxiety, and not just by those who had a personal or public interest in oil development. Conservationist opponents watched, too, with a scrutiny even more intense and with far greater awareness that the stage was being set for a political conflict of historic proportions. In the days before the Nixon inauguration Senator Gaylord Nelson (D-Wis.), a member of the Senate Interior Committee, spoke to this conflict and of the central rule to be played in it by the new interior secretary:

> There are a number of very, very tough confrontations that are going to occur, that are coming on us now . . . involving industry and other parts of the private sector pitted against government and conservationists and scientists and thoughtful citizens who are concerned about the degradation of the environment. There is in my judgment no way to compromise; that is, you either fight the battle and lose it or you win it, but there is not really a half-way house. The Interior Department is going to be deeply involved, directly or indirectly, in most if not all of these confrontations which are on the horizon.

Whether the new President proved as astute as Gaylord Nelson in recognizing that a watershed confrontation was at hand, whether he proved wiser than Nelson in seeing that a compromise *must* be found, whether he proved skillful enough in maneuver to establish a halfway house in which the needs of both resource development *and* environment could be reconciled would one day count more heavily than Vietnam or Watergate in fixing Richard Nixon's place in history. The first test would be his choice of an interior secretary.

When President-elect Nixon introduced to a national television audience the "men of that extra dimension" who would constitute his Cabinet, few of the viewers south of the forty-eighth parallel had ever heard of Wally Hickel—governor of Alaska, forty-nine years old, affable-appearing, even handsome in a stocky way—the nominee for secretary of the interior. "You can see that his eyes are looking over the horizon to things unseen," Nixon said of Hickel. "He's going to bring a new sense of excitement, a new sense of creativity, to that department."

Walter J. Hickel *did* have an extra dimension, all right, but there was a question whether that dimension jibed with the tenor of the times. He was a battler and a gambler better suited for an earlier, more robust era. The third oldest of ten children of a tenant farm family on the Kansas flatlands, schooled no farther than the high school at Claflin, where he attained, by his own account, "something less than a 'C' average," he first dreamed of escape, bucks, and glory in the boxing ring; he was the Golden Gloves welterweight champion of Kansas before he decided, at nineteen, that there must be a better route to fame and fortune. He chose Alaska, arriving there with 37 cents in his pocket and a knack for building houses and selling them, which he soon parlayed into multimillion-dollar holdings in residential developments, shopping centers, and hotels.

He retained the romantic's curiosity and the boomer's optimism that pointed to opportunity beyond every fjord, so he gambled on a bewildering profusion of dog-and-cat ventures, some of them in mining, oil leasing, and land speculation. Later this would arouse fevered suspicions that he was secretly in bed with this or that special interest, but what invariably panned out was only another glimpse into the psyche of the incorrigible plunger.

Nor did Hickel, on his quick eclectic run to riches, forsake the quest for fame and glory—or what passes for them in an unheroic age. He was a leader in the local epic—the fight for Alaskan statehood, a builder of his state's Republican party and, as national committeeman, its spokesman abroad, a victor in the electoral jousts, consequently governor of Alaska at forty-seven, and now, at midpoint in his term, a nominee to the President's Cabinet. En route he had gained a reputation that surfaced in the media accounts following his nomination—brash, aggressive, confident, outspoken, can-do. All in all, the classic American success story, complete with frontier backdrop.

The trouble was that by 1969 the classic American success story had lost its charm for many, particularly in circles that had influence on the confirmation of an interior secretary. To those familiar with the unhappy annals of business manipulation of government, the self-made millionaire as nominee—especially so ebullient a specimen as Hickel—invited suspicion and demanded investigation. To the legions to whom the "winning of the West" had come to mean the brutal despoliation of the Indian, the confirmation of a frontier speculator who dreamed of shopping centers where Eskimos had for centuries baited their traps presented an unhoped-for chance to ride against General Custer. And to conservationists, lately a potent political force, the song of Wally Hickel was a cacophony of alarming dissonances: the screech of factories, railroads, and mines, not the rustle of green pastures and the murmur of still waters; the *conquest* of nature, instead of its preservation; the celebration of decisiveness, when what was wanted was humble hesitancy before nature's complexity. Most to the point, his nomination seemed to personalize the inevitable confrontation that was coming over the future of Alaska. Hickel was known for his "obsession with progress"; as businessman and politician he had been a prophet of "develop or bust," of changing Alaska into a modern industrial and commercial society, whereas most conservationists wanted to keep Alaska as it was—America's "last great unspoiled wilderness."

President-elect Nixon understood the likelihood that new nominees, suddenly set upon by questioning reporters, will say the wrong things. Immediately after unveiling his Cabinet choices, he issued a friendly but pointed warning to them, for their good and his, to keep their mouths shut until they took office, except to answer senators' questions during the confirmation hearings. Any nominee who needlessly created a controversy for himself or the incoming administration "will not be in the Cabinet." If Nixon had stuck to his guns, the energy history of the 1970's might have been very different.

Within a fortnight of the announcement of his nomination, Governor Hickel, instead of lying low until safely confirmed, had made a large target of himself by setting off a series of minifurors. The brash outspokenness that had served him well enough on a parochial stage did not "play" on the national stage.

In an interview with the *Anchorage News* he seemed to say that he would use the Interior post to push through his program for the development of Alaska and was taking the job primarily with that purpose in mind. On alighting from a plane in Seattle at 3:00 A.M., Hickel was informed by a reporter that the departing interior secretary had just announced a freeze on all commercial and industrial activity within Alaska's federal domain—aimed particularly at oil development—until the century-old dispute over Eskimo and Indian land claims was resolved. Hickel shot back: "What Udall can do by executive order I can undo," a guileless warning to both Native sympathizers and conservationists to gird for battle. In Washington Hickel called a press conference at which he offhandedly disparaged existing Interior policy as one of "conservation for conservation's sake." The phrase was a red flag to all committed conservationists. On its face it was a rejection of the conservationist *raison d'être;* for fifty years it had been used by opponents of all the major conservation measures. The incident assured the instant retrieval of a Hickel statement of a decade earlier to *Time* magazine: "We're trying to make a Fifth Avenue out of the tundra," a boosterism almost bloodcurdling to those who knew that tundra—the thin, tender vegetative cover that insulates the Arctic earth in warm weather and keeps its large ice component from melting and causing general disintegration of the terrain—is so delicate that, to use a favorite illustration, the erosive scar caused by one truck that had driven over it twenty-five years before was still visible and unhealed.

My partner Drew Pearson now led the attack on Hickel with daily blasts. The *New York Daily News* wisecracked, "Foot-in-mouth disease has broken out in Nixon's Cabinet . . . before it can be sworn in." *The New York Times* editorialized, "President-elect Nixon's choice for Secretary of Interior has confirmed the worst fears. . . ." An unprecedented volume of wires and letters opposing Hickel's confirmation poured in on the Senate.

From the start, then, it was clear that in choosing Hickel, the President-elect had not grasped what Senator Nelson and others did: that the next interior secretary was bound to be at the center of a convulsive confrontation between the needs of development and the needs of environment. Nixon had chosen neither a reconciling figure who inspired *some* confidence in both camps nor a bland enigma who aroused dread in neither, but rather a clumsy provocateur who created and forearmed enemies while dismaying friends. Later on Nixon insiders would reveal how far removed the Hickel selection criteria were from any consideration of the significance of the post. The initial choice had been Rogers C. B. Morton, a congressman from Maryland's Eastern Shore. Savvy, likable, popular in Congress, conversant with national issues and interest group politics, Morton possessed considerable qualifications. But when all the Cabinet choices had been made, it was discovered with hack alarm that not one of them had the extra dimension of being a westerner. There must be someone from a western state in the Cabinet, Nixon insisted. Morton was yanked as interior secretary and named Republican national chairman instead; Hickel, governor of the westernmost state, replaced him. Outside of geography, why Hickel? He was favorably known to Nixon for three things. He and his Alaska delegation had been solidly for Nixon's nomination as presidential candidate; he had made himself available to the Nixon for President campaign as one of the "surrogate candidates"—that is, barnstormers who were booked into second- or third-level speaking engagements—and a possible third factor has been suggested by John Ehrlichman:

> Early in the 1968 campaign I was on a United Airlines plane returning to Seattle from a Chicago political strategy meeting while Governor Walter Hickel of Alaska was up in the first-class section returning home from the same meeting. After the meal had been served, Hickel began "working" the coach section of the

plane, slowly moving down the aisle shaking hands and talking to every passenger. He introduced himself, said a few words of praise for Richard Nixon and gave out Nixon buttons to anyone who would take one. By the end of the trip he'd shaken every hand on the plane.

When I reached Seattle I phoned Bob Haldeman to tell him about Hickel's performance, and Bob passed along my description to Richard Nixon. I don't know what impact that episode actually had on Hickel's political career, but I'm willing to bet it had some effect.

By the time Hickel's confirmation hearings opened, the process known as pig piling was at work. Each day a new accusation exploded in the press or in Congress—concealment of oil lease holdings, personal conflicts of interest, unfit appointees in Juneau, pollution atrocities, victimization of Natives and of the environment to enrich business allies. In the main, these charges were ill-founded, overblown, or oversimplified, but as they piled up, they created around Hickel an atmosphere of vulnerability.

During several mid-January days in the witness chair, Hickel fended off his senatorial inquisitors, led by Henry Jackson, Frank Church, George McGovern, Lee Metcalf, and Gaylord Nelson, like a wounded grizzly set upon by courteous but tenacious wolves.

Had he not, as governor, permitted the U.S. Plywood Champion Company to cut 8 million board feet of lumber and to set up its mills in disregard of environmental standards? He had not. "That contract was made with the Federal Government. . . . I have my ideas what to do but they won't listen . . . we were not involved in that at all."

Was he not at this very moment allowing the Collier Carbon and Chemical Company to dump three and a half tons of ammonia into Cook Inlet every day? He was not. "I just want to assure you, Senator, that this is not true . . . the plant has never been in operation."

But each time Hickel swatted down a new charge, he would be hit with an old quote. How could the governor who had for years made a career of denouncing Interior Department policies to protect Native lands and wilderness areas, the governor who had said to the federal government, "We can't live within your laws, go fly a kite," now be made the guardian of those policies, the enforcer of those laws? The disputes on which Governor Hickel was impaled day after day were the furniture and the backdrop used to set the stage for a power play that would have significant repercussions on U.S. oil security in the 1970's. Hickel's defense had been stout enough, the calumnies against him outrageous enough, the "positive" aspects of his governorship numerous enough that, though his confirmation was clearly in trouble, opponents could not be sure the Senate would in the end refuse it, so strong was the tradition that a President was entitled to his Cabinet choice except in the most extraordinary circumstance. A surer course was to try to force Hickel, in his extremity, to give guarantees on key issues, to trade away vital decision-making powers of his office, as the price of confirmation.

This the embattled Hickel began to do. To Senator Metcalf he promised not to permit oil development in an Arctic fish and game area the senator was concerned about, "unless I am instructed by Congress or by members of this Committee that I should do that." Naturally enough, Metcalf was back in a few minutes for another delegation of executive power, this time in the form of a committee veto over the pending Interior Department decision on the transfer of 3 million acres of federal land to the state of Alaska. Hickel promptly acceded: "Senator, you have made a good point. . . . If there is any conflict whatever, or if you think there is any, I will bring this specific case to this Committee *and let them make the decision* [italics added]."

In such an atmosphere the committee Democrats day by day orchestrated mounting pressure on Republican Hickel to surrender, by giving them a veto over, his power as interior secretary to alter the Alaska policy which the departing Democratic secretary had imposed by decree in his final days in office. That policy was the freeze—a ban on all transactions of any moment throughout 96 percent of Alaska until resolution of the Native claims issue, which might be a matter of years and which risked—no, assured—that the claims bill itself would be held hostage to conservationist and other demands. The committee Democrats wanted the Udall freeze maintained until such time as they themselves agreed to its lifting, partial or entire, an enormous and extralegal transfer of power. It would give them a lock on all significant executive decisions regarding Alaska, it would safeguard their varied interests in wilderness and wildlife preserves, it would give them power of approval over all individual development enterprises, and it would guarantee them a free hand to resolve the multibillion-dollar Native claims settlement in their own way and at their own pace, assured that no property in federal Alaska could change hands or change character while they dickered and awarded. Hickel's vulnerability offered them the opportunity to gain all this on the cheap, by a cozy agreement that did not entail the trouble of getting a law passed or the dilution of sharing these prerogatives with the entire Congress. The brash Hickel boast—"What Udall can do by executive order I can undo"—had come home to roost.

Hickel at first tried to evade commitment because he knew the dimensions of his abdication here: He would be surrendering the historic functions of the executive branch to decide whether or when the transfer of federal land to Alaska under the Statehood Act could go forward, whether or when any mineral or commercial or public development could take place anywhere in this vast area one-fifth the size of the lower forty-eight states, whether or when a pipeline could be built that would make usable the 10 billion barrels of oil now known to be waiting under Prudhoe Bay.

The pros went to work on a beleaguered office seeker unaccountably left to his own resorts by the Nixon high command, whose power it was that was being bartered away.

Senator McGovern had just concluded a long softening-up exercise, dragging Hickel through the errors of his yesterday's testimony about the Kuskokwim co-op, when he changed pace and turned to the Alaska land freeze:

SENATOR MCGOVERN: Let me just ask you what I think is a key question on this, Governor. If you are confirmed as Secretary of the Interior, is it your present inclination to lift the land freeze?

GOVERNOR HICKEL: Senator . . . it won't be lifted unless we collectively with the native people can agree on a plan by which they are protected, and I think we can be sitting around in a room and discussing how we might do it.

The committee majority was not to be diverted by such Pablum, particularly Pablum that had ignored a committee role, so Senator Jackson, with all his power as chairman, moved in on Hickel, while Ted Stevens, the new Republican senator from Alaska, at his first hearing, was caught between his fears of his seniors and his fear of what Hickel might trade away.

SENATOR JACKSON: Governor Hickel, I think Senator McGovern raises a very good point . . . I would just hope that as Secretary you would keep this freeze on until we resolve the legislative adjudication of the Alaska native problem. . . . How do you feel about my comments?

SENATOR STEVENS: Mr. Chairman.

SENATOR JACKSON: Senator, we are trying to be fair. I am talking to the Governor now. Can't he respond and then I will be glad—

SENATOR STEVENS: Very well.

GOVERNOR HICKEL: Mr. Chairman . . . I would give your suggestions to me the deepest and most thoughtful consideration . . . I will promise you to give you probably the benefit of what you say, because I value your opinion so much, and I will do that.

Jackson, in his day, had toughed out many a committee room extortion and was not to be fobbed off by general expressions of esteem. He had begun to refer ominously to "the feeling of the Committee." His voice and demeanor hardened, and his request assumed the character of a precise demand.

SENATOR JACKSON: You can give us the assurance that at least for the next Congress, that is this year and next year . . . that we maintain the status quo so that we can legislate without any unnecessary feeling of compulsion or acting in undue haste. . . . I think if you could give that assurance, Governor, I do not think it would be difficult for you to do it, we could legislate I think in a very fine climate. . . .

SENATOR STEVENS: Mr. Chairman.

SENATOR JACKSON: Let me get this assurance, Senator. We have been very liberal with you. You are just a new man on the committee, and I am trying to be as fair as I can, but I do want to get the questions from the witness and then I will yield.

GOVERNOR HICKEL: Senator, I would hope you would not make it an unequivocal yes or no, for the simple reason that problems arise. . . . One example is the Barrow Airport. . . . You would stop even those simple constructions if I said, "Yes, the freeze will stay on" unequivocally. There are areas where you really cannot stop the progress of the state. . . .

SENATOR JACKSON: I would agree to that kind of exemption, any proper *public* use that must be met, but I am talking about *private development* [e.g., an oil lease auction or an oil pipeline; italics added].

GOVERNOR HICKEL: I would abide by your wishes and I will do so.

Hickel has insisted to me that he made these decisions on his own, without permission from the President-elect or his staff. This was the point, then, at which Nixon had ample and obvious cause to withdraw a nomination that was from the start an ill-considered mistake. His nominee had turned out to be a bogeyman to influential interests, and to appease them, Hickel was giving up powers of decision without which the administration could not effectively pursue its responsibilities for ordering energy development. One can imagine the ire of a Franklin Roosevelt or a Lyndon Johnson at this. But Nixon saw the matter in the light of the surface political appearance of the moment, not the long-term payout. A slip-up in the Cabinet confirmation process would slightly mar inauguration week and give the press a Nixon embarrassment to chirp about. So he tolerated Hickel's performance.

Scoop Jackson was not the sort of general to defeat a foe and let him slip away in the night and regroup. On the basis of his colloquy with Hickel, he drew up an ironclad agreement and required Hickel's assent to it at the following day's hearing. The Udall freeze was to be inviolable for at least two years, unless a Native claims bill was passed earlier. The only exceptions that could even be considered during that period would be matters of public necessity, such as the repair of a washed-out road or the reestablishment of a Native village wiped out by a flood, and even in these cases the interior secretary would have to apply to the committee for its formal concurrence in "modifying" the freeze. By himself the secretary of the interior could not so much as okay the setting up of a portable toilet anywhere in federal Alaska, his former fiefdom.

Thoroughly whipped by now, Hickel abandoned the forceful opposition to the freeze he had repeatedly stated during the hearings and abjectly submitted: "I have no disagreement. I think it's a wonderful position to take and I support the chairman."

Even in his acceptance of capitulation, Jackson emphasized the severity of his terms, so there could be no misunderstanding of them: "Well, that is fine. . . . It does mean during this interim the postponement of leasing and opening of the federal domain in those areas that may have great promise of development. But I think this is the reasonable way of proceeding and I want to thank you, Governor, for that understanding that we now have."

The date was January 17, 1969. Three days before it took office, the Nixon administration had gained an interior secretary but had lost the authority to act decisively in assuring that the oil potential of Prudhoe Bay and Alaska's other Prudhoe Bays became realities. The Alaska option was not dead, but henceforth its life would hang upon the mercies of governmental authority unnaturally divided and informally spread among different groups of diverse purposes.

Five days after a chastened Wally Hickel, confirmed at last, escaped from the hot seat in his new garb as a born-again conservationist, he was thrust back on it by an explosive event that his critics held up as a test of the bona fides of his conversion.

On January 28 oil from a Union Oil Company well in the Santa Barbara Channel started leaking from the sea floor. Soon it had collected into a large slick which made its way toward the beaches of the famed resort area five and one-half miles distant. In due course the beachfront was fouled by a thick black goo. Thousands of birds were trapped in it, and the anguishing sight of their death throes became a nightly drama on 80 million television sets.

The new interior secretary flew to the scene. Quite properly he called for a temporary shutdown by the six oil companies operating in the channel, and on his return to Washington, in concert with the President, he set his department to work on tighter drilling regulations, which were soon ready.

But Hickel faced a mounting contention that offshore oil drilling was *inherently* unsafe and should be shut down permanently. This demand came not only from conservationists and a demonstrative segment of the public but also from within the administration itself. John Whitaker, by profession a geologist, by avocation a longtime Nixon campaign aide, at the time a White House staffer and one day to be undersecretary of the interior under Nixon and Ford, recalls the dispute: "As the immediate problems subsided, a sort of schizophrenia on Santa Barbara developed within the administration. The environmentalist sympathizers wanted the leases rescinded and a permanent ban placed on exploration and production. Those with an eye to energy shortages wanted to resume production." There can be little doubt on which side the old Hickel would have stood, the Hickel who had arrived in Washington saying things like "We can never say that we are not going to use or develop our resources." But the evolving Hickel was mindful of his confirmation ordeal; the means of his rescue from it had left their mark on him.

An interior secretary in calmer straits than those of mid-transformation to a new identity, an administration that understood the premier economic problem of its term and therefore sought a balance between the need for oil and the need for environmental protection would have faced the agitation with a sense of perspective and not taken drastic permanent action before the facts were in. Up until the Santa Barbara blowout, there had been only one spill of consequence during two decades in which thousands of wells had been drilled off the coasts of Florida, Louisiana, Texas, California, Oregon, and Washington. The claims of vast and permanent damage did not square with the accumulating evidence. In four months the beaches were clean. The channel fish catch was greater in a six-month period after the spill than during a comparable period the year before. Scientific studies would show that damage to the biota was not permanent, that damage to sea life was not widespread, that the bird population had not been decimated, that there were no deaths among seals or sea lions, that animals were reproducing with their accustomed felicity, and that the area was experiencing what promised to be a complete environmental recovery.

An independent survey conducted by forty scientists, sponsored by the Alan Hancock Foundation of the University of Southern California, would conclude: "It is not surprising that the studies after the Santa Barbara oil spill revealed such a small amount of damage."*

But instead of waiting for the studies of the channel, Hickel responded to the tumult, to worst-possible-case scenarios. On April 1 he made permanent his temporary shutdown order on sixty-seven of the

*The report recognized that, in this case as in all cases, one could imagine circumstances that might pose a hypothetical danger: "*Recurrent* spills at *frequent intervals* would *probably* result in large ecological changes."

seventy-one leases in the Santa Barbara Channel. Thereafter, with the full concurrence of the President, he moved doggedly to rescind these leases. (When the time came for the White House ceremony celebrating the cancellation of the Santa Barbara leases, Hickel was disinvited, and Nixon took the bows alone.) Nixon aide Whitaker deplores both the decision and the motive behind it:

> Because of the overwhelming popularity of the environmental issue, the administration had elected to abandon an area of potentially significant oil production. . . . What the administration should have done, after tightening drilling regulations . . . was to complete immediately an environmental impact statement and, if the study showed the environmental costs to be acceptable, to allow leaseholders to resume exploration and production. Instead, an environmental impact statement on the Santa Barbara Channel was not completed until the spring of 1976.

As if to emphasize the finality of its decision against offshore development, the administration rejected pleas that it undertake a program of environmental assessment of the frontier areas of the Atlantic and the Gulf of Alaska to locate the safest places for a revival of offshore oil.

These acts and omissions erased the offshore oil option from the national agenda for the decisive period. More than two years would pass before the administration even *sought* to revive the preliminaries of an offshore leasing program. By that time, other Nixon decisions had made offshore oil so vulnerable to litigation and procedural blockage (see chapter 5) as to kill it effectively for an entire decade.

Thus, before the new administration had held office for ten weeks, in its thrashing about to quell small fires it had managed to destroy the offshore option and to encumber the still-bright Alaskan option in a procedural quagmire of "freeze" and divided authority. If this signified merely a clumsy but proper sensitivity to environmental problems which would be compensated for by a strong administration drive to speed along an environmentally sound Alaskan oil pipeline or an intensified attention to the many energy alternatives that were compatible with sound environmentalism, all might yet be well. But if it signified, instead, that the clear but silent needs of energy security five years hence were to be subordinated willy-nilly to the uproars and political advantages of the here and now, shipwreck lay ahead.

* * *

"What might have been" if oil development had been allowed to resume off the shore of Southern California after the drilling regulations were tightened in 1969, instead of being hamstrung through the 1970's, is revealed by what has transpired since some of the wraps were taken off in 1981.

In May 1981 the Interior Department held a lease auction on eighty-one tracts in the Santa Barbara Basin, an area about forty miles west of Santa Barbara, which attracted $2 billion in high bids. Within eighteen months several oil strikes were made, constituting two giant discoveries, one by the Socal/Phillips group, the other by Texaco, which, according to *The New York Times,* "inspired visions of an oil pool beneath the sea bottom rivaling the one found beneath Alaska's Prudhoe Bay."

As of this writing, the exploration of the area is still in its beginning stage, but the initial discoveries alone are expected to result in production of more than 200,000 barrels a day by around 1985, and there is a growing consensus among experts that impending discoveries will double or triple that figure—in an area that represents only a small pocket of the vast outer continental shelf potential.

Time has not sustained the charges of inherent environmental incompatibility that stymied offshore oil development for twelve critical years. Where offshore drilling survived, no major oil spill has occurred since 1969. Assistant Secretary of the Interior J. Robinson West declared in August 1982 that some $250 million had been spent on studying the effects of oil spills and other environmental hazards attributable to offshore operations and that no significant impact on the marine or coastal environments had been found.

That this assessment is fundamentally sound can be seen in the muted rejoinders of once-apocalyptic critics. "We don't believe the effects of oil pollution, especially the subtle long-term effects, have been sufficiently studied," says Frances Beinicke, spokesperson for a group called the Natural Resources Defense Council. John Zierold, chief California lobbyist for the Sierra Club, has adopted a cautious stance: "We have to await the results of some tests. We're not going to shoot from the hip on this one." Barry Schuyler, an environmental studies professor at the University of California at Santa Barbara, says, "People are concerned, but the feeling is different now. People used to drive to meetings in a gasoline-powered car and say, 'Let's get all the oil out';

now they are saying, 'If it's going to happen, let's do it reasonably.' "

Thus far there have been no environmentalist efforts to block the resumed drilling, and the state of California is on the verge of approving drilling in the close-to-shore area that is within its three-mile jurisdiction.

But too late. Had what is now being done been allowed to go forward in 1969, instead of being interrupted by Nixon and Hickel, the progress achievable by 1973 could have averted that year's oil cataclysm. Instead, under that new book of rules effective after 1973, U.S. offshore oil will merely help us to live with a fundamental disaster that need not have occurred.

Chapter Four

The Rejection of Self-Renewal

What were they doing in 1969 when U. S. reserves of crude oil dropped by one billion barrels? . . . We have been standing still while our oil reserves decline, while our oil companies put their resources into developing foreign reserves. . . .

—Philip Sporn, 1974
Member, National Academy of Sciences
and National Academy of Engineers

From 1950 through 1970, U.S. crude oil production increased at a 2.9 percent compound annual rate. But from 1970 through 1975, the United States recorded a 2.8 percent annual decline rate. . . . This shift from expansion to decline by the premier world producer of crude oil is an important factor in explaining the approximately fourfold increase in the price of crude oil from 1970 to 1975.

—Walter J. Mead, 1978
Professor of economics,
University of California at Santa Barbara

The oil barons greeted the election of Richard Nixon with the controlled jubilation of inside investors who have just learned that twenty years of covert plunging is about to pay off. Michael L. Haider celebrated with the careful blandness befitting a chairman of the Standard Oil Company of New Jersey (soon to change its name to Exxon): "Under Mr. Nixon, we can look for improvement toward a balanced economy. Under a more balanced economy, business will prosper. If business prospers, the oil industry will prosper." Lesser oil chiefs were not so diplomatic and got right down to essentials. "One of the things we knew before the election," said W. W. Keeler, chairman of the Phillips Petroleum Company, "was that Mr. Nixon has always been for protecting the depletion allowance." This calculation was confidently endorsed by Continental Oil's president, A. W. Tarkington: "Mr. Nixon's record is clear in his support of the depletion allowance, and it will survive the 91st Congress."

When oilmen argued the case for the depletion allowance, it was usually in terms of freeing investment capital for finding the domestic oil that would "keep America secure." But the industry's postelection assessments showed instead its expectation that the new President would give them a free hand in using tax savings for *foreign* development—by repealing the restrictions recently imposed by Lyndon Johnson in order to shore up the country's lagging trade balance. "Oilmen generally believe," reported *The Wall Street Journal,* "Mr. Nixon will junk the Democratic program of investment controls abroad. 'These restrictions have been holding us back,' says one major oil company executive, 'and Mr. Nixon surely will scrap them.' "

O. P. Thomas, president of the Sinclair Oil Corporation, rejoiced in having a President who understood where the national interest lay: "Mr. Nixon recognizes the fact that an extended period of suppressed foreign investment will work in the long term as a detriment to the country."

That the industry counted on enjoying the best of all worlds from its patient nurturing of the Nixon career can be seen from this expectation: Not only would Nixon preserve their American tax write-offs and permit them to be transferred abroad, without restriction, for oil development, but he would also see to it that the oil developed abroad must be *sold abroad* and not returned to the United States by arriviste oil mavericks in amounts that would undermine the domestic price structure which government-industry arrangements kept 40 percent above the world price. "We hope we will have a period when we won't have all this chipping away at the import program as in the past," said J. E. Chester, president of the Cities Service Company. To Standard Oil's Haider, Nixon had come along in the nick of time: "The oil import program can remain intact if we don't grant any further exceptions."

San Diego producer O. B. Armour summed up the postelection élan of oilmen, as they raised their glasses in exclusive clubs in Houston, Dallas, New York, and London: "There is no question about it, we will fare better under Mr. Nixon."

That the preservation of their tax and price subsidies was still the central political objective of American oilmen as 1969 opened reveals their innocence of a dawning era in which it was their basic functions —to explore, drill, transport, refine, distribute, enforce contracts, control volumes and prices—that would regularly fall hostage to political decisions. But within their frame of reference the immediate steps taken by Richard Nixon to redeem his campaign promises were highly reassuring.

There were three matters pending before the new President that the American oil establishment regarded as tests of his fealty. First, they faced an attempt to resurrect the tax reform issue in the new Congress; the Johnson Treasury Department, needing the support of congressional liberals to pass the 1968 Vietnam War surtax, had offered as ransom an official study of tax loopholes, along with recommendations for closing those found unjustified, and Chairman Wilbur Mills of the House Ways and Means Committee had sweetened the offer with a cryptic promise to hold some kind of hearing when the Treasury study emerged. Secondly, Johnson was dumping in Nixon's lap the problem of a politically explosive application from oil maverick Armand Hammer for permission to import 300,000 barrels a day of Libyan crude on the condition that Hammer build a refinery for it at Machiasport, Maine, and sell heating oil to New Englanders at 10 percent below the going price—a frontal challenge to the quota system that protected domestic oil prices, couched in terms that roused consumers against the oil companies. Thirdly, President Johnson, in order to quell skepticism about his Oil links, had delegated to the secretary of the interior the President's authority to set oil import quotas and to decide *who got them;* the oil establishment wanted that authority back in the White House, directly reachable by *its* influence, not in the Interior Department, where it was subject to falling under the sway of bureaucratic procedures that could become legal precedents.

Nixon acted promptly and pleasingly on these matters. As early as a month before his inauguration he moved to squelch any revival of tax reform. In mid-December, Chairman Mills had called on the President-elect and his advisers at the Pierre Hotel "transition headquarters" in New York City to learn their wishes on his committee schedule for 1969. Among the lesser items discussed was the just released tax reform study of the outgoing Treasury Department; Mills came away from the Pierre satisfied that there was no Nixon interest whatever in tax reform, and on his return to Washington Mills scratched it from his agenda. (As one Nixon aide put it at about that time: "On any realistic list of our priorities, I would think that tax reform would be pretty close to dead last.")

And by the time President Nixon had been in office a month, he had snatched back from the Interior Department all powers of decision over oil imports and had consigned Armand Hammer's Machiasport scheme to a limbo from which it would never return by creating a Cabinet task force to undertake a study of all questions pertinent to oil import policy, which had the effect of putting all those questions on ice indefinitely.

In the field of foreign affairs, too, Nixon's early undertakings boded well to the international oilmen. Shortly after taking office, he received a delegation of them—which included the heads of Mobil, Gulf, and Arco, Oil's lawyer-in-chief, John McCloy, and its most prominent financier, David Rockefeller. Their problem was the mounting hostility of the Arab oil states to the United States because of its seeming identification with Israel; their solution was to placate the Arabs, to stop being pro-Israel. To the oilmen's delight, Nixon, flanked by counselor Henry Kissinger, radiated empathy and made the pointed comment that one group he didn't owe a thing to for his election was the Jewish voters.

The oilmen were temporarily pleased with this President who openly discussed oil security policy in terms of election payoffs, but they would soon enough find this propensity to be a two-edged sword.

Richard Nixon had scarcely taken possession of the White House when it was swamped with angry mail demanding, of all things, tax reform! The proximate cause was the simultaneous explosion of two delayed-action bombs left behind by Lyndon Johnson. Citizens at work on their income tax returns in mid-January were enraged to stumble over, at the end of their labors, that 7½ percent Vietnam War surtax, which had been timed so as to make its public impact after the election instead of before it, and while they simmered, a focus for their frustration materialized when the Treasury study of tax loopholes, which Johnson had initially intercepted and hushed up during the election campaign, finally surfaced as the Johnson administration was leaving town.

On his last working day, January 17, 1969, Treasury Secretary Joseph Barr appeared before the Joint Economic Committee of Congress, then a powerless study group, to present the gist of the tax reform study. Barr condemned loopholes long sacred which let "high-income recipients . . . pay little or no income tax" and cost $50 billion a year in lost revenue that had to be made up from ordinary taxpayers. "There is going to be a taxpayers' revolt in this country if we don't do something about it," Barr declaimed, and his punch line caught the media's fancy. The "taxpayers' revolt" of 1969 was on.

The White House was only one target for irate mail denouncing tax scams. Over at Treasury, more mail was received on tax abuses in the weeks following Barr's testimony than in all the previous year. The Ways and Means Committee was bombarded by more thousands of fuming citizens, prompting its ranking Republican, Representative John Byrnes of Wisconsin, to issue a demand on January 30 for tax reform and a challenge to the new team at Treasury to get cracking on it. Wilbur Mills, no doubt grousing cavernously about the inscrutability of the American voter, snatched back the lead by scheduling committee hearings to begin on February 18.

The announcement of hearings propelled a question to the White House: What would its posture be? Treasury Undersecretary Charls Walker urged the White House to get aboard rather than be run over. Nixon agreed, though it required an about-face. On February 3, conferring with Mills and Byrnes at the White House, he said he now favored tax reform and would be sending the committee a Nixon tax reform package, though it would take the new team some time to put it together. Within the week he put out a public announcement to that effect, silent on details but loud in staking his claim to the issue.

As the hoped-to-be leader of the tax reform cause, Nixon faced a bit of a dilemma. To the public, the most conspicuous abuse needing reform was the oil depletion allowance, the very "loophole" Nixon was sworn to preserve and had, in fact, made a career motif out of swearing to preserve. The contradiction did not faze him. He simply declared oil depletion not to be a loophole. There must be plenty of other things to reform.

One by one, the oil chiefs debouched from their corporate jets at Washington's Page Terminal—the president of the Gulf Oil Corporation, the chairman of the Humble Oil and Refining Company, independent operators from Texas, Oklahoma, Ohio, Pennsylvania, Michigan —for their appearances at Wilbur Mills's tax hearings. Theoretically billions were at stake for them, but they were inflated with the unclouded exhilaration of participants in a fixed fight. Every so many years they gathered for these ritualistic exercises at which their arguments before the high bank of deferential faces always won admiring approval. They could remember Sam Rayburn casting off his aloof austerity to escort their spokesman of the year, General Ernest Thompson of the Texas Railroad Commission, to and from the hearing room.

True, the press carried tales of "tax revolt," but the revolt was not against *them.* Some of their colleagues—for instance, John Shaheen—had been assured by Nixon himself of his total commitment to oil depletion, an assurance confirmed by their observers over at Treasury, who reported strenuous activity on tax reform proposals but none of it directed at any of Oil's major concerns. *Business Week* magazine captured the ambience: "Ranged before the House Ways and Means Committee, the oilmen comfortably expected that their traditional arguments would get traditionally friendly reception on Capitol Hill. Never in the 43 years since the depletion allowance was first written into the tax code had the Ways and Means Committee made the slightest threatening gesture toward it."

The oilmen could remember having to postpone their testimony one time while thirty-three congressmen came on, one after the other, to testify on Oil's side, some of them commissioned to speak for entire state delegations. Another year there were twenty-eight. Not one congressman had shown up on the other side. In fact, so far as the House was concerned, there *was* no other side.

Reveries of yesteryear's glories suddenly became a cause of disquiet. This year only six members of Congress appeared to testify in Oil's behalf, and—something unheard of—five stepped forward to *oppose* oil depletion. When the executives took their turn at the witness table, they noticed a lot of new faces on the panel and an unbecoming note of distanced objectivity coming from some old standbys. Committeemen who had never before thrown a fastball now felt constrained to ask them real questions. How is it, Representative John Byrnes wanted to know, that Atlantic Richfield had had a book income of over $400 million between 1962 and 1968 yet had paid no income taxes? The industry spokesmen politely evaded the question with such certified tactics as: "I can't speak conclusively to this particular subject on some of these corporate structures; I only know how it affects me personally."

But Byrnes did not show his usual understanding: "It is incumbent upon representatives of the oil industry, who appear here . . . that they show us how these results occur and *justify* them." Byrnes was worried about those angry letters, and the oilmen were not helping him answer them. "Frankly, I no longer know what to write to my constituents," he said. "They write to me that they have read some place that an oil company made $450 million in four or five years and that the company didn't pay a cent in taxes to Uncle Sam."

It was an indication of the pampered insulation Oil had long enjoyed that such plaintive maunderings from troubled friends seemed hateful and disloyal. "I don't think I have ever been so startled," said the vice-president of one of the majors. "There was Johnny Byrnes acting just like Bill Proxmire."

The oilmen sought reassurances from Wilbur Mills, but another rung splintered beneath them. "Depletion has become a symbol," Mills told them. The people were demanding tax reform, and if the committee's bill were to have credibility in the House, it had to make some reduction in the oil depletion allowance. Mills's advice to Oil was to cooperate in a mild reduction lest a fight precipitate a far harsher cut. Frank Ikard, who seven years before had left his seat on the Ways and Means Committee to become Oil's Washington spokesman as head of the

American Petroleum Institute, acknowledged "a sharp change in the political environment facing the oil industry."

Since the early 1960's—with the deaths of Sam Rayburn and Bob Kerr and with Lyndon Johnson's forsaking of the Senate to pursue a constituency wider than Texas—the political machinery of oil supremacy had not received its methodical maintenance and had fallen into disrepair. "Never again will we put our trust in a few congressional leaders," vowed one oil lobbyist, adding a rich exhibit to the "what-have-you-done-for-me-lately" annals.

Eighteen of the twenty-five members of the 1969 Ways and Means Committee had joined it after Rayburn's death; that meant they had not been rigorously screened for loyalty to 27½ percent depletion. Demographic changes in the country meant that more and more congressmen had to pay court to encroaching suburban constituencies, among which it was not a plus to be seen as pro-Oil on taxes. Mills was right. Oil depletion *had* become a symbol. Or, as a less tactful committee member put it, failure to reform oil depletion would "stick out like a sore thumb."

Such comments were testimony to the impact of the battle begun by Hubert Humphrey, kept alive by Paul Douglas, and still waged by William Proxmire. "Several of the people [on the Ways and Means Committee] that I interviewed," wrote Professor Bruce Oppenheimer of Brandeis University, "credited Paul Douglas and later William Proxmire with keeping the issue alive for years when there was no chance of changing depletion. Their doing so, the respondents felt, had educated others in the Congress and in the press to the inequities of the depletion allowance. Even if people did not understand the depletion allowance, many could identify it as a tax break for the oil industry."

Oil was now paying for its many past steamrollerings of a responsible opposition that would have accepted compromise but wouldn't just quit. Overkill had built up over decades, a visceral animosity when a statesmanlike compromise would have lanced the boil. The animosity was intensified by the frustration of East Coast consumers with the import barriers to foreign oil, the uproar of beachgoers everywhere over the Santa Barbara oil spill. "It's a new group down there," said one lobbyist, of the Capitol Hill scene, "and they couldn't care less about the oil industry."

The oil industry, from the great majors down to the wildcatters, had heretofore been united in a monolithic phalanx for its confrontations with government. For years Douglas had tried to break it up by persuading the smaller companies that their interests diverged from Big Oil's. Why support 27½ percent depletion, he had argued, when the economics of Small Oil dictated it could never qualify for more than 16 or 18 percent? Why not serve the public interest and their own, too, by supporting a reduction in excessive benefits for Big Oil, perhaps in exchange for an easing of their own taxes? But Small Oil would never budge, feeling that there was strength in solidarity and that since they often acted as, in effect, subcontractors for Big Oil or as "pilot fish," finding deposits which the majors would then take over from them, anything that took money away from Big Oil would probably take it away from them.

But suddenly the pressure of opposition from without was causing cracks to appear in the unity of Oil. If there was a real danger of tax changes, various groups within the oil family, fearful of being damaged in a general debacle, would try to extricate themselves and let the devil take the big companies. President Nixon had received a backsliding letter from Nelson A. Steed, president of the Texas Independent Producers and Royalty Owners Association, saying that if oil had to be the object of tax reform, "some reduction in the 27½ percent depletion factor might be maintained without irreparable injury." And word was making the rounds that the Kansas Independent Oil and Gas Association, which claimed 1,400 members, was going to support Senator Proxmire's proposal to cut the oil depletion benefits for the big companies while leaving them untouched so far as smaller companies were concerned.

In the month that Richard Nixon took office, it was revealed that U.S. crude oil production *capacity* was down from the level of the previous year by 234,000 barrels a day. Here was one of those definitive shocks, like a first slight heart attack, which casts all that lies ahead in a new light. As this drop was the consequence of a two-year decline in proved reserves, it was bound to repeat each year thereafter unless the annual slippage in reserves was turned around. Without that turnaround, by 1974 the nation's productive capacity would be down 1.5 million barrels a day below the level of the year that had just ended. "Capacity" is the operative word here. Production itself, instead of falling, could continue to edge upward in 1969 and 1970 and then remain almost stationary for the next few years, but only through the expedients of (1) draining the excess capacity of the shut-in wells that had previously constituted the nation's short-term oil security and (2) accelerating the stripping of the proved reserves that had constituted our long-term security.

The news of falling capacity was promptly put into grimmer perspective by news of rising consumption. Extending the 1960's oil consumption trend to the 1970's, the American Petroleum Institute reported that there would be a yearly rise in demand of between 400,000 and 600,000 barrels a day. Alaska's Senator Ted Stevens pointed out in a Senate speech the elementary math here: Even at the lower figure the nation's oil consumption by 1974 would be up 2 million barrels a day over current demand. The spreading gap to be created by both falling production and rising consumption, if permitted to develop, must obviously be filled by imports, and unless national oil policy made some modest adjustments to alter the two trends, the import level must rise over the coming five years by more than 3 million barrels a day. Here was a leap from safety to peril—from a safe 3-million-barrel-a-day-import level, 85 percent of it coming from neighbors in the Western Hemisphere, to a 6-million-barrel-a-day-dependence, 2 million barrels of which must come daily from the Middle East, forbidden alike by public rhetoric and secret study. The putative danger warned of by Senator Stevens would accelerate if one used the high 600,000 figure in the API projection. The average yearly increase in oil consumption in the 1963–68 period had been 427,000 barrels a day, which offered hope that an increase rate limited to the 400,000 area could be sustained in the 1970's, though even this would lead to the gap Stevens was pointing to. But the annual consumption increase figure for the more recent past, 1966–68, was 540,000 barrels, indicating a rising trend—which, if unchecked, would advance the moment of reckless dependence on OPEC oil to 1973, instead of 1974.

Fortunately, the mathematics of the danger was clear in 1969, leaving five or six years to avert it. The looming retreat from essential self-sufficiency—under which we imported only as much oil as could be replaced from our shut-in capacity or could be temporarily done without—challenged governmental oil policy down to its roots. For what purpose was the public subsidizing the oil industry to the tune of several billion dollars each year in tax breaks and price subsidies if the day was in sight when it would not purchase the promised return—secure oil supply at a stable price?

In May 1969 the Senate Antitrust Subcommittee, under Chairman Philip Hart (D-Mich.), held hearings to explore this embarrassment—

and to publicize it. Testimony was received from economist E. Wayles Browne, who introduced a series of charts that traced interlocking trends in newfound oil, proved reserves, and production. The trend of oil production in the "coterminous forty-eight states" was slipping. The charts showed that the rates at which new oil was being found and added to pumpable reserves were so much lower than the rates at which it was being pumped and consumed that unless there was a prompt recovery in exploration activity, production would soon start to drop in absolute terms and would henceforth fall behind demand by a wider and wider margin with each passing year. Representatives of Exxon, the industry's premier data collector and analyst, were called in and confronted with Browne's findings on the future decline of oil production in the continental United States. Vice-president M. A. Wright responded with an exhibit that stated "the downtrend [as depicted by Browne] is probably a reasonable expectation and does not differ significantly from Humble's outlook." (Humble was then the name for Exxon U.S.A.) To the subcommittee's chief economist, Dr. John Blair, the conclusion was inescapable: "The extrapolation of these trends clearly pointed to the development of a domestic shortage within the near future."

Wright, however, did not concede that essential self-sufficiency was necessarily lost or that the tax breaks and import barriers had failed their purpose, explaining that the oil companies would meet America's need for new supplies by offshore drilling and the development of Alaskan oil. This response ought to have caused a chill in the White House, assuming its left hand knew what its right was doing, for the Nixon administration was launched along a path that would block both Alaskan and offshore oil development for many years to come. So long as it persisted in thwarting the solution offered by the oil industry, the alternative hopes for continued self-sufficiency—which the President never stopped proclaiming as his policy—were to slow down the rise in consumption through elementary conservation and to beef up oil and gas activity within our continental borders.

The question arises: Were routine steps toward conservation and stepped-up domestic exploration possible in the world of the early Nixon years, and would they have been sufficient to keep imports from increasing to catastrophic levels?

Of the projected import increase of more than 3 million barrels a day, safely half could be counted upon from increased production in Canada and Latin America, with a small assist from Asia and Oceania—all of them anxious to have the United States as the customer for their rising output. The United States, with its consumer price for oil artificially propped way above the world price, was the optimal market, often the indispensable one, for oil exporters whose production costs were too high for them to compete on the world market with Middle East exports. (The imports from these areas would, in fact, fulfill expectations, rising from 2.2 million barrels a day in 1968 to 4 million in 1973.) It was the remaining 1.5 million barrels—which had to come from the Middle East if it could not be supplied domestically or obviated through lower consumption—that threatened to upset a world oil balance which for twenty years had kept supply ample and prices in gradual decline. If the United States could look to its own resources and disciplines *now* so as to avoid in the early 1970's a large-scale entry into the Middle East oil market, thus leaving undisturbed those 1.5 million daily barrels to preserve the surplus that long had stabilized the European and Japanese oil markets, the Nixon administration could maintain throughout its reign the two conditions necessary for security of supply and cheapness of price: (1) U.S. nondependency on the Middle East that rendered it impervious to Islamic designs and left it, as in 1956 and 1967, with a credible capacity to break up "oil actions" against its allies, and (2) a sufficient oversupply in Middle East oil production to keep the "producing states" too impressed with the hopelessness of cornering the market and too busy competing with one another for customers to brood together about embargoes, cutbacks, and extortion schemes.

But without the Alaskan and offshore options, which were already well on the way to being discarded, was there left in the United States a realistic "domestic option"? There was solid reason to assume so. A jack-up of only 7 or 8 percent in the projected U.S. production figure, combined with a mild slowing of the consumption rise—both achievable by minor adjustments in existing government regulations—would entirely eliminate the need for that 1.5 million daily barrels of Middle East oil.

Certainly sufficient oil was here in the "coterminous forty-eight" and in already approved offshore developments, waiting to be found or to be recovered from known fields by improved technology. The geological surveys of the time showed that the oil was there to support a short-term production increase until the governmental obstacles to Alaskan and offshore oil were resolved. Nixon and his interior secretaries attest that this was their operating assumption, and events have since proved it to be true. In the twelve-year period that began in 1969, more than 40 *billion* barrels of oil would be produced in the continental United States under circumstances which, for most of that period, were highly disadvantageous to domestic development. None can doubt that a modest extra effort would have added a couple of billion more for the early seventies, which would have made the difference. (In 1980, after eleven years of drawing on the American oil that lay waiting in 1969, U.S. reserves ceased their decade-long decline that had averaged more than a billion barrels a year and began to hold even, a dramatic turnaround owed in part to the proximity and realization of more hospitable government regulation.)

To be sure, oil was less easy to get at, and costlier to bring up, than in the days before the more spectacular and massive pools had been exploited. But this was the precise circumstance that the extraordinary tax deductions and the propped-up selling price of oil in America were supposed to meet. If, then, the oil was there to be found in 1969 for an 8 percent production increase, and if there was any basis whatever for the existing tax incentives and import quotas, a tightening up and sharper focusing of these subsidy-incentive programs could have accomplished this limited task.

The large margin for improvement that gaped in our domestic oil development was shown in early 1969, when the Treasury Department published a study done under its commission the previous year by the CONSAD research group. CONSAD found that each $10 of special tax incentive to the oil industry was resulting in only $1 worth of newfound oil*—a finding that supported the contention of many students of the oil industry, as summarized by Professor Oppenheimer, that as of the late 1960's "oil companies do not reinvest the depletion incentive in new exploration."

If there were enacted a stricter requirement of oil company performance which raised this scandalous return to a still-dismal $3 worth of oil for each $10 of taxes forgiven, 100 million barrels of oil a year would be added to U.S. production. Four years of this (plus an opening year for it to begin to take effect) would add 400 million barrels a year, more than 1 million barrels a day, to our projected 1974 output.†

*CONSAD reported that tax deductions beyond ordinary deductions for business was costing the Treasury $1.3 billion a year and was resulting in additions to the U.S. petroleum reserves worth *at most* $150 million per year.

†Why, instead of accepting so low a return in oil for lost tax dollars, did the government not just keep the taxes and buy $1-a-barrel oil abroad for stockpiling here against the potential day of shortage or price extortion attempt? By counting on one's fingers, one can see that a year's worth of Middle East imports at the projected mid-1970's level could have been purchased for one year's

The task of slowing the consumption rise slightly was also manageable without any great to-do. Among oil economists, it was the rule of thumb that each rise of 10 percent in the price of oil lowered consumption by 2 percent because some consumers would reduce their purchases. Now it was inescapable that the disciplined maintenance by President Nixon of quotas on oil imports would, as it tightened supply relative to demand, automatically cause a slight increase in the price from year to year. A price increase of 4 percent a year, which over five years would raise the price of a barrel of crude oil 20 percent—from $3.25 to $3.89—would by the economists' rule of thumb reduce oil consumption projected for 1974 by 4 percent, or more than 600,000 barrels a day. Hence the advice to the new President from the prestigious *Economist* magazine in London, in March 1969: "If adequate exploration is to continue, higher *domestic* prices are required at a time when *world* prices are still falling [italics added]. Only the government can decide what price America should pay for being independent of insecure Middle East oil supplies. . . . This is the question that will be faced now."

In undramatic increments, in minor tinkering with programs already in existence, with the necessary authority either wholly in the President's hands or available from a Congress bent on change, lay the means to eliminate that projected 1.5-million-barrel-a-day gap that, unless closed, had to cause a switch of the world balance of oil power to the ambitious sheikhdoms of the Middle East.

Such, in general, was the "domestic option" before President Nixon. In fact, it was being pressed upon him in its various aspects—by oilmen, who wanted import policy stiffened up and price controls on natural gas ended; by economists, who argued that tax incentives should be changed so they were tied not only strictly to exploration but to the especially risky kind of exploration that wouldn't be done without subsidy; by liberals, who were ready to increase incentives for the small independent companies as a trade-off for cutting them for the big international companies; by managerial-minded Cabinet members, who were appalled at the wasteful indirection of present oil subsidies and the small return from them; and by Treasury Department tax technicians, who, recognizing that Congress was determined to tamper with Oil's taxes in 1969, sought to make the change a constructive one by keeping

---

lost taxes and stockpiled indefinitely. In fact, President Nixon received official proposals for just such a policy from his Emergency Preparedness Committee and from his Cabinet Task Force on Oil Import Control. That will be explored later.

the depletion allowance, but only when it was "plowed back" into new exploration or recovery activity.

A historic convenience had smiled on Richard Nixon. Just as it became clear that something had to be done to spur Oil to pour more of its resources into domestic exploration, even if that meant lessening its higher-profit activity in the Eastern Hemisphere, Oil's power in Congress to snuff out such impertinences had paled. An oil industry that suddenly saw itself vulnerable to looting by congressional raiding parties was an oil industry no longer able to thwart a demand that it either find more oil at home or lose the massive tax and price subsidies given it for that purpose. Throughout the first Nixon year—as the tax code was overhauled, as the congressional revolt against Oil bubbled, as the warnings of a future U.S. oil shortage escalated—circumstance would repose in the President's hands the leverage to force through a step-up in domestic oil and gas activity, if he wanted to use it.

Happily the new President would not have to summon up the vision of a prophet or even the abrasiveness of the innovator, for the oil issues had already been framed by others, and he need only don the mantle of the national interest and join one side or the other, thereby giving that side a majority. Three choices were thrust upon him—by Congress and by events—early in his term: (1) whether or not to join the effort to strip Oil of those tax breaks it received for foreign oil development that helped make exploration abroad more attractive than at home; (2) whether to stand pat behind the domestic depletion allowance or to join those who would tighten it up so that the $1.5 billion it rebated to Oil each year must be plowed back into domestic oil activity; and (3) whether to reaffirm, do away with, or transform the existing quota system that limited oil imports to 20 percent of domestic consumption, a system mightily defended in oil country as the sustainer of high-cost American oil production, but under hot fire elsewhere because it cost American consumers an extra $5 billion a year to pay a propped-up domestic oil price that was exempt from world competition.

Though he had come into office generally on the side of Oil's traditional subsidies and was specifically pledged to defend its tax allowances, Richard Nixon always reserved to himself the statesman's prerogative of changing his mind. And if the bad numbers on oil production and consumption did not stimulate that faculty, the political realities in Congress might.

* * *

Nixon was confident that he could be at once the champion of tax reform and the defender of what was widely seen as the very symbol of tax abuse. I was pursuing this story for a column that was to appear on May 7, 1969, and I learned from a Nixon insider of the private instruction which the White House relayed to the drafter of the Nixon reform package—Edwin Cohen, who had left the University of Virginia law faculty to become Nixon's assistant secretary of the treasury for taxation. Cohen was told that he might range far and wide looking for loopholes to close (after all, the White House could intercept any Treasury reform proposals it thought unwise). But the President laid down one inflexible guideline: There must be no change in the oil depletion allowance.

Cohen proved a prodigious worker and, using as raw material the Johnson Treasury Department reform study, he assembled within a matter of weeks an omnibus tax package, the centerpiece of which was repeal of the Kennedy-Johnson investment credits for business. Cohen's superiors, Treasury Secretary David Kennedy and Under-secretary Charls Walker, approved the package and forwarded it to the White House. On its way to the President's desk in March, it fell among a posse comitatus of White House advisers, led by John Ehrlichman, which pulled some of its sharpest teeth, such as its restrictions on the use of capital gains and municipal bond shelters to avoid *all* taxation. In April, as the President leafed through the sanitized pages, his wary eye on the watch for the backslidings of underlings and the snares of a refractory bureaucracy, he pounced on a suspect-looking entry. "Now, this doesn't change depletion, does it?" he said challengingly. "I'm committed on depletion."

He was assured that it did not, that the depletion allowance remained as it had come from the hand of its congressional creators forty-three years before. The bill was cleared by the President and submitted to the House Ways and Means Committee.

From the moment Assistant Secretary Cohen began to acquaint the committee elders with the administration tax package, it was made clear to him that Nixon's fundamental posture was hopelessly unacceptable. There was in Congress and elsewhere what the Brookings Institution called "a rising tide of criticism" of the thin results of so many fat incentives given Oil. Even committee members who had long carried the flag for Oil dismissed out of hand the notion that the House would swallow a tax reform bill that was silent on the oil depletion allowance, the one abuse the public was widely exercised about. If the

President wanted a general tax reform bill—and by April the White House was touting this as its top legislative priority—they said he would have to accept a cut in oil depletion; he couldn't get one without the other.

Cohen saw that the administration, if it stuck with its doomed position, would be dismissed as irrelevant to the legislation and would lose what should be a commanding influence over the bill it was counting on to be the premier Nixon legislative achievement of the First Year. A fiasco impending, Cohen cast about for a device that would bring Nixon back into the legislative mainstream without requiring him to renege too baldly on his pact with Oil.

Lying all about were the ingredients with which to build a majority behind not just slashing the oil industry's tax incentives, as if there were no oil supply problem, but restructuring them to do honestly what they were supposed to do: increase domestic exploration and production. Hale Boggs, the longtime oil industry spokesman from Louisiana who was to prove a decisive figure in the Ways and Means deliberation, was preparing to offer a proposal that would cut the generous allowance to the big companies but liberalize it for the smaller, exploration-minded independents, which in 1968 had discovered 80 percent of the oil and natural gas that was found in the lower forty-eight. John Byrnes was proposing an amendment that would increase tax incentives for *exploratory* wells—the ones that sought to find *new* oil—while cutting incentives for *development* wells, which merely exploited oil fields already found and thus did not require further incentives.

In the Senate William Proxmire, who on this issue was the heir of Paul Douglas and could speak for what was acceptable to liberals, had a pair of proposals—one to increase incentives for independents while reducing them for the majors; the other to increase tax subsidies sharply for active exploration while shaving them for ordinary development. Finance Committee Chairman Russell Long, the grand klagon of the old guard of Oil protectors, was prepared, if he could not keep Big Oil's subsidies intact, to lead a move to increase them for the independents. And other legislators were pushing bills to reverse the tax preferability of exploring abroad over exploring at home.

Thus circumstance had spewed up the elements of a potential majority that lay waiting for a leader to assemble and focus it—Democrats and Republicans, liberals and conservatives, pro-Oil and anti-Oil legislators, a majority that was prepared, for diverse reasons, to support measures aimed at making production incentives work as opposed to just cutting them, thereby to punish Big Oil's faithlessness in a way that would meet the energy needs of the future rather than merely requite the resentments of the past.

Cohen and his Treasury aides now adopted the concept of saving Oil's tax forgiveness by reshaping the incentives to make them more result-oriented; they came up with a tough proposal by which the administration could hope to put itself at the head of this scattered but widespread movement yet, after a fashion, keep the Nixon pledge to the oil tycoons. It kept the oil depletion allowance at 27½ percent but required that all taxes thus forgiven must be plowed back into petroleum exploration and recovery. If adopted and implemented—making the assumption that Big Oil would rather spend the money on exploration than hand it back to the government in taxes—it should raise the minuscule performance rate the depletion incentive now had. As we have seen, a raise from the current return of $1 in new oil found for $10 of tax forgiveness to only $3 for $10 would by 1974 increase domestic production by more than 1 million barrels a day. And the Cohen plan would not only save Nixon from the out-and-out breaking of a campaign promise but raise him to truly presidential ground from which he could say to Oil and to the country: "I am keeping the promise I made to preserve the tax incentives which are necessary to encourage the domestic oil production vital to the nation's security; all the oil industry has to do is use those incentives for the intended purpose." This plowback proposal would also restore the administration from a posture of irrelevancy in the oil tax debate to one of leadership from which it could parley in the national interest with the congressional innovators.

The Treasury Department sent the plowback proposal to the White House in late June and waited expectantly for the green light that would enable it to start putting together a coalition. The Ways and Means Committee was scheduled to vote on the oil provisions in the third week of July. When no response came, Treasury raised the matter again on July 17 by sending the President an elaborate memorandum.

By now the oil industry itself had conceded that stubborn insistence on 27½ percent depletion would end in certain defeat even in its citadel, the Ways and Means Committee, and was authorizing its committee friends to propose small cuts in the depletion percentage so as to fend off large ones. For instance, Representative George Bush of Houston, an oilman who was considered "the most articulate spokesman for the industry position," would propose a cut to 23 percent. Representative

Boggs thought Oil was not bending far enough, that 20 percent was the magic number that would satisfy the demand to twist its tail yet leave depletion mostly intact.

But still the President would not budge. No response was made to the Treasury recommendation of July 17. The administration, instead of organizing the forces of productive change, lodged itself as an obstacle against those forces. Whatever the change being offered during the Ways and Means Committee voting, the administration was cited against it—against abolishing the foreign tax credit and the foreign depletion allowance, against using revenues gained from increasing taxes on the international companies to liberalize incentives for the domestic independents, against any alteration in the doomed 27½ percent formula.

So the proponents of rationalizing the incentives never united on a common approach. Each such proposal stood alone and was defeated alone. What emerged from the chaos was the part of the Hale Boggs bill that cut the depletion allowance to 20 percent; the part that would have awarded the reclaimed revenue to the wildcatters who did most of the domestic oil exploration was defeated.

On August 2 the full House of Representatives ratified the tax bill of its Ways and Means Committee and sent it over to the Senate. The President's position had gotten short shrift in the House, and the question was: Should it be altered before the Senate took up the bill? Custom required that the secretary of the treasury be the leadoff witness at the Senate Finance Committee hearings in early September, when he must propound the administration's stand on oil depletion and other matters in dispute. In late August, Secretary Kennedy flew to San Clemente with Walker and Cohen, to confer with the President and his economic advisers. Since it was now obvious even to the White House that depletion could not be preserved intact, the Treasury men had revived their plowback proposal as one of the three options for the President to choose among. The other two were: (1) fight the House action in the Senate, or (2) accept it.

The Treasury men looked forward to hashing this out face-to-face with the President, even though he would be flanked by John Ehrlichman, H. R. Haldeman, economic counselors Arthur Burns and Paul McCracken, and financial expert Peter Flanigan, but in the end they would find it more baffling than routing memorandums through Ehrlichman.

The President was unambiguous on one thing: He wanted no part of plowback. "I made a campaign promise, and I intend to keep it," he declared. Treasury might regard plowback as a way to preserve the 27½ percent allowance Nixon had pledged to keep unchanged, but at the proof level of money in the pocket, oilmen would see a big difference, as Nixon perceived, between being free to do what they wished with their tax break and being forced to reinvest the forgiven taxes in exploration. That was a change that would be blamed on its author, Nixon, whereas up to now the blame lay with the Democratic Congress.

All right. If the campaign promise was the standard, the President must want option number one, opposing the House cut and putting up a fight in the Senate to restore it. But this reading assumed that the President was prepared to do something active about depletion, something efficacious about his promise to Oil, when his intent was to be passive, to honor his promise only by not openly betraying it, to avoid becoming entangled in the issue in a way that would cost him his public identity with the tax reform bill of 1969. When the Treasury team gave Nixon its judgment that the House cut in depletion to 20 percent would be upheld by the Senate in principle, though probably lightened a bit, he appeared to abandon his thralldom to campaign promises and to jump all the way to the other side; that ought to have been a signal for caution in interpreting. "We might as well go with it," Nixon said of the House cut. "Let's just accept it and go on from there."

Kennedy and his entourage flew back to Washington, thinking this was their marching order, and on September 4 the treasury secretary testified to the Senate Finance Committee, "Although the Administration did not recommend a cut in domestic percentage depletion, we accept the House approach to increasing the share of the national tax burden borne by the petroleum industry."

How maddening to be President, to be daily spoken for by others and to see one's Florentine strategems blown by forthright simpletons! It was a literal enough rendering of what Nixon had said at San Clemente and a true statement of the policy he would in fact follow—the administration never did contest the House cut—but its directness rudely violated Nixon's known preference for poetry over prose when in tight corners. It was not politic or necessary to announce "we accept," because that smacked of "we agree." One could not, of course, say, "We don't like it, but we're not going to do anything about it"—the actual Nixon stance. No, this was the sort of thing that had to be elided over by shifting the subject. It was enough to point out that the President opposed the cut—not loudly enough to draw undue press attention, just audibly enough for the lobbyists to hear—but that if Congress was adamant, and he still hoped those *Democrats* wouldn't be, he had to consider the national stake in the entire bill. Something like that was what he wanted said.

The predictable furor that followed Secretary Kennedy's testimony was chronicled by those students of the tensions within the GOP, Rowland Evans and Robert Novak: "Big Oil and its allies in the Republican Party were apoplectic, the oilmen crying betrayal and threatening to abandon Nixon in 1972, the politicians demanding that the White House reverse Kennedy or face the loss of campaign contributors galore, not to mention the state of Texas."

The contrast between the President's reluctant obtuseness in confronting the *problems* of oil and his vigorous adeptness in rising to the *politics* of oil—a contrast not peculiar to Nixon but characteristic of a system that produces politicians who are oriented toward the immediate political fallout of issues rather than toward the substance of problems and who deal not in solutions but in learning the lines and taking the tacks that will keep them personally out of harm's way—shows in the orchestrated campaign that promptly commenced in the White House to disown Secretary Kennedy and separate Nixon from what was, and would continue to be, an accurate statement of his policy.

John Ehrlichman led off by phoning distraught Republican senators on the Finance Committee to assure them that Dave Kennedy had gotten confused and had misspoke himself. Why, Dick Nixon would *never* back a cut in the depletion allowance. For the oil constituency itself, "confusion" wasn't enough. Something more graphic was required to lance the swelling suspicion that treason was afoot, something with a little anger and blood in it—villains, denunciations, reprisals. Harry Dent, the nocturnal White House specialist at special-interest politics, opened up with a letter to the press wire services, nominally addressed to a protesting judge, Barbara G. Culver of the West Texas oil country, saying that Secretary Kennedy "is to be corrected very soon" and that "the President continues to stand by his campaign commitments." For the rough stuff there was needed a spokesman high enough to look official but not so high that his words must have consequences. The Republican National Committee was chosen, in the person of its deputy chairman, James Allison, a Texan. On the same day the Dent letter to Judge Culver was released in Washington, an Allison interview dominated the front page of the *Dallas Times Herald.* "Walker is the culprit," Allison proclaimed, "and I think they ought to fire him. . . . I think he has operated on his own without considering what the President said in his campaign. Up to now, nobody's stopped him."

It is indicative of the fine Nixon calibrations behind the public hysterics that the President did not at any point allow his people to go beyond *ad hominem* diversions and challenge the substance of Kennedy's statement, for instance, by committing the administration to action opposing the House depletion cut or by hinting at the use of the President's ultimate weapon, the veto. Nixon was alert to avoid arousing such expectations—this had been his purpose at San Clemente—for if he were to be seen engaging in a running battle with Congress against a major item of tax reform, it would cloud his claim to the tax reform bill of 1969 as a Nixon achievement. It was to keep attention away from the weak segment of his front that the Nixon men were maligning two appointees to whom they had entrusted the United States Treasury Department as, respectively, a bumbling dunce and an out-of-control adventurer.

The incident was to have repercussions on the administration careers of Kennedy and Walker. It added to the White House estimate of Kennedy as a weak secretary, and it deepened palace guard hostility toward Walker that would prevent him from replacing Kennedy as secretary when the latter departed at the midpoint of the first Nixon term.

On September 25 press secretary Ron Ziegler, eluding a clamorous White House press corps, said in substance the same thing that Kennedy had said, but in an anesthetic way: The President continued to oppose the cut in oil depletion, but of course, he "will abide by the judgment of Congress."

On the next day there was a presidential press conference, and as Nixon tiptoed around the inevitable depletion ruckus, he furnished a guided tour of what he did best.

> QUESTION: You told an audience in Houston last fall that you opposed reducing the oilmen's depletion allowance. Do you still oppose it?

THE PRESIDENT: As a matter of fact, I not only told the audience in Houston that, but that has been my position since I entered politics in California 22 years ago. It is still my position.

I believe that the depletion allowance is in the national interest because I believe it is essential to develop our resources when, as we look at the Mideast and other sections of the world, many of the oil supplies could be cut off in the event of a world conflict.

On the other hand, I am a political realist. I noted the action of the House of Representatives in reducing the depletion allowance. Also, my primary concern is to get tax reform —the tax reform which we submitted in April, which goes further than any tax reform in 25 years. We need that tax reform above everything else.

Some of the items that I recommended, the House did not follow my recommendations, and the same will be [true] in the Senate. When the bill comes to my desk, I intend to sign the bill, even though it does not follow all of my recommendations, provided it does not require a revenue shortfall. That is more than I believe the Nation can stand.

The statement was a tissue of painstakingly deceptive inferences. For example:

1. Thirteen years of unbroken decline in domestic drilling and the slide in home investment compared with foreign investment by U.S. oil companies were irrefutable proofs that the existing tax incentive system was not filling the role Nixon claimed for it and that to fill that role it needed to be overhauled, not reaffirmed.

2. The President's claim to authorship of the tax reform bill was spurious; not only was the administration about the last passenger to scramble aboard the tax reform bandwagon, but it was a lethargic passenger. The bill was almost entirely the handiwork of Congress and has since been cited by journalist and academic alike as a striking example of the lack of executive impact.

3. What the President hailed as a once-in-a-generation reform and as something "we need . . . above everything else" (and this

claim was not floated carelessly but was the heart of his message) was in fact a typical congressional combination of grab bag and juggling act, so watered down in its reforms and puffed up in its new tax exemptions as not to be significant reform at all (else the nation would not still be awaiting it).

4. Nixon was *not* prepared to veto the tax bill, as he said he was, if the final version gave away more taxes than it recouped and resulted in a revenue shortfall. A month before the September 26 press conference he had agreed at San Clemente to accept a $1.3 billion revenue loss in the bill and would eventually sign a measure containing a $2.5 billion loss.

But in terms of the President's objectives, the answer was a tour de force, a classic exercise in Nixonian ratiocination. Crossing the same treacherous terrain on which Secretary Kennedy had straightaway blown up, Nixon smoothly accomplished the most delicate disengaging maneuver, during which he moved from principled support of the oil lobby's position, on grounds of national necessity, to refusal to fight actively for that position, also on grounds of national necessity. But there is never a point at which a clean photo of that abandonment can be taken, nary a phrase or word that, even lifted out of context, could be placarded against him in oil country.

Nixon had to address two disparate groups simultaneously under a spotlight—the general citizenry and the oil constituency—and he sought to beam a different signal to each without shorting out. To the citizenry he presented himself, almost unobtrusively, as the protagonist of far-reaching but uncontroversial reforms, the large-minded statesman who wore the collar of no particular interest and was practical enough to accept disappointments in order to pilot *the* bill to safe passage. To the oil constituency he came forward as their devoted lifelong advocate who continued to be their advocate. He could not change the fact, of course, that the Democrats controlled Congress. If they were adamant against Big Oil's tax benefits, he could resist only up to the point where the life of the whole bill was endangered. He said nothing that would undermine Oil's chance to prevail in the Senate, as Secretary Kennedy had done by formally surrendering in advance. But he effectively, though gently, cut off any hope of a veto in its behalf—before that hope could get steam behind it and be the more damaging to deny, for down that road lay a swamp of bad imagery.

Oilmen could grumble that Nixon was too restrained in resisting the liberals, that he was more worried about general opinion than he was about them, that it was *results* that counted. But in the absence of active collaboration in the depletion cutback, Nixon's offense was not the kind to draw the sword over or close the purse. When payoff time came, in the campaign year of 1972, would oilmen turn against the well-disposed Nixon for being ineffectual, or would they blame the national Democrats? Nixon was confident that though he had to disappoint Big Oil from time to time, he could adequately position himself for that choice —as long as he avoided the gratuitous slight, the disloyal word, that embitters allies when mere injury will not. At the press conference of September 26 he made no such slip, and the uproar that followed the Kennedy testimony quickly subsided.

The House action cutting oil depletion to 20 percent, which was estimated to cost oil companies about $600 million a year, still had to run the gauntlet of Chairman Long's Senate Finance Committee.

Russell Long's grip on the affections of his committee colleagues and on their campaign kitties and pet legislative schemes was unoppressive but enduring. If his peers in the House—Mills and Boggs—were headed into alcoholic storms, Long had somehow survived amid those storms for decades and seemed to be headed out of them and into calm waters. Behind his bulbous nose, his infectious amiability, and his incorrigible penchant for depraved and ribald stories of Louisiana politics was perhaps the canniest logrolling talent left in the Senate. With a foot in both ideological camps and an amused tolerance of any man's political or personal hang-ups, this persuasive son of Huey and nephew of Ole Earl could find and keep common ground in any corner of the Congress. Nor did he deal only in transactional empathy; there was a bit of fear at work. "You may *think* you can beat him today," Wisconsin's Gaylord Nelson once said, "but you *know* he'll be back to beat you tomorrow."

Well did he know the transience of public indignation, and for months he had been waiting out the anger against tax loopholes. By late October, when he finally scheduled committee votes on the tax bill, the fury of the previous January had abated considerably; by December, when he would sit down with Mills and Boggs to compromise House and Senate differences, it should be paler still.

Long's October nose count of his committee showed he had the votes to restore the depletion allowance to 27½ percent; if the count held, he should be able to negotiate a compromise with the House at, say, 25 percent. I had learned from sources within the departed Johnson White House that Long felt he could count on the support of Eugene McCarthy and Vance Hartke, to whom over the years he had given much education as well as help in arranging Oil campaign contributions.

But on October 23, the day appointed for the votes on oil depletion, he lost both Hartke and McCarthy. Hartke voted against Big Oil outright. McCarthy's defection was more subtle; he just didn't show up to vote (a McCarthy resort Paul Douglas had experienced from the other side of the issue). Long's motion to change the House 20 percent figure back to 27½ ended up in a tie vote—8 to 8—and a tie meant the motion was defeated.

Long looked around the table at colleagues for whom he had put many a package under Christmas tree tax bills. "How about raising it to twenty-three percent so I'll have something to bargain with?" he wheedled appealingly. It was a personal favor he was asking, and he got it—at a saving of a hundred million or two for the oil companies.

In a moment he was back for more. After all, he did have half the committee. "Isn't there something we can do for the little guy?" he challenged, his voice wet with compassion. He proposed to use the tax revenue just taken from Big Oil to cut the taxes on Small Oil—the independents—thereby keeping the money in the same extended family. A skeptic, familiar with Longisms, asked what Long meant by the "little guy."

"An operation with less than five million net income," Long tried to slip by. But in vain. Didn't that mean a gross income of 25 to 30 million? This perked up John Williams of Delaware, a chicken farmer with the husbandman's antagonism toward favors for speculators, who started in on how huge an enterprise a $30 million grosser would be in chickens. When a prickly purist like Williams began to mix in, Long knew it was time to close as quickly as possible. He narrowed his definition to $3 million and got his amendment adopted.

Thus an eleventh-hour reprieve. A December conference would be held among the tax chiefs of Senate and House to reconcile the differences between their respective bills. These "conferees" were granted wide latitude; what they agreed on would be all but binding on the Senate and House, which could not amend the conference report but only accept or reject it *in toto.* From Russell Long's "little guy" amendment, the conferees could easily fashion a tax reform that was also an incentive to oil production—if the White House got behind it.

And by December the President and his advisers had added cause to get behind it, for they had been exposed to almost a full year of small but cumulative erosions of the energy abundance they had inherited, erosions left unrepaired because of their own actions in shutting down oil development offshore, where 43 percent of the remaining favorable locations for U.S. oil discovery were located, and in delaying the start of the Alaska pipeline, the beginning of which was now farther off than when Richard Nixon took office. More natural gas had been used up than found in 1969 for the second year running, turning an aberration into a trend. The construction timetable for nuclear power plants—and more than half of all new power plants on order were nuclear—had somehow slipped a year in 1969. Demand for electric power had risen by 8.1 percent instead of the expected 6.9 percent, increasing the need for coal, but a coal mine safety bill, which the President was about to sign, and an anti-coal-pollution bill, which was working its way through Congress with the President's support, would, however laudable their primary objectives, have the by-product of making coal less and less available and usable.

For all these shortfalls—of coal, natural gas, and nuclear power—the only resort was oil. Yet 1969 was ending as the first year in which capital exploration expenditures in the United States, which had long been declining in comparison with our overseas oil investments, declined in absolute terms. Expenditures for oil exploration by independents was down to half the $2.5 billion spent in 1956. The year-end figures showed the White House that in December 1969 U.S. oil reserves were 1 billion barrels below where they had been in January. And so a presumably heightened awareness of energy deterioration militated for tax action to revive the independent oil operators who did the bulk of exploration in the continental United States.

That the President still commanded great influence on the process, if he would use it, was shown when at his request the conference committee cut in half the deficit that then existed in the tax reform bill, from about $5 billion down to $2.5 billion. But the climax of the story lies in anticlimax, in what was *not* done. No administration intervention was made in behalf of domestic oil production, and with the House unyielding, the provision Long had salvaged was dropped in conference.

In the end only one significant change occurred in oil taxation:

Depletion was cut from 27½ percent to 22 percent. To the extent this affected oil exploration at all, it harmed it. The tragedy, however, was the lost opportunity to improve the situation. During the first three Nixon years domestic capital and exploration investment in oil, instead of rising to meet demand, would fall by 12 percent from its 1968 peak. The number of exploration wells drilled in the United States in 1970 would continue to decline from the 1956 high of 16,173, dropping to 7,693.

What happened to American oil exploration in the first Nixon term was illustrated when the University of Pittsburgh's petroleum engineering department, after flourishing for most of a century, had to close its doors in 1972 because only six students applied for admission in 1971.

From these events emerges the basic outline of the priority system that was to govern Nixon decision making in energy matters. Three competing claims had been sifted out and chosen among: the requirements of national energy security, the demands of the oil lobby, and the lure of the hour's political *cause célèbre* (for this hour, tax reform). The lowest priority was assigned to U.S. energy security. Typically it was not even addressed in its own right but was wafted in as background music to lend an aura of high purpose to Nixon's otherwise eyebrow-raising reiterations of lifelong fidelity to Oil's tax breaks. Just as the energy security issue was a disposable prop at the press conference, so it was deeply subordinated in the closed-door decision-making sessions. In these, whenever the need for real *result-oriented* incentives in place of the existing placebos was raised, as it was by the Treasury proposal and by the need to take positions on various congressional bills, Nixon cavalierly dismissed it with reminders of his campaign promises to Oil. This became the more revealing when he also made it clear that he did not intend to go to bat for those promises and was actually in the midst of disengaging from them because they stood in the way of something he valued more: being the signer of a tax reform bill.

But he wanted to disengage from his promises to Oil in the least visible way, the least culpable way in its eyes. That was the way of lowest profile—to do nothing active. If he would not fight Congress in Oil's behalf, neither would he join Congress and participate in writing the depletion tax change, though change was sure. This meant surrendering the administration's opportunity to make the change a help to the national oil security, not merely a twist of Oil's tail. Hence, every proposal to stiffen up Oil's tax privileges into real domestic exploration incentives—whether "plowback" or preferential deductions for exploratory wells, or reversal of the tax advantage given foreign oil activity over domestic, or increasing the tax break for wildcatters as a trade-off for decreasing Big Oil's—was rebuffed by the White House and opposed in Congress by administration spokesmen.

Nixon's press conference and campaign hustings solicitude for the oil industry was thus of a demonstrably higher order than that shown to energy policy. Oil is saluted as a worthy, longtime ally. Its tax loophole is invested with the "national interest." Big Oil is reminded of Nixon's two decades in harness to its cause. Continued fealty is pledged. And when the moment for decoupling arrives, he has the grace, as he reaches for his trousers, to blame circumstances beyond his control. "On the other hand, I am a political realist." For there is a still higher priority —the popular cause of the moment. In puffing up the tax bill as the nation's greatest need and in insinuating himself as its chief author, the President's hyperbole shows where his heart is.

A dichotomy of mischievous potential can be seen here. The President *opposed* tampering with Oil's tax privileges when the purpose was to step up domestic oil activity. But he *accepted* tampering with its tax privileges—a kind of tampering that hurt rather than helped domestic production—when the purpose was to smooth the passage of a general tax reform bill and thus to link himself with the minicause of the season.

By itself this juxtaposition is not an occasion for rending one's garments. Politics is politics, and the President still had many choices before him, any one of which could have recouped the opportunity lost by the failure to rationalize the oil industry's production incentives. What made it ominous, however, was that during the approximate time span that the President's trade-off on tax policy was taking shape, he made other choices of a similar kind, raising the probability that a *modus operandi* of unhappy portent was taking shape.

For one example, in 1969 the Office of Energy Preparedness proposed to the President a program for stockpiling up to a year's worth of oil imports, which could be done quite cheaply with spot oil prices in the Persian Gulf hovering around $1 a barrel. The program was opposed by the oil lobby, which, like other producers' lobbies, was averse to having a stored-away surplus overhanging its market. Nixon went along with the oil merchants and took no action on the proposal—then the cheapest, shortest, and most direct route available to short-term energy security. But though Nixon would defer to Oil when the issue was the constituentless one of energy security, the Santa Barbara spill in 1969 showed how he would slip away from Oil when its adversary was a cause with popular oomph.

This pattern was reinforced when the Cabinet Task Force on Oil Import Control, which the President had created in February 1969, made its report to him in early 1970. Subsequent events made it clear that Nixon had expected the task force to reaffirm the existing quota system, but he had made the mistake of appointing as its head a professor of economics, Dr. George Shultz, the secretary of labor, and Shultz had staffed it with solid economists who brought in a scholars' report. It concluded that the system for limiting oil imports was an inadequate safeguard of U.S. oil security, though it was costing American consumers up to $5 billion a year in higher prices for oil.

The importance of this report lay not in its tariff proposal as a substitute for quotas, or its oil projections (which would in time be rendered obsolete by the contrary thrust of Nixon energy decisions at home and abroad), but in its official warning to the President from his Cabinet that a grave problem had begun to unfold, a problem that could be solved all right but only by a coherent policy that included *some* hard choices.

The task force confirmed that the gap between domestic oil production and consumption would widen steadily in the 1970's. It established the danger point above which Middle Eastern oil imports should not be allowed to rise—5 percent of consumption in the 1970's, 10 percent thereafter when new oil sources around the globe would ease the security factor. And the report recommended that while phasing out the present quota system, we should phase *in* a series of alternative measures the President could choose among that were aimed at protecting U.S. energy security at a fraction of the cost of the import quota system.

Among these carefully wrought measures were: various methods of stockpiling oil for emergencies; making a distinction between secure and insecure sources of imports and acting on that distinction by arranging energy security pacts with Western Hemisphere nations while placing permanent caps on Middle East imports; completing the Trans-Alaska Pipeline no later than 1973; pressing ahead with development of U.S. offshore oil; putting in place a standby rationing program under which the American economy could absorb the loss of substantial oil imports in time of cutoff yet function effectively; establishing with tariff revenues a synthetic fuels program to produce oil from coal and shale rock; and putting in place a system for continuously reevaluating the effectiveness of whatever mix was adopted to assure that imports were not rising too fast or domestic production too slowly.

The oil lobby was dismayed by the report, for the oil import quota system was as valued a source of government-guaranteed profit as the tax privileges. Nixon sided with Oil, rejected his Cabinet task force report, and declined to implement a single one of the above measures. In a parallel of Nixon's statement to his tax expert, Edwin Cohen, rejecting plowback ("I've made a campaign promise and I intend to keep it"), the President's political spokesman in the Cabinet, Attorney General John Mitchell, appeared before the task force, after it was known to be turning against the quota system, with a pointed, if coded, warning: "Don't put the President in a box on this."

Yet, as in Nixon's response on the tax issue, when the impetus for scrapping import barriers came not from long-range energy security calculations but from immediate political ones, his solidarity with the oil lobby evaporated. When the consideration became heading off local uproars over temporarily tight supplies (which had to be stoutly endured if import quotas were to play their assigned role of forcing domestic readjustments in either consumption or production), or quieting demands from industries—like petrochemicals—for more imported oil, or holding down artificially the politically potent consumer price index by using a flood of Middle East oil to head off domestic oil price increases (which were essential if domestic oil activity was to increase), Nixon again and again scooted from the side of the oil lobby, opening one breach after another in the system of import barriers by letting in great floods of imported oil whenever there was prospective demand for it. This was a policy, or rather a lack of policy, which had to end in destroying the existing energy security system, not in the eyes-open manner of the task force, but in the Dr. Feelgood sense of prescribing uppers without admitting what was being done, of building up an addiction without coming to terms with the malady.

A common thread ran through these decisions in the early Nixon period. Energy security always lost out to a more politically potent competitor, despite the dread statistics that were steadily accumulating. To appease the oil lobby, Nixon thwarted any schemes for domestic oil expansion that trespassed on Oil's established governmental privileges. To appease other lobbies, he thwarted Oil's own programs for domestic oil expansion. Had the President been either Oil's consistent captive or Oil's consistent foe, energy needs would have been served *part* of the time; as it was, Nixon's early performance recalled the terse scouting report on a baseball prospect: "He is a weak hitter, but he is also a poor fielder."

On April 30, 1972, President Nixon kicked off the oil phase of his reelection campaign at the Floresville, Texas, ranch of John Connally, his second treasury secretary. (Connally had recently left that office to head the national Democrats for Nixon drive.) To the assembled millionaires from Big Oil and other walks of life, Nixon repeated his ritual pledge of support for the depletion allowance. He had lately reassured the oil barons through White House statements of his continued support of the oil import quota system.

Had there been ironists present, they might have taken the renewed pledges as omens that the tax depletion and quota systems were in for further trouble. Indeed, during Nixon's second term, oil depletion would be altogether wiped out for large companies, in the wake of another wave of popular indignation, though this time Nixon would not be around at the finish for the bill-signing ceremony. And within a year of the Floresville gathering, the oil import quota system would be dismantled by Nixon, though almost to the end he would continue to indicate his support of it.*

That Nixon was heartily cheered by Connally's guests, that they queued up to press his hand and would fill his campaign coffers with sums of unprecedented size, that Texas would give him 66.2 percent of its presidential vote were testimony that he had all along gauged the oil crowd's mentality with flawless exactitude. He could act against oilmen's legitimate interests in a half dozen different contexts yet keep them tolerably in line behind him by playing on their attachment to their government tax and quota bonanzas. He didn't even have to *deliver* on these narrow promises; as long as the national Democrats sounded menacing to Oil, as he counted on them to do, it sufficed for him to sound supportive, though in the breach he was lukewarm and ready to defect.

Oil continued to accept Nixon's presentation of himself as one who limited the losses it would otherwise suffer. Following the 1969 cut of oil depletion in the House, the President was praised at a Treasury-hosted gathering of oil executives and lobbyists for his staunch support

*On February 27, 1973, John Ehrlichman publicly denied that the White House was even considering the total suspension of oil import controls; on April 12, 1973, they were totally suspended.

and for the access to his Treasury Department which the industry enjoyed. One executive said, "If a Democrat were in power, we would not have this power."

The reward for Nixon's acuity was now pouring in. In the 1972 fund-raising season, senior oil executives and stockholders would contribute more than $5 million to Nixon's reelection ($4,981,840 was *recorded*). The same compilation, which includes only gifts of $500 or more, showed *no* contribution to the Democratic opponent, George McGovern. One industry, therefore, provided at least one-tenth of all the financing of the entire Nixon campaign, still the most expensive in history.

Oil would pay for this overkill, in predictable and unpredictable ways —not only in intensified hostility from the soon-to-be-ascendant Democrats but in the guilty embarrassment such largess caused to the Nixon administration. A year or so after the election the catastrophic harvest of its oil policy would force Nixon's White House advisers to recognize that Nixon's hold-down of oil prices since 1971 via price controls was in several ways exacerbating U.S. dependence on Middle East oil imports and that the administration must reverse its field and take the cap off domestic oil prices in order to encourage both domestic production and conservation in oil use (for even as the U.S. consumer complained of gas lines, he was filling up his tank, whenever he could, at half the world price). But though the White House advisers came to this conclusion, they had to abandon it, for they realized that Nixon had taken so much money from Big Oil that to permit a rise in oil prices would appear to be a return payoff. So the price lid stayed on and on, making an effective national oil policy impossible.

Chapter Five

Paralysis: The Botching of the Trans-Alaska Pipeline

Technologically, energy independence is well within our reach. Politically, it may not be within our reach at all.

—Lester C. Thurow, *The Zero Sum Society*

Had offshore drilling and producing continued as foreseen and had North Slope oil come into operation, these two sources could have contributed about two million barrels per day by now . . . roughly the equivalent of direct and indirect imports from Middle East sources.

—Hans Landsberg, 1974

PART ONE

As one by one the options for maintaining self-sufficiency in energy were nullified by government action or default—the blockage of offshore oil development, the rejection of tax incentives to revive oil development on *land,* the regulatory straitjacket placed on coal in 1969 and 1970, the refusal to combat the archaic price control system that was shriveling natural gas production, the spurning of proposals to stockpile oil for emergencies—the importance of the gigantic Alaskan oil discovery of early 1968 would become more and more magnified. From being one of many solutions, it would become the one last hope.

Here, too, a shadow was cast, across the tundra, by Nixon policy. As long as the land freeze continued, there could be no drilling to find the new Prudhoe Bays that geological formations indicated were there somewhere, nor could the oil from the original Prudhoe Bay be brought up and moved across Alaska to market. But oilmen were confident that once they announced their specific plans, once they made clear the potential of North Slope oil and the nearness of its realization, the land freeze would thaw under the rays of reality.

So in February 1969 the three major owners of North Slope oil—Atlantic Richfield, Humble Oil (the Exxon division to which Arco had turned for capital), and British Petroleum (long the leading leaseholder on the North Slope)—announced plans to spend about $1 billion to build a pipeline that would cross Alaska from top to bottom, carrying oil from the edge of the ice-closed Arctic Ocean almost 800 miles south to Prince William Sound, on the ice-free Pacific. There, where the fishing village of Valdez slumbered, except for the furtive scurrying of realtors, a great depot and port were planned, able to service fleets of 200,000-ton tankers that would move up to 2 million barrels of oil each day to Seattle, San Francisco, and Los Angeles.

Natural obstacles made the Trans-Alaska Pipeline one of history's most dismaying undertakings. It was not so much the 789-mile length or that the pipeline would have to cross three steep mountain ranges, twenty-six rivers, and several earthquake zones, three of them active, or that most of the route was over a wilderness without roads or indigenous supplies and facilities of any kind. It was the weather and the nature of the terrain that gave pause. In the winter months of sunless semidark, life groped brittlely forward as through a frozen fog. At 50°F below zero—and it was often much colder—the lubricant on the helicopters froze, truck tires came apart from their wheel rims, and the jetted steam used to clean drilling bits instantly turned to ice. A continuing bitter wind drove fine snow everywhere and created the equivalent impact of minus 100° F on the human body. Where the ground ought to be there was permafrost, a loose association of gravel, sand, and ice, all frozen together and extending down 1,200 feet, in some places 2,000 feet. Attempts to drive steel piling into the winter permafrost merely crumpled the pipe. Wastes could not be buried, nor would they decompose. Dirty plates left twenty-five years before by the navy after its last melancholy mess at nearby Umiat remained unaltered as though suspended in a time warp.

But the men who moiled for oil were not to be deterred. Working areas could be enclosed and to some degree heated. Ways were found to penetrate the earth, and the cold could even be taken advantage of, for instance, by pouring heated water around pilings, which instantly froze into a solid foundation good for several months. Human wastes, lest they become immortal, were sealed up and shipped out to where they would properly disintegrate. Aircraft of unusual hardiness were found and modified for the Arctic. Vehicles were kept running continuously from fall through winter lest once stopped, their workings would become enfrosted and unrevivable. Men, however, were not so easily kept serviceable. After a month of all this they tended to lose zest and turn sour. So, for two weeks out of every six, they were flown out to Fairbanks, that northern Babylon of wondrous and warming amenities for the snowblind and frostbitten. Duly debauched and depressurized, they trooped back to the Arctic gloom rejuvenated, almost jocular.

It remained for summer to deliver the lowest blow. Even as its perpetual half day of gray mist, its average temperature of 40° F, and the sudden arrival of millions of mosquitoes buoyed the spirit with their comparative agreeability, the permafrost's crust—an eighteen-inch layer of grass, moss, and lichen—thawed and became incredibly fragile. Vehicles running over it soon created an oozy quagmire into which they sank inextricably; worse, once this insulating mantle was disturbed, the permafrost below thawed, too, turning last season's supply of gravel into a fifty-six-foot ravine. A heavy pipeline four feet in diameter, laid on such a surface however tenderly, would erode the crust, creating underneath it a gully that would soon enough be a roaring river of thawed ice that would wash out the line and flow on destructively, perhaps for eternity. So traffic had to be halted from April to October, and engineers faced problems they had not before dreamed of. But summer could be "scheduled around." Techniques were developed to build summer roads by piling gravel atop the tundra, and ways were found to secure the gravel safely. As for laying the pipe, none doubted that whether it had to be buried underground or elevated aboveground or surrounded with layers of liquids or of gases, it would somehow be laid from Prudhoe to Valdez—and payday.

Thus the exploitative drive of the oilmen did not falter, nor had the most appalling obstacles caused any notable lag in their timetable. The oil strikes of unknown extent that Arco had made in the winter and spring of 1968 had by July been authoritatively measured. By August engineers from Houston's Humble Pipeline Company were scattered across Alaska, preparing feasibility studies for a pipeline that could go in a number of directions. Before autumn the billion dollars had been raised and Exxon and British Petroleum had been welcomed aboard as bankrollers and coowners; by year's end a series of harrowing decisions had been made—on the route, the construction mode, the financing package, the strategy for beating the weather and terrain. By February 1969 the first petitions were before the Interior Department, 800 miles of extrawide pipe had been ordered from Japan, and oil company recruiters were rounding up men and machines from around the globe for a start on construction in September, when the pipe would begin arriving and the permits would presumably have cleared Washington. Completion was scheduled for mid-1972.

* * *

But what of the political obstacles? There were essentially two. One had to do with social justice: securing for the Natives of Alaska—the Eskimos, the Indians, and the Aleuts—their fair share of the Great Land before it was parceled out to mineral speculators, pipeline builders, and entrepreneurs of every stripe who would now flock there. The need for justice was reenforced by the need for legal certainty; without a definitive settlement by Congress, landownership in federal Alaska would be subject to litigation that would hold every enterprise in jeopardy. The other impediment was a question of public policy: How much disruption of the wilderness and its wildlife, treasures in short supply in the world of the late twentieth century, was to be traded for the economic and strategic boons of Alaskan oil?

Both problems were inherently solvable; events would prove that. The critical question had to do with timeliness, with the capacity of the American political process not only to "do the right thing" but to do it within a time frame that exhibited some sense of priority, order, purpose, and cohesion. The national need—for a large new source of oil by the mid-1970's—allowed for a year, perhaps two years of delay, to settle the Native claims and work out the environmental safeguards. So long as the three-year work of active construction began by early 1971, all could yet come right.

Four American institutions would be tested by this challenge of timeliness: the oil industry, the Alaskan officialdom, the Congress, and the Nixon administration. Happily none of them was called on to make any sacrifice of its interest to the national interest, and each of them had its own peculiar incentive to clear away its portion of the political hindrances so that pipeline construction could start by the fall of 1969.

Just as the travail of the Trans-Alaska Pipeline is a metaphor for the American energy crisis, so the roles played by these groups introduce the question of how little the energy crisis has to do with external events and how much it has flowed from a pervasive incompetence and faithlessness in the performance of routine tasks and trusts Americans once thought they could do better than anyone in the world.

The oil company consortium building the pipeline wanted to speed the process because it stood to suffer heavy financial losses each day the start was delayed beyond September. It moved quickly, after its February announcement, to disentangle its enterprise from the land freeze and from potential land title litigation. Representatives went to the

Native tribes along the pipeline route and, by offering construction jobs and proposing a no-lose arrangement which would preserve the value of any Native property rights that might be later validated by Congress or the courts, secured formal waivers permitting the pipeline to pass through. Having achieved this, the consortium presumed that the Interior Department and the Interior committees should be able by July to process whatever approvals they had to give. Making allowance for two months beyond July for bureaucratic and legislative inefficiency, the oilmen planned on a September 1969 construction start at the latest and were spending hundreds of millions of dollars to have men, matériel, and equipment in place at that time.

The same disposition toward an early start was shared by the three *public* institutions. Each envisioned disposing quickly of its attendant task. Within none of them was there visible a predominant or even significant opposition to the pipeline per se, such as that which was overwhelming offshore oil development.

For the Alaskan establishment, the task was simply not to make waves. It would be splendid, of course, if Alaskan leaders could actively aid Congress and the administration by impressing conservationists with shows of regard for the environment and by helping their neighbors, the unschooled Natives, to unite on and present their claims to Congress. But it would be enough if Alaskans could merely forgo spectacles of indifference toward the environment and niggardliness toward the Natives that would alarm conservationists and alienate congressmen. The motivation of Alaska's political and business establishment was the highest; nothing more need be said than that oil development and pipeline construction would create thousands of jobs and hundreds of businesses, would yield more than enough taxes to remove the overhanging threat of state bankruptcy, and was worth many thousands of dollars in revenue to each and every Alaskan. Hoping that the pipeline would be reassuringly on its way by September, the state of Alaska was then planning to hold a great oil lease auction of state lands adjoining Prudhoe Bay, which was expected to bring the state treasury almost $1 billion—eight times the annual state budget.

In Congress, by common consent, the Senate Interior Committee would set the pace on the Native claims bill, having held hearings on it in Alaska a year earlier. After locking up 96 percent of Alaska in a freeze so that it could proceed undistractedly with this legislation, the committee recognized a special responsibility for prompt action, an obligation acknowledged by Chairman Henry Jackson when he was tightening the screws on Governor Walter Hickel:

> I would just hope that as Secretary you would keep this freeze on until we resolve the legislative adjudication of the Alaska native problem. I can assure you there is no reason, with two Senators from Alaska that are on the committee now, why we cannot act on that legislation *at this session of Congress* [italics added]. I think this would be the feeling of the Committee. . . . Otherwise the Congress is guilty of negligence in delaying this problem as long as it has delayed it, and I think I speak for most of my colleagues here on the Senate Interior Committee that we will act at this session [i.e., before the end of fall].

As for the administration, whether or not it had the vision to see that its foreclosure of offshore oil created a void that made Alaskan oil the more urgent, it had other reasons for handling its pipeline responsibilities with a brisk efficiency. An Alaskan oil boom was the longtime dream of the new interior secretary, and Hickel's success in pushing it along was essential if he was to preserve his political base. Richard Nixon's 1968 campaign team had wooed industrialist contributors with promises that a Nixon administration would be more sympathetic in moving business problems through the toils of Big Government; here was a maximum-profile opportunity to deliver on those promises. And the pipeline, the greatest commercial construction project in history, was the kind of achievement the new President wanted to be identified with, the kind of visible triumph presidential terms are celebrated for in schoolbooks and popular histories. When it was proposed that the pipeline route should go through Canada so that the oil would arrive in the Midwest, where the need would be greater than on the West Coast, Nixon was adamantly opposed. This, he insisted, was to be an "all-American" achievement.

By the same token, the Nixon men, who had already twice felt the power of the conservationists—in the furors over the Hickel nomination and the Santa Barbara Channel blowout—and who were now saddled with Hickel's agreement to get permission from the appropriate congressional committees before proceeding with any project in federal Alaska, well knew the capacity of the environmental issue to rise up and make a fiasco of the Alaska pipeline, a fiasco that would be laid on the

White House doorstep. They recognized the importance of presidential leadership and seemed prepared to furnish it.

Soon after the oil consortium had announced its pipeline plan, the White House stepped in publicly and set in motion a mechanism designed to assure the protection of the Arctic ecology, allay the fears of doubters, and involve the President directly and prominently in the timetable, thus to emphasize the importance of moving the project ahead not only swiftly but safely. As an earnest of his administration's determination to protect the environment, the President had named a conservationist luminary, Russell E. Train, as undersecretary of the interior. Train was now put in charge of a "departmental task force" to oversee North Slope oil development. In a further accommodation of environmentalists, President Nixon in April personally announced that he was beefing up the task force by adding a "conservation/industry ad hoc committee," whose roster included reassuring conservationist names. At the same time Nixon set a four-month deadline—September 15, 1969—for the task force to report directly to him on the precautions that oil company engineers would be required to observe before permission was given for a construction start.

This was a statesmanlike beginning that promised to face squarely the questions of substance, cut through the bottlenecks of procedure, conciliate the principled opposition, and give to the enterprise the stature of a national undertaking.

It should have been elementary to backers of the Trans-Alaska Pipeline that what must be avoided above all was the kind of environmental atrocity in Alaska that would galvanize the opposition. But in its impatience to begin making money from North Slope development, the Alaskan establishment rushed in and created one. The temptation that led to it was the booming airfreight traffic to the roadless North Slope that was supplying the needs of a spreading complex of oil installations and pipeline preparations. At the peak of North Slope activity, air carriers flew more tonnage to Prudhoe Bay in twelve months than had been delivered to Berlin in the eighteen-month airlift of 1948–49. And it was done without cost to the taxpayer or damage to the terrain.

But it was also done without profit to Alaska's truckers and their teamsters union allies, a potent force in Alaskan commerce and politics, determined and able to claim tribute from almost anything that passed through the largest American state. The thought of money to be had from oil activity, but of outsiders getting it, is repugnant to the Alaskan and intolerable to the trucker. All that needed doing to get the truckers in on the freight profits was for the taxpayer to build a 400-mile "winter road" across the trackless, frozen wilderness from the Fairbanks area, where roads stopped, to Sagwon on the North Slope. With the winter of 1968–69 already begun, the state of Alaska would have to move quickly if the truckers were to get some of the action before the return of "warm" weather stopped vehicular traffic in April.

Governor Hickel (serving his last weeks before Nixon tapped him for the Cabinet) polled the legislature, and amid breezy talk of competitive bids and low costs and trucker tolls that would "more than pay the cost," the road to Sagwon was authorized. As it happened, the competitive bids were turned down because it would take private contractors ten precious days to gear up; the job was turned over to the state's Department of Highways; the true cost was twice Hickel's estimate; and the tolls never materialized.

Alas, the consequences went beyond those commonplaces of political construction; the state, to meet the truckers' need for haste, threw away the ecological rulebook. The technique for building a safe winter road was known—heaping snow on the tundra and compacting it into a raised berm which would make a tolerable road until it began to melt in April—but it took time. Instead of piling up the snow, the state turned loose its bulldozers to scrape it away and to gouge a road into the tundra. Completed in March, the tundra-denuded roadbed disintegrated into a canal come the April thaw, a gaping scar across the Alaskan wilderness. It was, to quote a University of Alaska professor, "the biggest screw-up in the history of mankind in the Arctic." To conservationists throughout the lower forty-eight, it furnished a needed rallying point, an instant monument to everything they feared in the Alaskan boomer mentality, a "proof that Alaskans could not be trusted with the last remaining wilderness."

Even in its own terms, the winter road was a dud. Truckers got in only one month's hauling and delivered but seven and one-half tons of freight, a tonnage that three Hercules air transport flights could have accommodated. And the eroded-away road would have to be built again during the winter of 1969–70.

The episode was not without its element of unintended mirth. Back when winter still concealed the carnage, when teamsters were preparing to roll their rigs north, and few were questioning the runaway costs or the stillbirth of the touted toll system, the new governor of Alaska, Keith Miller, staged a christening ceremony, assuring with fulsome grandiloquence that the creator of the Walter J. Hickel Highway should not beat the rap: "This impossible road shall be known by the name of the man whose courage, foresight and faith in the Great Land gave Alaska what surely will become one of its greatest assets."

By summer the Hickel Highway debacle bared by the spring thaw had percolated through the conservationist community and allied academic enclaves and had helped catalyze a decidedly negative attitude toward the pipeline. This became apparent when the Twentieth Alaska Science Conference convened at Fairbanks in August 1969, attended by scientists, conservationists, oil company spokesmen, politicians, and students of public affairs from around the country.

The conference presented advocates of Alaskan oil development with an opportunity to reassure, or at least to divert with food for thought, potential opponents in intellectual circles who had shown their puissance in influencing official policy. To do so, the proponents would have to make a good case that appearances were deceiving, that Alaskans treasured the Alaskan environment as much as outsiders could, that oil revenue was desperately needed by the poor of Alaska, white and Native. If the proponents were to keep their eyes on the ball, they must be able to overlook the smug condescensions of outside ideologues who were having a good time disparaging Alaskans for overattraction to the Golden Calf and insufficient ardor for nature.

But the pipeline advocates were too obtuse or thin-skinned to stoop to the occasion. James H. Galloway, who as a vice-president of Exxon assumed he need jolly no one, was obviously affronted to hear conferees debating whether or not Alaska's oil should be developed, as if there were a choice. And he took it on himself to set them straight: "The oil is going to be extracted and some of the country hitherto unmolested is going to be torn up in the process. Let's not fool ourselves. This activity is already far past the point of return."

Alaska's Senator Ted Stevens, on his better days a perceptive conciliator, gave way to umbrage at anti-Alaskan barbs and played to his constituents instead of his audience. "I am up to here with the people who tell us how to develop our country," he said, and went on to denounce the initial federal regulations to govern the pipeline construction, drawn up with conservationist participation, as "stupid, absolutely stupid" interferences that reeked of federal contempt for Alaskans' capacity to manage their own land. As tactical error tightened its intoxicating grip, Stevens was emboldened to take on the *cause célèbre* venerated by all studious conservationists: the 10,000 used oil drums abandoned at Umiat, in the Alaskan Arctic, by the U.S. Navy when it moved out in the early 1950's, drums that, impervious to the passing decades, still defaced the frozen tundra. Senator Stevens saw those old oil barrels as assets. Looked at realistically, they were an "enhancement" of the area, he argued, later explaining that the Eskimos could have retrieved them and sold them for $7 to $10 apiece. The market was not identified.

The imagery Stevens evoked—of environmental protection and economic uplift carried out by mean scavenging—was equally an affront to the professional environmentalist, who thought in grandiose terms of continental land freezes and eternal prohibitions, and to the rude Native who, coming to see himself as the owner of oil wells and endless vistas of timber, did not take kindly to being declassed to junkman of the Arctic. To many in the Fairbanks conclave, the arrogance of the oilmen and the provincial chauvinism of the politicians were a double-barreled confirmation of their worst fears. Conferees who had come to Alaska with suspicions to sort out went home with a cause to prosecute.

As helpmate in bringing about the second prerequisite to a pipeline start—passage of a Native claims bill—white Alaska got off to an even worse start than it had at inspiring outsider confidence in its reliability on the environmental front.

First, the Alaskan establishment—the predominant element of politicians, lawyers, editors, and businessmen—defaulted on its responsibility to help the Natives, divided as they were into numerous factions of Eskimos, Indians, and Aleuts, to unite on one claims proposal and present it effectively to a distant and incomprehensible Congress. So the Alaskan Federation of Natives was not ready with *any* proposal until June 1969, losing more than half a year and frustrating Senator Jackson's plan to get the claims bill passed by the Senate in early 1969.

Second, when a New York-Washington legal squad, headed by former Supreme Court Justice Arthur Goldberg and former Interior Department Solicitor Edward Weinberg, stepped into the void and helped the Natives come up with a formidable brief, the Alaskan establishment fought the result in an ignorant and suicidal manner: They contemptuously dubbed the Native proposal the Goldberg Bill and suggested, as one observer aptly summarized the innuendos, that it was "a Jewish scheme foisted on dumb Indians." Just as a year earlier influential Alaskans could not bear to see out-of-state airlines getting to carry the freight to the North Slope, so now they bridled at the thought of Natives getting acres they might want for themselves or siphoning off one forty-fifth of the future oil revenues.

The Alaska bar presented an especially sorry performance. Its members, apprehending that they had so ignored the Natives over the years as to have missed the boat on the greatest legal fee bonanza a bush lawyer could dream of, began to have fits of remorse. As Goldberg and Weinberg gathered in the soon-to-be-rich tribes, the provincial lawyers filled the air with strange-to-the-ear talk about ethics and with demands that the New York interlopers withdraw. It did not pacify the Alaskan lawyers to learn that the impeccable Goldberg toiled for free—"as a public service," he said—but rather it enraged them the more. Some sent letters to Goldberg that accused him of hustling the clients of Yukon nonentities and interfering with their "contracts."

Such impertinences had serious consequences. Mr. Justice Goldberg proved a bit stuffy and flighty for an old labor mouthpiece. "These communications are entirely lacking in the respect owing one who has served our country in three of its highest offices," he protested, and he informed the Alaskan Federation of Natives that he was resigning as its counsel rather than endure further indignities. But then he wired the AFN an offer to reconsider his resignation. The AFN, apparently indifferent to Goldberg's services to the Republic, held a vote in Anchorage on whether to accept the resignation—and came up with a tie. On a second vote they mustered a majority for giving Goldberg another chance to make up his mind.

Several weeks passed after this thin expression of confidence while Goldberg wrestled with the imperatives of dignity. Finally he agreed to come back in, after arrangements had been made to write a schedule of attorneys' fees into the claims settlement. But critical months had been wasted and future delays born, for the Natives acquired a distrust for lawyers—Alaskan and outsider alike—that would henceforth dog the claims process.

In mid-November 1969 an edifying tableau was on display in the Senate Interior Committee conference room. U.S. senators do not gladly expend their time and equanimity haggling over abstruse property settlements for invisible tribes in somebody else's state. Yet here they were around the long table, under the lead of Chairman Jackson, "marking up" the Native claims bill—which is to say, agreeing on the legislation clause by clause. A few days of this should produce a finished bill. Recommended by the prestigious committee without partisan division, a bill of so technical a character was sure to pass the entire Senate all but unanimously, and it seemed likely that the House would follow the Senate's lead.

There was a general air of optimism around the conference table as the markup began. Over the past couple of years—since the imminence of oil had begun to concentrate the mind—a rough agreement had developed over the parameters of the Native claims settlement, agreement among the Natives, the Department of the Interior, the state of Alaska (as represented by Governor Hickel's 1967 task force), the Alaskan congressional delegation, and the chairmen of the Senate and House Interior committees.

Among these, it was widely accepted that about 40 million acres should be reserved, in some fashion, for the Natives—some of it where they lived, some of it in far-flung areas of economic value; some of it under full Native control, some of it for limited use only. In addition, there should be a large cash payment, distributed over a period of years —between $500 million and $1 billion—some of it paid by the federal government, some of it by the state of Alaska out of a minute portion of its coming royalties from oil and other minerals. Important differences existed over just how the "some-of-its" should be spelled out, but with a consensus almost in hand on the broad outlines, the give and take of the legislative process should resolve the rest. The committee hoped to produce a bill within a week that would pass the Senate within a month.

Legislation in the closing weeks of a congressional session, however, is vulnerable to the defection or the ineptitude of any part of the consensual coalition, and now blows of both kinds fell from white Alaska on the Jackson effort. As its opinion jelled, it had become clear that white Alaska had no vision of walking into the oil-rich future hand in hand with the original inhabitants. It was insisting on taking for itself the federal lands of suspected mineral value before the Natives got anything, and it was demanding that any cash settlement for the Natives must be paid entirely out of federal revenues. If the Natives were to get the 2 percent of oil revenues legislators bandied about, let it come out of the feds' 10 percent, not Alaska's 90 percent.

So went the talk on platforms and across barstools in Anchorage, Juneau, and Fairbanks. Vox populi welled up in Chamber of Commerce resolutions and newspaper broadsides that in due course were synthesized into a message from Alaska's Republican Governor Keith Miller which on November 19 was dumped on the Senate Interior Committee with the effect of a bucketful of sand on a humming machine.

The gist of the governor's letter was that the committee must not be taken in by the grand larceny being attempted by the Natives and their out-of-state Edgar Bergens and should hearken to a "fresh approach"; that the committee was contemplating giving the Natives far too much land and money; that the 40-million-acre formula was "completely unacceptable and unrealistic" and that half would be more than enough; that the Natives must not be given any land the state wanted for itself, nor should they have the right to *choose* any land, but rather should accept what was allotted to them; that the rights to any minerals found under land that the Natives did get must belong to the state, not the Natives; that the $1 billion cash settlement was twice what was called for, and even $500 million would be tolerable only if the federal government paid it all; that "we are unalterably opposed" to paying a 2 percent share or any other share of mineral royalties to the Natives. The governor closed by regretting the "false hopes that have been raised among our native peoples by the eastern attorneys representing them."

The governor's letter was the less magnanimous against a historical background with which the senators were intimate. Ten years earlier, upon Alaska's becoming a state, Congress had granted the new state *government* 102 million acres out of the federal domain, which it would have years to select; this amounted to almost a third of the entire territory and five times the proportion of land given to the other western state governments upon admission. Beyond that, Congress had granted the Alaskan government 90 percent of all future mineral royalties from the public domain, almost three times the percentage allowed to the other western states.

Even if the senators were disposed to overlook the tone of *lèse majesté* inherent in such Miller presumptions as "completely unacceptable" and "unalterably opposed"—and they were unlikely to do so, for this was a prickly body, self-consciously in command of the bunghole on the honey barrel—many of them could not but be offended by the governor's stinginess toward the Native and ingratitude to the nation, for they were in the business of divvying up vast benefactions and they knew that in politics the saving grace of greed is the willingness to give a little to get much.

One member threatened to offer to the claims bill a spoiler amendment that would give his own state the 90 percent of federal mineral revenues that Alaska was getting, thus to invite a stampede of home state chauvinisms. Another member was for dropping the legislative approach altogether and letting the Natives and the land sharks go at it in the courts—a route sure to paralyze the Alaskan economy for decades. Mischief was loose, and the consensus that had all but jelled began to dissolve, a process that was encouraged when the committee's two Alaskan members, the Democrat Gravel and the Republican Stevens, came onstage not as healers but as adversaries and monopolized the committee's waning time and tolerance with an ongoing debate.

Chairman Jackson extended himself in an attempt to rescue the claims bill from the returning boomerangs so recklessly tossed by Alaskans. Under his prodding, Gravel and Stevens, who to their credit saw that Miller's anti-Native line and the intra-Alaska donnybrook were inducing in Congress a spirit of malign neglect toward the Great Land, compromised their differences. A subsequent parley involving Jackson, ranking Republican Gordon Allott, and the two Alaskan senators produced a new consensus that closely resembled the package Governor Miller had condemned: 5 to 10 million acres to be given outright to the Natives, who would have full control and mineral rights; 40 million acres the Natives could use for subsistence purposes; $500 million out of congressional appropriations; another $500 million to accrue from a 2 percent sharing in state mineral revenues. A December date was set for committee adoption of the compromise.

But then another boomerang circled back to shatter any hope of the bill, and of a pipeline start, in 1969. Senator Gravel, apparently eager to be seen back home as the Henry Clay behind the looming legislative victory, reached clumsily for the role of committee spokesman and spoke to the press about plans and agreements that, while deliverable in the deft hands of a Jackson working behind the scenes, were endangered by the spotlight of publicity. Senior senators do not like to read in the morning paper that some freshman has committed them to meet every day on a bill until it is passed or to hear that a newspaper in Alaska has printed the details of a compromise reached in their name before they have had a chance to deliberate on it. So the December meeting of the Interior Committee, which might have produced a bill, brought only acrimony and a statement from Chairman Jackson that all previous understandings were off, that a year's work had washed out, and that the committee would start from scratch next year, when it would have to consider each element of the issue separately and put together a brand-new bill. The only positive step the committee took that day, before adjourning for the year, was to make Jackson its "sole spokesman" henceforth on the claims issue, a pointed rebuke to Gravel.

It was the operating plan of the House Interior Committee to sit back, let its Senate counterpart take the lead in thrashing out pipeline-related issues, and then work its will on the Senate's handiwork. So while the Senate deliberated, the House committee kept an eye on the business, held a meeting or two, and sent its Subcommittee on Indian Affairs on a field trip to Alaska. All the while its members were forming impressions according to their singular ways of sizing up public issues.

These impressions were to be hurtful to the Native claims legislation and to clearance for the pipeline. The unnecessary harm was a reflection not only on the capriciousness of congressmen but also on the failure of Alaska representatives and oil consortium agents to tend to their knitting or even to understand that there was knitting to be done.

A progression of miscalculations led the Trans-Alaska Pipeline System (TAPS) to make a bitter-end enemy of the ranking Republican on the House Interior Committee, John P. Saylor of Pennsylvania. Overestimating its engineering readiness and underestimating Interior's seriousness about permit qualifications, TAPS geared its purchasing of 800 miles' worth of steel pipe to a September 1969 start that was illusory. The premature deadline for delivery had the effect of excluding the American steel industry, which did not currently manufacture the unusual four-foot diameter pipe required and would need time to create new production facilities. United States Steel was anxious to convert a Texas plant for this order. Kaiser Industries proposed to build a new plant in Alaska. Others vied for a share of this steel contract of historic dimensions. If TAPS had studied its engineering problems with a tad of humility and its political problems with a hint of intelligence, it would have perceived that time was something it had plenty of and that allies in American industry, in organized labor, and in Congress were something it had great need for; instead, it peremptorily ordered the entire lot of pipe from three Japanese firms which, as one might suspect, were ready to meet the specifications and the deadline with no ifs, ands, or buts. Not even the pipe scheduled for use in the second and third year of construction would be American-made.

Enter John P. Saylor. Being a patriot of the old stripe, he believed in "buy American," the more so in the case of a national project that fed off the federal domain and needed federal permits. Being a Pennsylvanian, it seemed to him worthy that Bethlehem Steel and its workers and stockholders should have some of the gravy. And being a congressman, he was affronted by the spectacle of a bullion train passing through his sphere unmolested and shamed that Japanese were to get all the booty and Pennsylvanians none.

By September, when the oil consortium appeared before Saylor's committee, shipments of pipe were arriving in Alaska marked with the name Nippon Kokan Kabushiki Kaisha. Saylor was a blunt man. Why had they not ordered their pipe from Bethlehem Steel? he wanted to know. Waving aside their answers about the four-foot pipe and the deadline as so much humbug, he passed sentence: "You've just ruined yourselves as far as I'm concerned. I'm going to fight you every way I know how, and the only way I know how is conservation."

Years later, whenever oil executives passed through their marshaling yards at Prudhoe or Fairbanks or Valdez and observed workmen applying rust preventive to endless stacks of timely received but as yet unlaid Japanese pipe, they would remember Saylor.

In October the House Indian Affairs Subcommittee, headed by Representative James A. Haley (D-Fla.), known to be sensitive about recognition of his importance, arrived in Alaska to tour Native villages and to make an inspection flight over the route designated for the pipeline. The disgruntled Saylor, still fuming over the alien pipe, accompanied the tour in the skeptic's seat. Considering the significance of these gentlemen to TAPS and to Alaska, one would have expected Alaskan boosters and the vaunted oil lobbyists to attend on them discreetly, to fabricate an occasional demonstration of local esteem for the visitors that they might "think positive" about Alaska, to be ready to smooth away misconceptions planted by cranks along the way, and to inflate the touring fact finders with as much helium as was seemly about "the greatest private construction project in American history" and about the Great Land's reverence for nature.

Instead, the congressmen were suffered to arrive at their forlorn stops unheralded and to depart unnoticed, in demeaning contrast with the triumphal passage over the same route a short time before by one Ted Kennedy—who could claim none of their legislative omnipotence. And when, perhaps soured by the indifference of their hosts, they flew over the pipeline route, which they assumed was "frozen" of all activity while awaiting *their* approval, and saw cropped hilltops, a swath of destruction extending through the wilderness for tens of miles, and recurring scenes of what appeared to be drilling, no interpreter was at their ears to remind them that they had specifically sanctioned all this, that the raped hilltops were in fact only "gravel borrow pits" for an approved road, while the gash through the forest and the drilling sites were all part of the authorized program to collect core samples of the soil, a precaution so that the environment could be truly protected when the real construction began.

The fact finders returned to Washington assuming they had witnessed an outrage, if not an insurrection, and when an unbriefed Undersecretary Russell Train appeared before them—as luck would have it to seek their permission to lift the freeze—they rounded on him uproariously. A gleeful Saylor loosed a bewildering mix of metaphors. Crying "denuded hilltops," he accused Train of "coming to us after the horse had been stolen" and of permitting devastations that "were enough to raise the hackles on the back of even a bald head!" Train was sent away, rebuffed and empty-handed. Upon finding out what had hit him, he wrote to the committee explaining, with attempted tact, that what members had seen was only what they had specifically authorized, but a grievance once clutched to the bosom is not easily given up.

It might be argued that the lobbyist cannot be held responsible for the misunderstanding by congressmen of their own handiwork. But it would be a specious argument. It is *ever* to be assumed by the lobbyist, is in fact his reason for being, that in general the Congress does *not* know its craft; that its operating milieu is ignorance of the details and consequences of its enactments; that the lobbyist's task is either to dispel that ignorance or to enlarge it, depending on the circumstance, but never to ignore it.

The hostile reception by the House Interior Committee to Undersecretary Train's request that it concur in lifting the pipeline freeze coincided with a hesitation by the Senate committee, which declined to act until it held hearings. The Senate turndown was traceable to mounting opposition to the pipeline from the conservationist movement, dormant only a few months before but lately roused to action by the misadventures and laxities. Train had argued to senators the need for action within a "time frame" that took into account the oil companies' construction schedule and investment losses from delay. But senators weren't impressed.

"How long are we going to permit the private sector to decide these issues for us?" asked Gaylord Nelson.

Lee Metcalf agreed: "Much as we need oil from the North Slope and would like to accommodate these people who've invested billions of dollars there, we should proceed cautiously."

These congressional rebuffs brought home to all concerned parties the gravity of the abdications wrung from Hickel during the confirmation hearings. Up to now this reality had been masked. The committees had treated the pipeline, which had *formally* come along after their pact with Hickel, as one of those public necessities for which the freeze could be "modified," and on two occasions they had promptly given their concurrence to Hickel on modifications to facilitate pipeline preliminaries—first a fifty-mile stretch of road, then the "geological and engineering investigations" along the proposed route, the menacing aspect of which from the air had so disturbed the touring House members.

For its part, the Interior Department had gotten away to a promising start, considering the unprecedented problems presented by so mammoth a project crossing so vulnerable a landscape under the suddenly intense searchlight of the environmental age. Interior had promptly identified the basic environmental questions, had insisted that the oil consortium answer them while at the same time seeking answers from conservationists and its own experts, had issued tough construction stipulations as the answers came in, and was setting up a monitoring system to assure that they were carried out. Interior had not lagged behind TAPS but had kept ahead.

But now, just when the disparate horses Hickel was driving were turning balky or unruly, the committees had asserted their new rights and had taken from him both his lash and his sugar by making it clear that the secretary's orders and promises were illusory, mere recommendations to be debated and disposed of by two committees of Congress.

This was an unhappy moment for such a demonstration, for Hickel had reached a crisis point in his stewardship over the Trans-Alaska Pipeline. He had embarked upon the pipeline enterprise as one who by ingrained instinct and working philosophy was *for* it. He had the frontier entrepreneur's feeling for big projects that created business and jobs. He believed that mineral resources locked away in pristine wildernesses were for people to *use* and that ways could, and must, be found to keep environmental disruptions to acceptable levels. He had almost unbounded faith in the capacity of technology to overcome both the engineering and the ecological obstacles that this gigantic project comprehended. He would insist that technology did in fact solve them, yet he was sympathetic to the arguments of the builders that they should not at the outset be expected to produce theoretical, definitive answers to all the riddles that lay in wait 500 miles and two years down the line, that the pragmatic thing was to solve them as they were encountered, drawing on the know-how learned along the way. And he was constantly pricked by reminders that the home folks in Alaska, a continuing factor in the life scheme that stretched ahead, were counting on him —with increasing testiness—to deliver on the permit; if he did not, a host of businesses formed in anticipation of oil boom commerce would go bankrupt, as would the state government itself if the anticipated oil revenues did not materialize.

He had come to this task as one who found it tempting to succumb to the mystique of Big Oil omnicompetence. "I have always had a healthy respect for America's oilmen," Hickel has said. "The men at the top are some of the most talented in the country. The men in the second and third echelons can be some of the roughest and toughest men in the world. I like them and respect them." Hickel insisted, however, that he was not "awed by their power and influence." In important respects, the industry had enhanced its image on the North Slope—in its surmounting of the dread Arctic to find and bring up ice-locked petroleum, in the ingenious innovations of its cold-conquering complexes that were spreading across the tundra and that would be capable, on demand, of pumping enough oil to equal thrice our Middle East imports, and in the vast scope of its plan to pipe that oil across a spectacularly forbidding subcontinent.

The oil consortium's initial approach to Hickel's federal jurisdiction was somewhat lacking in suavity but was full of the presumption to be expected from titans. "The first time they came in was in April 1969," recalls a senior Hickel aide, "and the message I got was they had to have the permit by the first of July or there would be a disaster."

After being informed of the intended pipeline route, two teams of Interior Department experts went to Alaska, one in April, another in May, for an on-the-scene examination. Between them they traveled the entire route, compared notes with resident experts on the area's ecology, and came back with the same conclusions: Most of the 789-mile pipeline would have to be elevated several feet above ground, primarily because the oil would go through the pipe at a temperature of 160° F (it came out of the ground at 125° F, and the friction of the pumping process would raise it higher)—a heat that, if the pipe were underground, would melt the surrounding permafrost, gouging out an ever-widening channel of erosion that would be murderous to the environment and destructive of the pipeline itself. But here the enigma began. TAPS was proposing a conventional *underground* pipeline, to be buried for all but 40 miles. TAPS seemed overly mindful that elevated construction cost a good deal more than conventional and oblivious to what permafrost could do to the pipeline as well as the ecology. "They approached the whole thing as if it were routine, just a matter of digging a hole and putting the pipe in," Russell Train later recalled. For Interior officials who were extending themselves to bring off a quick, smooth permit process, this was a confidence shaker.

And it was only the first. Promptly upon receiving TAPS's formal request for a pipeline permit on June 6, the Interior Department shot back a list of inquiries designed "to indicate the kind of questions to which satisfactory answers will be required before permits can be given for the use of public lands." The department asked for answers within a month, so it could act in "anticipation of the industry's timetable." TAPS replied in ten days but with a broad description of its plans rather than precise answers to the questions. This rubberiness was to be a continuing policy, as demonstrated at a subsequent meeting in Alaska when Undersecretary Train pressed for specifics on where TAPS intended to elevate the pipeline. The TAPS spokesman replied sanguinely: "We'll elevate the pipeline where we have to."

And again: Interior was concerned about an especially hazardous area of permafrost in the Copper River valley. The earth there contained an unusually large proportion of ice that would be melted by the heat from a buried pipeline. When it melted, Interior maintained, the pipe would perforce bend, perhaps break. Yet TAPS insisted it could bury the pipe even there, though it would not say how it could do so safely.

This pattern of unresponsiveness threw a monkey wrench into an approval process that Hickel and Train wanted to be a model of efficient speed. For a time Interior hoped it was due merely to the crotchets and defense mechanisms of Big Oil—the congenital secretiveness long nurtured within the oil companies; the unconscious arrogance fed by decades of getting their way on their terms; the disdain for mere bureaucrats and their ineffectual paper shuffling felt by men who had manipulated senators and even chiefs of state. Train sensed in them the professional's resistance to meddling by amateurs, a tone of "We know how to do it; now leave us alone."

Well, Interior was prepared to swallow a lot of that. The Lilliputians, knowing the strength of their rope, could forgive Gulliver's ill-mannered underestimation of them. They hoped that departmental persistence would soon bring the giant into conformity, and they reported to the White House, according to a careful student of its dispatches, that they were "moving toward approval of the project on the theory that TAPS would provide the necessary technical information if prodded sufficiently."

By October Hickel was thoroughly stymied by the oil consortium's unresponsiveness, which he saw playing into the hands of a rising conservationist opposition. To meet these threats, he needed the undiminished authority of his office and administrative flexibility, so that both sides not only would know that they must reasonably come to terms with him but also could see piecemeal evidence that they could gain their ends by doing so. For, as Hickel reported to the President: "We have been guided by the view that oil development and environmental protection are not inconsistent. We have attempted in our proceedings to reach an equitable balance between the just concerns—and timetables—of industry, and the just concerns—and timetables—of the public."

Hickel worked out a scheme to lead both sides into cooperation. He would focus first on those segments of the route with the least environmental problems, granting permits section by section as TAPS came up with the data requested on each. Construction could start right away in some zones. He would put teams of government inspectors in the field to work alongside the construction crews, inspectors who would enforce regulations and would wield the power to shut down construction whenever they were violated or when unforeseen dangers arose or during periods of special hazard to wildlife—nesting, spawning, migration. Thus could Hickel prove month by month his reliability to the conservationists, satisfy the demands of his Alaskans for action, and draw the oil consortium into full cooperation while yielding to its basic complaint that it should not have to answer today a question it would not face in actual construction for two years.

It was with this scheme in mind that Hickel had sent Train before the Senate and House Interior committees in October to request a further modification of the Alaska land freeze. Their refusals to grant it, pending hearings and the like, destroyed the plan by undermining his authority. For the oil companies, it would be gamble enough to build half a pipeline in the faith that a secretary of the interior would one day let the other half be built, but to add two congressional committees to the game made the odds too great. Nor need the conservationists bargain with Hickel for half a loaf so long as their friends on the Interior committees were of a mind to prolong the freeze indefinitely.

Wally Hickel thought at first that the oil consortium's noncompliance was rooted in a supercilious ignorance of the uniqueness of Alaska. "It was like the companies wanted an open hunting license," he said later. "They had no engineering plan, no criteria on how to build a pipeline, no nothing; but they wanted a permit."

The companies acted, to the Interior Department at least, as if an Alaska pipeline were no different from the pipelines they had built in the lower forty-eight and as if the department should okay it in the perfunctory manner of the past. "They turned in proposals like they were going to build a pipeline from Lubbock, Texas, to the Gulf, as if they already knew the terrain they were going to confront. I knew that Alaska is just not that way."

When enough time had passed so that ignorance of Alaska could no longer be a tenable explanation, Hickel began to question first their competence and the effectiveness of their organization and then their interest. "It was really a loose organization, and I couldn't get any information from them. It seemed like they did not have an interest in building the pipeline."

This suspicion grew and came to center on Exxon.

During an interview with us in 1981, former Secretary Hickel voiced suspicions that Exxon was purposely "dragging its feet" from mid-1969 to late 1970, that it "was not enthused about the pipeline" because it didn't have as much "production" on the North Slope as its partners Arco and BP did and because Exxon *did* have large holdings in Canada's Mackenzie River valley and therefore wanted the Arctic pipeline routed east and down through Canada to the American Midwest rather than south through Alaska to the Pacific. Hickel blamed Exxon for a series of delays in meeting Interior's permit timetable. Each partner in the pipeline consortium had a veto power over any submission to Interior, Hickel pointed out to us, "and the Exxon-Humble people did a lot of vetoing." Moreover, occasional reports surfaced of foot-dragging accusations against Exxon by British Petroleum officials, even of a lawsuit allegedly threatened by BP's chairman, Sir Eric Drake.

But if there was a rift between Exxon and BP over deliberate delaying tactics, it was contained and resolved within the Oil family, and journalistic probers have not been able to sink their teeth into it. BP officials subsequently soft-pedaled the matter, and Hickel, when pressed for more details, referred me to former subordinates, who were unable or unwilling to provide them. Certainly some oil companies had reason to fear the too rapid influx of Alaskan oil. As *Business Week* put it on February 1, 1969, "Oil producers are fretting over the possibility that when cheap Arctic oil begins invading the market in 1971 or 1972 the present price structure and the state production controls that help maintain it may crack." And certainly there was a lack of push by the oil combine in much of 1969–70, in sharp contrast with its all-conquering drive before and after—a lack that tolerated an organizational shambles so bad as to make deliberate delaying tactics superfluous. "We were faced with 'decision by committee,' " Hickel explained. "By the time all the principals could be telephoned in Texas, New York, California, and London it could take weeks just to get an agreement on what color to paint the toilets in the construction camps."

Faced with so laggard a performance, and one which the interior secretary suspected was deliberate, why did not the secretary—and the President, who had set September 15, 1969, as the deadline for substantial progress—start knocking oilmen's heads together when the deadline passed without progress? It was these *officials,* not oil executives pursuing differing estimates of corporate advantage, who bore the responsibility for national oil security and who, holding the keys to the vast federal oil domain and to the investigative apparatus and to tax and price subsidies, wielded ample power to persuade and compel.

But according to those who should know, it was just not in the President's makeup to involve himself in such functions of personal leadership as herd riding and arm twisting. The present Senate majority leader, Howard Baker, judges that Nixon lacked the follow-through and give-and-take capacities for getting-things-done leadership: "Nixon never paid enough attention to it to be successful." John Ehrlichman, Nixon's chief of staff for domestic affairs, says that while Nixon supervised the posture taking on charged issues freighted with political gain or danger, such as aid to parochial schools, abortion, and crime, he remained aloof from substantive involvement in *governing:* "Indeed, Nixon insisted I not bother him about many things. He preferred to stay away from the environment issues, for example—free to criticize, taking the bows when things went well, disowning me when EPA closed a plant or ordered expensive retrofit for a factory."

Just so with the Alaskan pipeline when it turned out to be more than an occasion for political ribbon cuttings and began to offer nettlesome problems. The President faded away from the pipeline scene and over the next four years would pop up only for a rare offstage PR directive.

If when the President withdrew from the pipeline stallout he had left behind an interior secretary with the undoubted power and presidential backing to whip things back on schedule, all might yet have come right, but by late 1969 Nixon was also withdrawing from Hickel.

The President had decided, after a few Cabinet meetings, that his interior secretary was a lightweight. Hickel was something of a garrulous butter-in who regarded Cabinet sessions as forums for airing his pet nostrums. Back when candidate Nixon was shagging delegates for the upcoming Republican National Convention of 1968, he had listened attentively to the then Governor Hickel, in an Oregon motel room, expounding on the same notions: the boons of lower interest rates; the panaceas to be achieved through bond financing. Then Nixon had heard out Hickel with feigned admiration, occasionally interjecting sweet nothings ("Wally, you really belong in the United States Senate," or—turning to aide Robert Ellsworth while motioning toward Hickel—"Bob, the top of the ladder starts here"). But by now Nixon had taken to wincing when Hickel spoke up in the Cabinet and would wave him down or cut him off with "We've gone all through that," and he became less and less accessible to Hickel and the problems of the Interior Department. Hickel could not get in to see Nixon because, according to then Nixon aide William Safire, Hickel was not seen as *un homme sérieux* (a standard that does not seem to have been rigidly applied).

For reasons entwined in the Nixon personality and method of politicking, the President would neither work with Hickel nor replace him but was keeping him on in a power vacuum while the effectiveness of the department in all things requiring clout and decisiveness was crumbling. This was partly because Nixon was squeamish about disciplining or firing people face-to-face and partly, as shall be seen, because he didn't want a troublemaking, embittered Hickel on the loose until the next election—the midterm election of November 1970—was safely past.

By late 1969 the word was out among Washington lobbyists and government higher-ups that Hickel no longer enjoyed the confidence of the President, and speculation hummed that Hickel's authority in sensitive areas (i.e. business regulation) would soon be shared with one Fred Russell, known to be close to Nixon and well regarded by the regulated business interests. By early February 1970 Hickel was having daily discussions with his staff about what he termed "the cool winds blowing our way from the White House" and the hostility Interior was encountering there on its budget requests and on Hickel projects in general. So it was that over a long period of time, dating from the first Cabinet room humiliations beginning in April 1969, the buffeted Hickel gradually became too unsure of his muscle to take on the oil pipeline consortium frontally. There then developed a stalemate, during which Interior continued to block the construction start but failed to raise the kind of hell that could have broken the bottleneck.

In the latter months of 1969—while Secretary Hickel waited impotently for signs of compliance from the mysterious TAPS that would justify him in again asking permission from his congressional overlords to raise the freeze, while the Senate Interior Committee floundered toward its first but not last impasse over Native claims legislation, and while Alaskans made ready to try again for a piece of the North Slope freight traffic by rebuilding the Hickel Highway so as to bypass the long stretches of terrain destroyed by the first attempt—a minor, uncontroversial piece of cosmetic-appearing environmental legislation was making its routine way toward uncontested passage by Congress. Unrecognized amid its platitudes was a sleeper clause—102 (2) (c)—that would magnify incomparably the obstacles before the Trans-Alaska Pipeline.

The purpose of the bill was laudable enough: to ensure that the adverse environmental consequences of government action would be anticipated and weighed, to make federal bureaucrats more environment-conscious, more mindful of the damage that could flow unintended from their acts. The device for making them more mindful—and here is the sleeper—was to require them, when making decisions on any new project involving federal property, money, or permits, to draw up an "impact statement"—a list of all the possible adverse effects, along with a list of all the alternatives they had considered before making their decision.

The idea for this "action-forcing mechanism" had come from Professor Lynton Caldwell, a sometime consultant to the Senate Interior Committee. The Senate author of the bill, Chairman Henry Jackson, saw the clause not as a new grant of power, positive or negative, but as a sort of management prod for "internalizing environmental values." The members who cleared the House version in committee thought they had kept out anything controversial. The bill had been modeled after the Full Employment Act of 1946—a toothless expression of benign sentiments—and staffers said that the subcommittee "wanted pretty much to leave it that way."

When a House legislator, more anxious for something toothy, proposed that agencies be empowered to issue stop orders on environmental grounds, he was voted down on the theory that such a clause would prevent this modest bill from passing, and when it was pointed out that one of the rhetorical clauses in Jackson's bill, about people's "environmental rights," might somehow be misconstrued to bring the courts into the act, the clause was stricken from the bill.

So the National Environmental Policy Act moved to passage, without receiving real news notices, without arousing lobbying activity, without exercising the executive branch, without discussion in interest-group newsletters beyond mention of its name and legislative status. Such argument as it engendered concerned only matters of in-government housekeeping and jurisdiction. At key legislative stages, NEPA was not even considered worthy of roll call votes.

Once enacted, NEPA became the responsibility of the President, advised by his administration. If the President were apprised of a serious reason to veto it, and did so, that would be the end of it; its authors had recognized that it was not the kind of a bill to override a veto when they removed from it any controversy they were aware of.

In the Interior Department, Solicitor Mitchell Melich took a look at NEPA in December 1969 and saw something that worried him. Though he did not foresee the litigatory swamp that yawned, he did see that all those impact statements could cause a lot of paper work for the department and could lead to any manner of complications. He advised Secretary Hickel to urge the President to veto it. But the beleaguered Hickel, gun-shy over repeated gestures of White House displeasure, did not consider Melich's objection serious enough to risk another rebuff. "I don't believe anyone in government even studied that bill," Melich recollects. "It looked simple, so legislators passed it."

The officer in immediate charge of pipeline clearance, Undersecretary Russell Train, saw no occasion for alarm, though he would head the agency created by NEPA: "When NEPA was enacted and signed into law no one appreciated, and I mean no one, not even the congressmen who passed it, the implications of the impact statements or avenues for litigation that were implicit and explicit in the Act. No one really discussed it. It wouldn't have occurred to anyone to exempt the pipeline or any other project from NEPA. . . ."

Congressional staff members involved with the legislation confirm Melich and Train. "If Congress had appreciated what the law would do," said one, "it would not have passed. They would have seen it as screwing public works." Concurred another: "If Congress had known what it was doing, it would not have passed the law." But did the President and his advisers know what *they* were doing?

On New Year's Day 1970, Mr. Nixon was at San Clemente. It was a slow news day in the middle of a holiday weekend, and the President, wanting to take advantage of the news lull and as well to start off the new decade with something portentously appropriate, sifted through the slim pickings on his desk and came to the National Environmental Policy Act, awaiting his signature.

The bill had passed through the supposed gauntlet of concerned agencies and presidential counselors without encountering serious objection or even causing much discussion, recalls the White House aide on environmental matters, John Whitaker. No one had spotted the sleeper. "No consideration was given . . . to how the new law could be used as an instrument of litigation to delay or stop federal projects that required federal permits."

So the President called in the newsmen and signed the bill into law with a flourish that made front pages across the country: "It is particularly fitting that my first official act of the new decade is to approve the National Environmental Policy Act—the 1970's absolutely must be the years when America pays its debt to the past by reclaiming the purity of the air, its waters and our living environment."

The failure of Congress and the White House to see NEPA as a revolutionary act was due not to the inscrutability of the bill but to their own lack of diligence and craftsmanship. The absence of concern, the lack of interest must be laid not to the simplicity or blandness of the bill but to the lack of care, competence, and imagination among those who should have seen in it a vehicle of revolutionary change. For once it was on the books, conservationists, lawyers, and judges quickly enough pounced upon it, seeing in it a new, fundamental grant of power that was to become the environmentalists' major tool for turning around public policy in the 1970's.

The environmental impact statement required by NEPA for all undertakings was, creatively seen, a threat to almost every project in the land. Once a statement was required, the question of whether that statement was adequately rendered, whether it fulfilled the intent of the law, could be challenged. What if the impact statement was laxly prepared? What if it did not list *all* the potentially adverse effects or did not discuss them thoroughly enough? What if it did not explore all possible alternatives? The theoretical possibilities for challenging an impact statement were immense.

Take the Trans-Alaska Pipeline. The potential adverse effects upon every kind of animal, bird, fish, plant, insect, soil, and water could be argued back and forth for a decade. And the government would not only have to defend the adequacy of its statement on all this but would also have to prove that the agreed-upon pipeline route was both defensible in itself and superior to any other imaginable route or means of transport on land, on sea, or in the air and why the others had been found wanting.

Two findings remained to transform the potential effects of NEPA into actual effects. Should not any citizen be able to challenge impact statements? And were not the courts the place to challenge them? The courts promptly agreed to both propositions.

On the day the courts did this, the near certainty of years of litigious delay—a delay that would be far more important than the ultimate decision—descended upon every developmental undertaking of consequence in America. Had this come about as the result of a conscious national choice, openly argued, with all the penalties as well as the gains understood, it could have been called a noble experiment. But to adopt, after presumed deliberation by House, Senate, and White House, with a host of interest groups looking on, so consequence-laden a measure *by accident,* in ignorance of its lurking but unnoticed content, is a grotesquerie that calls into question the basic capacity of our political system.

By mid-April 1970 the mistakes and negligences of a year had suddenly claimed their combined price. The ignored Alaskan Natives, charging that TAPS had reneged on its promise of construction contracts and jobs and angered that it refused even to treat with them on this small matter, had sued for breach of contract and obtained a court injunction against the pipeline's crossing of their claimed lands. The conservationists, contending that Interior had failed to produce the impact statements required by NEPA, had obtained yet another and more serious court injunction forbidding any permit until the litigation was resolved. The inability of Congress to pass a Native claims bill had caused a federal judge to despair of allowing the pipeline to go forward. And Secretary Hickel, hoping against hope for progress from TAPS, had instead received a fell report from U.S. Geological Survey chief

William T. Pecora: "The collective judgment of the works group is that TAPS has not demonstrated acceptable fundamental design criteria for below ground construction in permafrost of a hot-oil pipeline that would be reasonably safe from failure."

To make matters worse, Secretary Hickel was scheduled to face his fellow Alaskans on April 23 at an Earth Day rally at the University of Alaska. While he was mulling over how to deal with the reality that the Trans-Alaska Pipeline was hopelessly bogged down with no light visible at the end of any tunnel, he got a directive from President Nixon that ended this first year on an appropriate note of circus and futility. Obeying it, Hickel departed from his Earth Day speech to make an awkward interpolation: "I am announcing tonight that I will issue the permit for the pipeline right-of-way."

But only God knew when.

PART TWO

By the spring of 1970 a year had passed since TAPS's first visit to the Interior Department to request prompt action on a Trans-Alaska Pipeline. Three hundred million dollars' worth of Japanese pipe stood idle, stacked in seemingly infinite rows, at Prudhoe, Fairbanks, and Valdez, in mute accusation that some fantastic mistake had been made. Not a foot of it had been laid, and the obstacles to a start had multiplied, not diminished. Now, under NEPA, the environmental criteria had been widened and made more nebulous, and the solutions had to be found acceptable not just to one department but to a gauntlet of successive courts. The Native claims bill of Senator Henry Jackson not only had to start its process anew but faced opposition, instead of mere inertia or ineptitude as in the past, opposition from the newly alert conservationist bloc in Congress, which saw it as a hostage to use against the pipeline, and from some of the Native factions, which, angered by the slights of the oil companies and the slurs of white Alaska, were demanding more than was apt to be given and were getting caught up in the air of contention that brews aboard a vessel long becalmed.

Yet, from the vantage point of the national need for North Slope oil —and of the consuming nations' need for continued "looseness" in the world oil market—there was still enough time to work out both problems. Even if it took another year to pass a claims bill and satisfy the demands of NEPA, so long as construction started by 1971, the pipeline could still be finished by early 1974. That was soon enough to protect the United States from that extra margin of oil dependency that could stir Arab-Iranian hopes of risklessly gaining control of the international oil market. The pipeline need not be complete to have its effect. Once it was well under way and seemed to be proceeding on schedule, the mere imminence of a new oil flood would so strengthen the U.S. government and so unnerve at least some of the diverse group of would-be conspirators as to postpone any reverse in the balance of oil power for a number of years.

There were two reasons for assuming, as most observers did, that during the second year the incoherent lethargy of the first year would be turned around. First, American dependence on foreign oil would visibly widen, transforming a theoretical future problem into a clear and present danger; during this second year U.S. oil imports would rise to 4 million barrels a day, a figure close to the limit that would be available from secure sources in the 1970's. *Since oil consumption would continue to rise, it could be seen with mathematical certainty that without Alaskan oil by 1974, we were on a track that would place us at the mercy of OPEC.*

Secondly, a turnaround in the performance of the pipeline consortium buoyed hopes that its impasse with government regulators and overseers was on the way to resolution. There were a number of corporate actors in this reversal, and they were moved by various motives: Lately liberated by the falling projections of U.S. oil production from any fear that Alaskan oil would cause a price-depressing glut, disabused of the cocky presumption that they could bulldoze their way past Hickel or that he would soon be replaced, humbled by accumulating proofs that Oil's political predominance was under fierce challenge by the new mass phenomenon of environmentalism, painfully initiated into the multiple levels of delay now built into the American system, uneasy that an almost $2 billion investment was tied up fruitlessly and that incomparably more in expected profits was cloaked in deepening uncertainty, the oil companies of the North Slope had shaped up their act.

In the summer and fall of 1970 TAPS—the loosely organized, internally divided, and frequently incompetent hodgepodge so despaired of by Secretary Hickel—was reorganized into the Alyeska Pipeline Service Company. Alyeska was 80 percent held by Humble Oil (Exxon), Arco, and British Petroleum, with the remaining interest divided up among other oil producers on the North Slope. The new consortium was tightly knit and forcefully led by Edward Patton, formerly of Humble Oil. Immediately it began to heal its relations with the Native tribes, making solid offers of construction jobs, lining up Alaska's white establishment behind a claims bill acceptable to Natives, and throwing Oil's Washington lobbying resources behind passage of such a bill.

Suddenly Oil was again conducting its affairs with a skill and flair that matched its international renown. Prudhoe Bay was now a showcase for environmental responsibility. "Spotless" was the word used by returning journalists. Used oil drums—that old bête noire—were nowhere to be seen, for they were systematically rounded up and hustled off to Fairbanks, where they were sold for scrap metal. Ordinary refuse was collected every day and either flown out or "treated" on the spot. Sewage was rendered blameless by the most modern of disposal plants.

Whatever the oil roughnecks might be thinking in their hearts about Alaska's 400,000 caribou, so beloved of conservationists, the newly benign oilmen gave them the run of the place and treated them like woodland nymphs. "They play, graze, sleep and mate under and around oil rigs and long sections of a test pipeline," wrote an observer. "On windy days they huddle against it for warmth. When the wind shifts they go to the other side, using ramps the oil companies built for them, or walking under the line where it has been raised to allow their passage."

The new discipline produced prodigies of sensitivity. When a marauding polar bear lumbered into a workers' dormitory yard, those who in yestermonth would have reached for a rifle instead called for the Nytol unit. The rampaging beast was harmlessly put to sleep with a tranquilizer gun, trundled onto a helicopter, flown back to its habitat under nervous eyes, and gently rolled out onto the tundra, there to be watched over from a distance until it roused itself and began romping about in full vigor.

The stark North Slope was transformed into a center for visiting scientists flown in from around the world by the oil companies. There were more than sixty such eminences there in the summer of 1971. European ecologists had painstakingly designed those caribou ramps to be tempting as well as serviceable, and they would return regularly to supervise their installation. Archaeologists, with Oil footing the bill, directed the excavation of ancient campsites where thousands of years ago the "first Americans" had rested during their migrations from Siberia across the Bering Sea. Enlightened hospitality transformed the oilmen's gruff camps into a riot of exotic scholarship. Wildlife specialists wandered about, one dispatch reported, "studying Alaskan birds such as the scoter, green-winged teal, mallard, ptarmigan, greater scaup, black brandt, eider, pintail, whitefronted goose and old squaw. They're also studying the habits of Arctic fish, of polar bears and grizzlies, of wolves and foxes, even of moose and lemmings."

Once their consciousness had been raised, the oil companies pressed on with a Prussian extravagance. In sensitive areas firearms were banned to prevent backsliding; workers were forbidden even to *walk* on the tundra, except in winter; mischief was foreclosed by banishing whiskey and beer from the BP and Arco camps and drilling areas. Two hundred varieties of grass were tested in order to select three ideal strains for reseeding earth displaced by construction; so that not a moment would be lost, plans required the reseeders to follow just a few yards behind the construction crews—that is, if construction ever got started.

The prescriptions of the environmentalists were agreed to by the hundreds with as much enthusiasm as the oilmen could muster. If a quart of crankcase oil was spilled on the ground, construction had to stop until it was cleaned up, and numerous "spill reports" were to be filed on each incident. A strange network of roads, built on top of five feet of gravel spread over the tundra, was taking shape. Quite properly the works of road builders would wait upon the habits of fish. Alyeska undertook to postpone removing gravel from streams until well after the salmon seasons were completed, so as not to disturb the salmon on the way to and from their spawning grounds. One series of stipulations aimed at protecting fishing would result in the new port of Valdez's having the most advanced antipollution system in the entire world. Tankers preparing to take on oil were to discharge water from their ballast tanks into special treatment plants, from which it must emerge pure enough to drink. Waters in the harbor were to be tested every day and must be certified as oil-free by fishery experts.

These routines were but as frosting on the cake compared to the precautions being taken to safeguard the main engineering process. The oil consortium agreed to scrap its original plan for a pipeline that was 90 percent buried. It was now test-drilling along the entire 789-mile route in order to analyze the permafrost content. To protect the tundra during these tests, it was using methods perfected in offshore drilling operations. Wherever the permafrost analysis dictated, according to agreed-upon criteria, the pipeline would be elevated 4 feet above the tundra (10 feet at several hundred sites known to be moose crossings), and it now looked as if elevation would be necessary for at least half the route. At the University of Alaska, Alyeska had buried a 600-foot section of pipe in permafrost, and under the direction of the Institute of Arctic Biology faculty, the effects of circulating fluids through the pipe at 160° F were being measured.

The earthquake hazard—with its specter of broken pipeline and hot oil pouring out—would be overcome by building the line in a zigzag crawl across earthquake regions, with the pipe mounted on skids and specially welded so that it could roll with the blows, as it were, enduring a shifting of the ground without bending or breaking the line. Where there was danger of landslides, protective walls would be built. Pipelines built with these precautions could survive an earthquake—twisting, straining, but never breaking. But what if the best that one could do against nature proved not to be good enough and the line broke after all? A system of cutoff valves and special electronic devices monitored by computers would divide the line, in effect, into separate compartments that could be closed off within minutes of a break, and cleanup crews would be promptly dispatched to the scene.

In early July Alyeska delivered to the Interior Department sixteen boxes of data, standing five feet high; another shipment would be made shortly, concluding the consortium's responses to all the questions the combined curiosities of the U.S. government and the environmental movement could devise. So by mid-1971 the oil consortium had demonstrated, with unprecedented specificity, that it not only had swallowed the draft of environmentalism but had mastered the science necessary to the safe construction of the most ambitious venture of its kind ever undertaken. Alaska's most prominent crusader for land use planning, Dr. Victor Fischer, director of the University of Alaska's Institute of Social, Economic and Governmental Research, pronounced his blessing: "The basic environmental questions have been faced and engineering can solve them."

Under the goad of profit the pipeline consortium had recast itself so as to meet the changed temper of the times and the new challenges of government. If that achievement were to bear fruit, government, too—the officialdom of Alaska, the Interior committees of Congress, the relevant arms of the administration—must disenthrall itself and rise above the default of duty and the self-serving fatuity that had marked the first year of the process. Instead, the second year was to be a reprise of the first, confirming the adage that when history repeats itself, it does so as farce.

Governor Keith Miller led off in April 1970 with a bold stroke. Up for reelection, convinced that he must be seen to do something to revive the stalled bonanza, he searched the musty statute books for a sign that he *could* do something. He found an 1866 federal law which vested in the state the rights-of-way over public lands "not reserved for public use" when some universal boon such as a road was needed. Grandly booting aside the federal triumvirate of court, Congress, and administration, Governor Miller "authorized" the haul road himself, to yahoo cheers all across the state.

The haul road was by itself a great commercial enterprise, estimated at $120 million—almost as much as the previous year's state budget. If Miller could get the road built while Washington wrangled over the pipeline, it would revive both the economy and the spirit of Alaska, reverse the spiraling unemployment among stranded construction workers, and enable swifter completion of the pipeline once it got started. It might even strike a spark that would shock the federal octopus to action.

Under Miller's scheme, the state would build the road and temporarily pay for it, taking the $120 million out of the $900 million it had received at the September 1969 oil lease auction; the oil companies would repay the state in use charges when the pipeline was built alongside the road. The companies agreed to the Miller plan, with the reasonable provisos that if the pipeline permit never came through, they would be off the hook or if the permit did come through but forced a change in the pipeline route, the state would be recompensed only for that portion of the road that was parallel. Fair enough, agreed the governor, pointing out to taxpayers that the road would be a solid investment in Alaska's future even if the pipeline were checkmated in Washington since the road would open up the north and make possible the development of vast deposits of mineral wealth such as the now unreachable copper known to abound in the remote Brooks Range.

But here the comic opera aspect of Alaskan statecraft rolled across the stage, basso buffo. At Juneau the state legislators, so far shrunken from the pioneer stock that they refused to risk a piece of their unearned billions for a grubstake or to recognize that the gravy train is not on a one-way track, approved the road plan only if the oil companies were to foot the bill in full, even if fate never permitted them to use the road.

The oil companies straightaway pulled out of the deal, causing consternation among the legislators, who claimed not to have known of the oil company conditions. The governor responded that he had indeed warned the legislature that the consortium might reject its repayment demands, and he called a special session to repair the mess. By the time it met, however, Governor Miller had decided that to raise the issue would only invite more circus and make the state a national joke rather than a rallying symbol. So he let the plan die. Inquiry into the chemistry at work in the legislature had shown that there was more behind its performance than smallness of spirit. Envy had played its part. The representatives from the Anchorage and Juneau-Ketchikan areas were reluctant to vote for a project that would dispense disproportionate benefits to their unworthy cousins in the Fairbanks area, which geography had made the supply base for northern road construction.

James E. Bylin, observing the performance for *The Wall Street Journal,* summed it up as a small-bore failure of magnanimity and vision: "Alaskans jealously guard their stake in Alaska's development and bristle if they think the other guy may be getting too much. . . . Some Alaskans appear more concerned with divvying up the current spoils than in using the money to bring in fresh resources that could be used to develop the state and alleviate its social problems."

In any event, the haul road was doomed to that limbo wherein the pipeline waited. The bracing spectacle of an aggrieved province's rising to shame the national government into some semblance of cohesion was not to be. Henceforth Alaska would nurse its internal grudges and wait, sullenly but supinely, for Washington to mete out rewards and punishments.

Revolution having petered out, the focus of hope returned to the seat of government. In the spring of 1970 Secretary Hickel had said that the permit would be approved, the federal district court had enjoined the issuance of any permit pending its deliberations, and Senator Jackson was pushing a Native claims bill off to a quick start that would enable it to pass his committee in April and the entire Senate in July.

All three branches of the federal government were thus enmeshed in the permit and claims processes. If their combined action were to "clear" the pipeline within a year so that construction could start in the summer of 1971, two modest achievements had to transpire: The House of Representatives had to produce a Native claims bill compatible with the Senate's, and the Interior Department had to conclude the permit process on terms quickly acceptable to the federal court.

The initial obstacle to be surmounted was that the House is only infrequently at work on the public business in the second half of an election year. When the Senate-passed Native claims bill reached the House in July, there were scarce three weeks left before the annual congressional vacation that consumed August and spilled over into September—not enough time to work in any markup sessions. After Labor Day there were but three weeks before the mass evacuation of Capitol Hill in late September to contest the November congressional elections. November and December would be junketing time. So that was it for the year. Since this was the second session of the Ninety-first Congress, any bill that did not make it to final passage in both houses, no matter how long its genesis or urgent its substance or numerous its sponsors, would at the moment of adjournment revert, like Cinderella's coach, to a pumpkin—less than a pumpkin, for it would become a nullity.

The second obstacle was the strain which a few consecutive days of steady, workmanlike effort placed on the human resources available. When legislative time is rationed so sparingly, a premium is placed on the cooperative attributes of humankind; the negative crotchets and abrasive vanities that seemed to abound in Congress, or at least in the House Interior Committee, assumed a distressing power to obstruct.

Representative John Saylor, who had turned sour on TAPS because of its failure to buy its pipe from Bethlehem Steel, was, inconveniently, the ranking Republican on the House Interior subcommittee that was to handle the Senate bill. Enter Representative James A. Haley, chairman of that subcommittee, formerly the president of Ringling Brothers and Barnum & Bailey Circus. Now seventy-one, white-haired and angular, Haley had for more than a year been suffering a slow burn over the amazing failure of anyone to solicit his aid or even to sound out his views on the claims bill. Not the Natives, or their celebrated New York lawyers, or the white Alaskans, or the representatives of Oil, or the administration had paid him the court he felt due or had even exhibited the slightest tremor over what his position might be. Neglect where congressional vanity is concerned has its consequences; like a long-spurned suitor who holds the homestead mortgage, Haley awaited his day, and when, after Senate passage, the interested parties remembered him and came around, he refused even to see them.

"The lobbyists will write this bill over my dead body," he vowed. In fact, as far as Haley was concerned, *no one* was going to write it, at least in 1970. When his subcommittee received the Senate bill, he declared, to Saylor's delight, that it might be months before it could be acted on. Then he buried it deeper, adding that there were many "Indian bills" with higher priorities.

Alaska had a man on the House Interior Committee, Edward Pollock, a Republican in his second term. It was his duty to see to it that the committee was alert, or at least accessible, to the most important measure in a decade for Alaskans. But Pollock was representative of that American tradition of officeholders who campaign for another office or another term, instead of governing in the one they have been elected to. For most of the year he had absented himself, as he would again in the all-important July weeks, to campaign in Alaska for the Republican nomination for governor. "It was an ill-concealed secret," wrote claims bill historian Mary Clay Berry, "that the older members of the House Committee resented Representative Pollock's frequent and prolonged absences and felt he had not been doing his share of the work." When present, Pollock had been a bust at the game of ingratiating himself with the seniors, especially his leader, the bileful Saylor, and his chairman, the alienated Haley, and he had failed to build up a collection of due notes he could now call in. His need for a legislative trophy to brandish in his flagging gubernatorial quest made it all the more satisfying for Haley and Saylor to pigeonhole the claims bill, and his nonimpact on the committee made it easier to do so.

But high above Haley and Saylor and a dozen other subcommittee chairmen and ranking Republicans loomed, in grandeur indescribable, the Interior Committee's seventy-five-year-old boss of bosses, Wayne Aspinall of Colorado, overseer, along with Henry Jackson, of that third of the nation that was owned by the federal government. When the timing was right, Aspinall had the personal power, as he would demonstrate almost two years hence, to jot down a multibillion-dollar Native claims bill on the back of an envelope and get it passed by his committee. And though he was ambivalent toward the Natives' designs on the many millions of *his* acres, he was responsive to the needs of the oil industry, a major force in his home district, which had lately roused itself to the danger of drift and begun lobbying for any kind of a claims bill that would get the issue out of the way. Representatives of several oil companies now contacted Aspinall and kindled his sympathies; the bill would at least be scheduled for consideration. This was left to Haley, who set aside four days, but the obdurate Haley chose the last four on the committee's legislative calendar, in the third week of September, close to the moment when the Grand Chairman would go home to Colorado to campaign, always the signal for the committee to fold up for the year.

Came the third week of September, and the Haley-Saylor subcommittee considered the Native claims bill, after its fashion, rejecting several formulas, reaching tacit understandings on parameters, but agreeing to nothing on paper. And on September 22 it adjourned for the year. When reporters asked Congressman Saylor about the status of the claims bill, he quipped: "The only thing we haven't had is the funeral."

The legislative process, like the judicial, has its courts of appeal, and there would be a final chance to revive the claims bill. September 25 was the day fixed for Chairman Aspinall's departure for his biennial campaign safari through Colorado, and on the twenty-fourth he would preside over a last meeting of the full Interior Committee. Pollock, back in Washington after his defeat in the gubernatorial primary, planned to appeal for action, if he could get recognized. His influence, however, was at an even lower ebb than before, now that the voters had clipped his wings and he was a lame duck serving out his last days.

Chairman Aspinall had given some thought to the claims issue. He and an aide had in fact sketched out the Aspinall solution—the usual billion dollars in cash for the Natives to be paid over a period of years from a variety of sources; the 40 million acres, but only for "use" in hunting, fishing, nut gathering, and the like, with ownership and control remaining with the Interior Department (and thereby, of course, himself); and maybe 3 million acres to be given outright to the Natives, land around their villages. But pushing this through would entail protests from dissatisfied Indians and a lot of back-and-forth with the Senate, so Aspinall had decided to let it lie until next year. He had more important fish to fry: in October, his campaign; in December, his honeymoon—for the old widower had taken a wife that summer. After election day, having dutifully deferred his bliss for reasons of state and career, he would set off with his bride on an around-the-world tour, an event for which American embassies and territorial governments across the globe were even now making preparation.

Besides, the reappearance on Capitol Hill of young Pollock had already given fresh cause for umbrage. Had he not accused Aspinall's committee of "just playing games" while Alaska burned? And—can it be believed?—someone had leaked to the press premature speculation about the committee's intended claims bill, and suspicion had fallen on Pollock and his staff. It was an inopportune moment for a replay of the apparently irrepressible conflict between the weakness of Alaskans for press accounts of achievements that won't materialize and the fury of committee elders at half revelations of semiagreements that are going nowhere.

Such were the undercurrents when the windup meeting convened. Pollock tried and failed repeatedly to gain recognition, and his cause might never have been raised had not Representative Ed Edmondson, the respected Oklahoma Democrat, raised it when the group was on the point of disbanding: "Mr. Chairman. . . . We have pending a matter of extreme urgency in the case of the Alaska Native land claims bill."

Aspinall looked silently down the long table and up to the clock which hung over his own oil portrait. Then the oracle spoke: "The Chair wants it known that he's going home." Although this would doubtless have been quite sufficient in that lingering twilight of baronial power, the "Chair" favored his lieges with some reflections on the Alaska matter. He was displeased with the Natives for overreaching and for not at least uniting in their overreaching. He had a minimal regard for the lobbyists on the case. He wanted it understood that the chair could not be stampeded by state leaders and their cries of woe or by Native leaders and their lawyers. As his eyes circled the table, they paused at Pollock. The breach of trust involved in leaks to the newspapers, he went on, was to be deplored. And he philosophized: "People have a tendency to foul their own nests."

Pollock, finally obtaining permission to speak, made a stout appeal that the bill not be dropped now, so close to completion, with so much of the year left to run, and at such cost to Alaska's economy and the nation's oil security. If the chairman would order a resumption of deliberations, the remaining differences could be resolved *"in a few hours."*

All those around the table well knew that the alternative to a few hours taken now, or even a few days, was another year of delay. A bill that died now would have to be born again next year and undergo all the growing pains that had brought it to this point. The ensuing scene says a lot about the motives and priorities of congressmen as against the needs and urgencies of the nation.

The chair had detected a note of reproach in Pollock's plea. "Don't point the finger about working," he said, bristling. "The gentleman was home to campaign. *I'm* going home to campaign."

Pollock continued to agitate, and he was supported by Edmondson. There is something in the nature of a discussion around a table, though it be between hinds and princes, that is both democratic and volatile. Two voices raised in behalf of such combustibles as justice denied and crisis unmet might soon be ten.

Or so thought the wary Saylor. Sitting at the chairman's right hand, he divided his attention between baiting the plaintiffs with diversionary charges and watching the clock. It was almost time for the House to go into session on the floor, and once it did, the committee could not remain sitting if any member objected. At the instant of twelve noon, with Edmondson ably pressing Aspinall for a reprieve, Saylor rose. "I call a point of order. The House is in session." The committee's legislative year had ended.

On his way to the door, Saylor encountered a seething Pollock and, lest the junior miss the personal reproof in what had transpired, rubbed it in:

SAYLOR: If you'd tried to help me anywhere along the line during the four years, you'd have gotten my help.

POLLOCK: You haven't been any leader as far as I'm concerned.

SAYLOR: Nuts to you. I'll be around next year, and you won't.

Six months later, on March 16, 1971, Chairman Henry Jackson of the Senate Interior Committee was about to open a hearing he had been trying to get under way since January, when he had for the third time presented his Native claims bill. This time he had protected his rear by dragooning as cosponsors the two Alaskan senators, Stevens and Gravel, who were present. His plan was to get the hearings on this dog-eared question quickly out of the way with one day of testimony from the other principals: the administration and the Natives. Then, having satisfied the formalities, he could push the bill through his committee and get it over to Aspinall so early that there could be no excuse for inaction. It had taken two months to get the major spokesmen lined up: the secretary of the interior, to give the Administration's position; Senator Fred Harris (D-Okla.), who had emerged as the Senate champion of the Eskimos; and various of the Native leaders.

But the expeditious sendoff was not to be. The interior secretary had sent word he could not appear and asked for a raincheck until the administration had firmed up its formula. Fred Harris was also absent, having gone to New York for the funeral of Whitney Young, the civil rights leader. The Native leaders were on hand, but they refused to dignify such a rump affair by their participation.

Jackson canceled the hearing, making a note to try something for April. He doubted now that he could get this star-crossed bill to the House before summer. It would not escape, after all, the cycle of vacation, logjam, and committee filibuster. Maybe by Christmas . . .

While the Congress thus alternated between frivolity and futility, how had the executive branch been performing? As we have seen, the interposition of NEPA and of the attendant court reviews compounded the obstacles facing the administration's permit process in the spring of 1970. Still, the effort and experience of a year could be built on, and outside events were creating that atmosphere of calamity which, it is our peculiar boast, makes American democracy work.

We know now that a year was left to get the Trans-Alaska Pipeline started if it was to be completed in time to ward off the great oil upheaval that was brewing. Our decision makers in 1970–71 could not know this with precision, of course, yet so many energy alarms were suddenly sounding all around them that they *did* know, or at least they proclaimed, in internal memorandum and public statement alike, that the creation of new, safe sources of oil was an urgent national priority.

Foresight was no longer required, nor even elementary mathematics, to spot the onrushing energy crisis. The occasional brownouts during the summer heat of 1969 had by the summer of 1970 progressed to widespread power breakdowns, forcing a cutback of services throughout the eastern seaboard. Shortages of fuel oil were involved, and suddenly it was noticed that American refining of fuel oil had so declined that the East Coast was dependent on foreign sources for 93.7 percent of its supply.

A winter fuel oil crisis was officially predicted. The White House announced it had alerted its Emergency Petroleum Supply Committee to be ready for a possible world oil emergency. Newspapers the country over front-paged warnings of shortages, higher fuel prices, plant closings, and worker layoffs. The natural gas shortage that had loomed in projections actually arrived here and there, causing gas companies to refuse new customers and notify existing ones—industries and residential developers—that they could not provide the increased supplies that had been counted on to fuel expansion.

Washington appeared to recognize that something serious was afoot. Congress held "emergency" hearings in 1970, while the State Department was warning the White House that the United States was rushing headlong into a dependence on Eastern Hemisphere oil far beyond what was safe, in terms of either security or price. The President formed a White House executive committee charged with producing "new or revised energy policies to alleviate the acute shortage of clean fuels . . . and to insure an adequate fuel supply for the next five years." Secretary Hickel formed yet another committee to plot the oil future fifteen and thirty years hence. Assistant Interior Secretary Hollis M. Dole, speaking at Stanford University in January 1971, indicated a department and an administration conscious of the impending danger:

> Beyond this winter loom the peak load demands for electricity of next summer; the steadily diminishing capability of the gas industry to meet demands upon it; the ever-increasing dependency upon foreign oil, particularly along the East Coast; the tenuous supply of fission fuels available at reasonable cost for burner reactors; and the mounting constraints upon the use of coal in metropolitan areas—we shall feel—for the first time—the full force of foreign oil supply interruptions directly as they occur. This will be something totally new to our experience; to be dependent upon foreign oil sources not only for a substantial and growing part of our normal oil supply, but for *all* our emergency supply.

In all these warnings and inquiries and new committees, oil journalist Ruth Sheldon Knowles saw a barnyard analogy: "The problem wasn't that of locking the barn door after the horse was stolen. It was trying to find the barn to see if we had a horse." In fact, we had a number of horses, and the most ready for the starting gate was the one in Alaska. Its advocates breathed new life from the hubbub about oil shortage that surfaced in so many places and was continually recharged by untoward events during this second year of the pipeline permit process.

The conduct of the administration now takes on a significance of deeper hue. Up to 1970 what transpired was primarily a lack of foresight about a looming danger and a politics-as-usual complacency, lamentable in hindsight, but not damning. Foresight in matters of state, after all, is the gift that separates the great from the merely adequate, and all history chastens us that we have no right to expect greatness.

But by mid-1970 the vanguard of the danger had *arrived,* the invader's advance units were already in the suburbs, and the question that would dominate the new decade and beyond had become one of basic governmental competence and coherence. Can our leaders *act* on a certified and acknowledged menace, or are they mesmerized by obstacles and diverted by routine ambitions? Does our governmental structure retain the capacity for follow-through?

The Nixon-Hickel pledge to that University of Alaska audience on April 23, 1970, could have signified the emergence of an administration policy or might have been just talk to create a momentary appearance of action. Nixon could not deliver on the pipeline permit unless he was prepared to exert strenuous personal leadership. Only if he would mobilize the White House and the administration-at-large behind the goal of 2 million barrels a day of Alaskan oil by 1974, only if he would line up his congressional allies, shock the oil consortium into swifter compliance, spur the bureaus to greater urgency, and impress upon the courts the need for accelerated review within the context of a national emergency could he redeem the words he had put in Hickel's mouth.

No such follow-through was made or even begun. Far from involving himself, the President never once sat down with Secretary Hickel to thrash out the Alaska opportunity. Nor were the truncheons of H. R. Haldeman, John Ehrlichman, and company utilized in clearing the way. The top-echelon members of the Hickel Interior Department we interviewed can recall no instance of White House take-charge intervention. The most important Republican on pipeline matters in the House, Representative Saylor, continued to be the pipeline's most obdurate opponent. Far from rallying its resources behind Hickel's lonely, slowly succeeding effort to convince the oil companies that the only route to a permit was to solve, not evade, the environmental problems, the White House weakened those efforts by progressively undermining Hickel throughout 1970.

Two weeks after his University of Alaska speech Hickel—seeing a linkage between his exclusion from direct communication with the President and the alienation of anti-Vietnam War students who felt excluded from "the process"—wrote a letter to Nixon warning that the President was suffering from a failure to communicate with the young and even with his own Cabinet: "I believe this Administration finds itself today embracing a philosophy which appears to lack appropriate concern for the attitude of a great mass of Americans—our young people." Then to the nub: ". . . permit me to suggest that you consider meeting on an individual and conservational basis with members of your Cabinet."

Now struck the Alaskan nemesis—the weakness for premature publicity. The Hickel letter was slipped to the Washington *Star* and was on the streets before it got to the White House on May 6, 1970. The resulting publicity feast, picturing the President as a figure so out of touch that he was a stranger to his own Cabinet, damaged Nixon as only adverse testimony from one's own ranks can. Nixon aide John Whitaker called Hickel aide Pat Ryan in a dudgeon.

WHITAKER: You find the son-of-a-bitch who leaked that letter and fire him.

RYAN: You find the son-of-a-bitch who wouldn't let Hickel see the President and fire *him*.

The air was now filled with predictions of Hickel's ouster, and if he had been dismissed and promptly replaced by a well-supported figure capable of leading the fight for domestic energy revival, the deterioration of Interior's drive and capacity for decision could have been arrested and turned around. Instead, Nixon chose the worst possible path so far as the public business was concerned. Chief of staff H. R. Haldeman announced to a group of White House insiders that included speech writer William Safire that (1) the decision to fire Hickel had been made, but (2) its execution was to be deferred for six months, until November—that is, until after the midterm elections.

The reasons were obvious to politicians, if appalling to administrators. A fired, embittered Hickel would be a loose cannon rolling around in a campaign season. He might mount a harmful anti-Nixon campaign; already he was being introduced by such as John Oakes, editorial page editor of *The New York Times,* as "the man who was once known as the bad guy but who is now the good guy of the Administration." Or he might return destructively to Alaska politics, upsetting the reelection prospects of Nixon-supported Governor Keith Miller and Senator Ted Stevens. Better to keep Hickel on, as a semicaptive under partial wraps, while continuing to cut him down anonymously in terms of prestige and influence—a course made possible by Hickel's obtuseness and his determination to cling to his office for as long as possible.

Each month thereafter had its petty Hickel humiliations and its White House tips to the press so they could be publicly elaborated: Hickel disinvited from a White House prayer breakfast; Hickel disinvited from a White House press briefing; "Hickel is through," said departing White House aide James Keough; "It's only a matter of time before Hickel leaves," leaked Nixon press secretary Ron Ziegler in a "backgrounder." He had become the butt of Washington jokes. In late June, humor columnist Arthur Hoppe summed them up by caricaturing Hickel arriving at his office in the morning:

MR. HICKEL: What is my appointments schedule today?

MISS PANGLOSS: 8:45 A.M. Arrive at office.
6:15 P.M. Leave for day.

At one point Hickel went as high as he could go—to Attorney General John Mitchell—to protest what this was doing to his department: "John, all these rumors that I might get fired or relieved are upsetting my department something terrible. You know yourself how difficult it is to run a department when your people, especially the heads of bureaus, think that their boss is going to get fired at any moment. All sorts of political games get going."

Mitchell claimed to know nothing.

Hickel at last seemed to be giving up; he spent much of August and September traveling—in the Arctic, Norway, Sweden—as if to get as far away from Washington as possible.

As the fall campaign approached, small-bore White House treachery took a new twist. It was decided that Hickel could be of use, before he was fired, in the campaign, and he was booked for a grueling fourteen-state tour. The President called him to San Clemente for a little temporary public rehabilitation before he set off on the campaign trail. The old jollity was restored, photographers were called in, and after they had left, the heavy-burdened interior secretary was warmed, at last, with presidential rapport and advice: "Wally, the thing to do is to get into a city, make a speech or something, and then find a way to get on television. For every person you reach in a speech, you can reach thousands more over television."

Hickel recalled with a pang that this was the exact message he had gotten from candidate Nixon near the start of their relationship two years before. Maybe it was his *only* message.

As soon as the campaign ended, the White House "Hickel is through" press leaks started up again. Mitchell now told him to resign: "Your relationship with the administration is just not compatible." Hickel refused to leave unless fired by Nixon personally, and on November 27 Nixon finally called him in and fired him. "There is a mutual lack of confidence," Nixon said. Hickel was told he could not wait for the arrival of his successor but must clear out by day's end. The office of interior secretary would be vacant for about two months.

For an entire year, then—the critical year, the last one that could have made a difference in domestic oil production for 1973–74—the drive for the Alaskan oil that lay waiting was stalled while one interior secretary was being dismantled and another was undergoing confirmation and basic training.

The arrival in January 1971 of Rogers C. B. Morton as interior secretary was to demonstrate how little thought the White House had given to Alaskan oil, for it ushered in not a new approach to the pipeline but rather several months of anarchy before a return to the old approach. Morton, who had been doubling as Republican national chairman and congressman from Maryland, was a huge, white-thatched, amiable, outspoken fellow who at times could cause as much accidental carnage as a playful polar bear at a picnic. Though two years of department deliberation had passed since the oil consortium's public announcement of its pipeline plan, Morton seemed to regard Alaskan oil as a new and mysterious question he would have to investigate from scratch.

In January Morton declined either to embrace or to disown the Hickel-Nixon commitment to the Alaskan pipeline, whether or not it met environmental safeguards, and he refused to speculate about any sort of timetable for granting the permit. In February he halted the recent momentum the pipeline had acquired from Oil's revived push by declaring that he was "a long way from deciding that this pipeline was the way to do it." No decision, he said, could be expected before *summer!*

In March he began to explore a radically different pipeline route, the Mackenzie River valley route, running down through Canada to the American Midwest, and at his insistence, oil company executives flew to Ottawa to talk it over with the Canadian government. It was an interesting idea, but by 1971 it was two years late. The Canadian route was twice as long as the trans-Alaska route and would therefore take about three years longer to construct. Moreover, the Canadian government had not even begun *its* processes of evaluation and approval, which promised to be as cumbersome as our own.

By early June Morton was publicly doubting the advisability of *any* pipeline. "I don't favor the damn thing much at all," he said in Seattle. "I don't think you need all that oil out here. I think it's an awful lot of money to spend for thirty or forty million barrels of oil."

That in June 1971, two years after the brownouts of 1969, the secretary of interior should be venting preposterously idiosyncratic views about the American need for oil, the economics of oil, and the amount of oil on the North Slope (at least 300 times 40 million barrels) aptly revealed the degree of coherence in administration policy.

It was a period when every concerned agency of the administration was recommending a different policy on the pipeline—a ferment that might possibly have been helpful two years before but that now added to the paralysis of decision making.

The Environmental Protection Agency suggested further study—the bureaucratic death sentence. The Federal Power Commission urged a prompt go-ahead on the pipeline. The President's Council on Environmental Quality wanted to take "a very good look at" transporting oil from the North Slope "via submarine and at various pipeline routes across Canada." The Emergency Preparedness Office, differing from Secretary Morton, found the U.S. oil situation so "tight" generally and so "particularly critical" west of the Rockies that it favored immediate permit approval combined with strict enforcement of environmental safeguards. The Defense Department found this procedure inadequate; while mute on its supposed specialty—national security implications—it was voluble in insisting on a veto power in pipeline matters for its Army Corps of Engineers.

In the wake of this Babel, President Nixon sent to Congress on June 4, 1971, a message on the development of U.S. energy resources, a response to the rising fears of energy shortage. Advocates of the Alaska pipeline plunged through the message with a mixture of eagerness and trepidation. But there was not a word anywhere about Alaskan oil or the Trans-Alaska Pipeline. At a time when the pipeline consortium was making a maximum effort to qualify, when the environmental and technical problems had been solved, when one last concentrated push could have gotten Alaskan oil over the top in time to make a crucial difference to the world, it had amazingly become a nonissue to the

White House, the missing plank in a platform whose other planks, which shall be examined in a later chapter, would have no impact whatever, singly or all together, on the energy crisis of the 1970's.

But though removed from the presidential agenda, the Trans-Alaska Pipeline permit application still had its place on something more important—the bureaucratic calendar—and was due for decision in the summer of 1971. In July and August the Interior Department received the final installments of Alyeska responses, installments that now filled thirty boxes with documents. Meanwhile, the flirtations with "alternate routes" and submarines and the like had cooled of their own unlikelihood, the remoteness of shale oil had sunk in, and by process of elimination the original pipeline proposal had returned to favor.

Too late. Man, having weighed and doubted and improved on the proposal for more than two years, was ready to approve; but Procedure had long since taken over from human choice, and Delay would render moot what passed for human decision. To "process" all the Alyeska responses to all the hypothetical agency questions through all the places they must be referred to would take at least until the end of 1971. Secretary Morton set "early 1972" as the new decision date, while cautioning any optimists who still survived that "whatever we decide will go straight into the courts."

Nineteen hundred seventy-one was the third and last year of grace. Its squandering passed an invisible sentence upon the West—dependence on the sheikhs and dictators of the Middle East, whose schemings toward oil control had begun again to wax as the oil sufficiency of the United States waned.

Two years too late to have sprung loose the start of pipeline construction before NEPA and the litigators and the disarray at Interior descended upon it, the Native claims bill was enacted, in December 1971. How needless and artificial the obstacles had been was shown by their too late evaporation. Suddenly all the heretofore unmeshable pieces came together and all the bystanders put their shoulders to the wheel. The crusty Aspinall compromised with the industrious Jackson; the discordant entries from Alaska, Stevens and Gravel, now ran in tandem, with Stevens playing a statesmanlike part in the outcome and Gravel leaving his mark on it. Alaska's replacement in Congress for the unfortunate Pollock, Nick Begich, proved just the tonic needed to warm up those elders who could be won, while isolating the captious

Saylor, who cast the lone "Nay!" in the conference committee. And strong White House participation tipped the balance toward generosity for the Natives and helped whip the claims bill down the home stretch.

Construction of the Trans-Alaska Pipeline would not begin until February 1975, in the aftermath of the oil crisis that the pipeline could have averted had it gotten under way, say, only two years behind schedule instead of five and a half. Once the political obstacles were lifted by the joint, summary actions of the administration and Congress which cleared away the judicial and bureaucratic labyrinths, an epic task of construction was accomplished in little more than two years. Though there were short-lived furors over environmental violations, the great feat was carried through without significant damage. The pipeline opened its valves on June 20, 1977, and was soon delivering 1.5 million barrels a day to the port of Valdez.

Had it come, or been clearly on the way, in 1973, this onrush of new oil would have preserved the world oversupply, wiped out totally U.S. dependence on Middle Eastern imports, and thus deterred the oil revolution that was concocted out of a contrived temporary "shortage" and U.S. vulnerability to it and intimidation by it. But coming as it did in the late 1970's, Alaskan oil entered an upside-down oil scene in which the OPEC revolution exercised enough control to negate new Western oil by cutting back production of Middle Eastern oil, which it could now easily afford to do because of the price explosion, a scene, moreover, in which U.S. leadership had become too conditioned to vulnerability and intimidation to take advantage of new strength.

Chapter Six

The Dragon and the Spider

Oil is for the Arabs! Why do you not exploit your lost wealth which is being plundered by aliens? Remember that the oil which flows under your land is seized by your enemy! Remember oil, your lost wealth! Oil is for the Arabs!

—A Nasser slogan on
Radio Cairo in the 1960's

For our part, we do not want the [oil] majors to lose their power and be forced to abandon their roles as a buffer element between the producers and the consumers. We want the present set-up to continue as long as possible and at all costs to avoid any disastrous clash of interests which would shake the foundations of the oil business.

—King Faisal, spoken by
his oil minister, Sheikh Yamani, 1969

By the spring of 1971 the United States had all but forfeited, act by act and omission by omission, the energy independence that was the backstop of the West's oil security. What was now to be its recourse?

The Nixon prescription—never proclaimed as policy and, in fact, occasionally denied—could be discerned in his periodic relaxations of the statutory limits on oil imports, especially from the Middle East, which between 1970 and 1973 were to rise by 500 percent. Henceforth the administration was to rely on the oil states of Islam to make good the mounting shortfalls of all the major domestic energy sources.

For more than twenty years Richard Nixon had been running for office on policy planks that promised to prevent the very dependence on Middle Eastern oil he was now silently embracing. He believed he could risk spurning the collective wisdom of the past because of the ease with which the U.S.-led oil giants had broken up the various attempts of the Arab states to gain control of the oil market in the year before he was elected President.

"Nothing was further from Nixon's mind than an energy crisis . . ." Henry Kissinger tells us. "[OPEC] . . . was not perceived as a serious cartel," he explains. "Conventional wisdom [by which Kissinger may mean his own and Nixon's] was that the oil producers could not force the price up by restricting production because the necessary cutback would have to be so steep as to bankrupt them before it would affect the industrial democracies. The fiasco of the Arab oil embargo of 1967, which collapsed in the face of uninterrupted Iranian, increased American and other Western Hemisphere production, seemed to confirm this."

Indeed, if the President were prepared to act as vigorously in defense of Western oil dominance as had his four predecessors, his gamble might have succeeded indefinitely. The caveat is all-important, however, for beneath the obvious weakness of the Islamic states in the oil world Nixon inherited, their hunger for self-determination and their craving to give the West its comeuppance had a depth and resilience which demanded constant watchfulness by Nixon and which invites a look back into its proximate origins and its potentialities.

Whether or not the oil-endowed nations were getting less than they deserved for oil, they ardently believed that to be the case. As has been seen,* the isolated efforts of individual states to enforce their will on the multinational oil industry had failed repeatedly during the 1950's and 1960's because of Too Much Oil in Too Many Hands. Each failure raised anew the question now to be explored: Why do not the oil-blessed states act *together* to reduce production, to create the scarcity that enables extortion, and to prevent the oil industry from playing off one of them against the other?

Attempts at collective action had also failed because of the inability of the Arab states to achieve a working unity. Yet there were two hard-to-measure forces on the loose, either of which had a potential, should it catch fire, to override temporarily the enmity of Syrian for Iraqi and Saudi for Egyptian and Ba'athist for monarchist, blast open their parochial horizons, and unite them in a common cause. The first was the magnetic appeal of one man, Nasser of Egypt; the second, hatred for one enemy—Israel. In the middle of 1967 there was to be a joint igniting of both these combustibles, an explosion that was both a warning of the suppressed danger ever threatening the oil status quo and an illumination of what must be done, and continue to be done, if that danger were to stay repressed.

Gamal Abdel Nasser, the plebiscitary dictator of Egypt, was the greatest crowd drawer ever heard of in Arabdom—not just in Cairo and Alexandria but in Algiers and Damascus, Riyadh and Beirut. The look of him standing in an open car—from afar tall, firm-jawed, and big-

*See pages 5 to 16.

shouldered, from closer up of caring face with sensitive eyes, contagious smile, and outreaching good humor—so confirmed in the flesh the legends of his defiant deeds and regenerative blueprints as frequently to provoke a mass hysteria that dislocated a capital city for days and brought home to the local despots at Nasser's elbow how thin was their writ.

Eldest of ten children of an Egyptian postal clerk, intermittent school dropout, strapping boy of the streets and alleys of Cairo and Alexandria, Nasser was admitted to army officers' school when the bar to commoners was let down, one day to emerge the leader of a conspiracy that would become a revolution.

He was an agitating force sweeping across the walled enclaves of the Arab world, the amplified voice of its frustrations and yearnings, the hounder of its foreign exploiters and native quislings, the hero of deeds that projected to kindred millions electrifying pictures of common shames avenged, of the foreigner expelled, of Muslims operating modern technology on their own, of an Arab nation no longer the pawn of outsiders but standing independent and aloof from *all* the great powers, making them bid against each other for the friendship of the Arab heartland.

To a rising, restive generation from Rabat on the Atlantic to Kuwait on the Persian Gulf, the major events of Nasser's career were a succession of vicarious retributions and fulfillments. He had overthrown the British puppet, King Farouk, to emerge as the first native ruler of an independent Egypt in 2,500 years. He had ended the foreign military occupation of his country, forcing the withdrawal of 80,000 British troops. He had dared reverse the familiar roles and play manipulator of the great powers, crushing the Communist party of Egypt but winning massive arms aid from the Kremlin, breaking up the Anglo-American plan to align the Middle East against Russia but luring huge Anglo-American aid projects aimed at countering Moscow's aid. He had nationalized the Suez Canal and successfully operated it, and when Egypt was invaded in retaliation by Britain, France, and Israel, he weathered the storm, and in the end his Suez takeover prevailed. He had lofted the dream of the Aswan High Dam—"seventeen times larger than the greatest pyramid," a TVA for the 1,000-mile-long Nile Valley —and when American and British financing was canceled, he fulfilled the dream nonetheless by means of Soviet aid and the expropriated earnings from the Suez Canal. With pitiable resources available for combating Egypt's desperate and pervasive poverty, he had somehow kept going a momentum of hope, an atmosphere of schools opening everywhere, of factories being built, of rural estates being divided up among the landless fellahin, of a native middle class emerging, of the breaking of the ancient closed circle of futility, of the stirrings of opportunity on a wide front.

That he also ran a one-party dictatorship, handpicked the major candidates for office, maintained a concentration camp in the desert for political foes, censored the press, and employed all the ugly arts of police state surveillance did not discredit him in the Muslim world. Nor, for that matter, did it alienate American and British attachés and journalists in Cairo, who in the main felt that Egypt was in no way ready for parliamentary democracy and that Nasser's authoritarian rule had wide public approval and was about the best that the Egypt of the times could hope for—as the CIA's Miles Copeland, for years the U.S. go-between with Nasser, maintains in *The Game of Nations.*

To the great oil companies—whose control over the international oil flow was made easier by the chronic divisiveness of the Middle East oil states—Nasser's overriding themes of Pan-Arab unity and the expulsion of the Western "exploiters" sounded an increasingly disturbing note. For though the odds against Middle East unity were almost insuperable given the tangle of overlapping animosities—Iraq for Egypt, Syria for Saudi Arabia, Kuwait for Iraq, Jordan for Syria, the Arab oil states for Iran—Nasser showed alarming signs of possessing an overarching appeal.

It was an appeal that reached out on three distinct levels. On the level of personal charisma he again and again showed the capacity to incite a worshipful hysteria among the Arab masses, however remote from his own power center. The ultimate proof of this magnetic pull was his 1956 reception in Riyadh, Saudi Arabia, a controlled fundamentalist enclave presumably far outside his orbit and impervious to his secular charms. Robert Lacey has described the event in *The Kingdom:*

> Tens of thousands of spectators rushed forward, cheering wildly, breaking through the police barriers, desperate to touch the god-like figure. The Army had to force a passage for Nasser's car along the road into the capital, and, whenever the Egyptian leader appeared in public in the course of his brief stay, pandemonium broke loose.

Riyadh had never seen anything like it. No member of the House of Saud had ever inspired such spontaneous display of passion.

For a long time thereafter the reactionary feudal monarchy of Saudi Arabia was the compliant paymaster for Nasser's attacks and maneuvers against reactionary feudal monarchies.

On a second plane, over the years since 1952 Nasser had proved a durable rallying point for key groups in most Arab states—the younger army officers, the bureaucrats, the university complex, intellectuals and artists, industrial laborers such as the oil and dock workers, the emerging middle class.

He turned Cairo's cultural supremacy in the Arab world into an ideological weapon. One Arab country after another came to have influential enclaves of Egyptian teachers, technicians, professionals, military instructors, and experts on government administration who, on the side, were natural propagators of the Nasser line. To develop an Arab managerial class that would fill the functions of the Westerners he wanted expelled, Nasser ran the only substantial industrial management school in the Middle East. Its faculty included instructors from the Harvard School of Business Administration, and the school went on the road, conducting classes in Baghdad, Tripoli, Beirut, even Khartoum in the Sudan. Thus he not only was preparing Egypt to become, as a *Fortune* article put it, "the industrial mecca of a modern Arab empire," but was making his revolution a beacon to the abler and more enterprising class across the Arab world.

Cairo had a natural ideological resource unmatched in the Arab world—its diversity of writers, directors, actors, singers, and technicians—and Nasser mobilized it, building up Radio Cairo and its "Voice of the Arabs" into the preeminent air attraction from Morocco to Iraq. Its mix of lively entertainment interspersed with dramatized propaganda and Nasser's version of the news gained it vast audiences. A 1956 CIA study of Radio Cairo concluded that "its prolonged effect was a conditioning of Arab attitudes much as is achieved under hypnosis. Even listeners too educated to accept the content of the broadcasts intellectually were beginning to assume certain attitudes without consciously knowing why."

Finally, Nasser was an unsurpassed operator at the subterranean level of politics. Though upright enough in many respects, he was amoral when it came to the use of subversion and even terrorism, regarding these as his counterforce against the established regimes and the great powers that bolstered them. He made himself the master of the labyrinth of fanatical groups that infested the Arab netherworld.

El Rais—"the boss," as he was sometimes called in the Arab corridors of power—could thus attack from various directions and via many proxies. The case of King Hussein of Jordan, in Nasser's early years, is illustrative.

The government of neighboring Jordan was a monarchy established after World War I by Great Britain, in which a British general still commanded the armed forces. In January 1956 King Hussein let it be known that he was on the verge of joining the Baghdad Pact, a British-spawned alliance of Arabs and Islamic nations which violated Nasser's fundamental precept that the Arab states should stand clear from great power alliances. Through inflammatory broadcasts over Radio Cairo and by activating his subversion network of bravos in Jordan, Nasser quickly fomented rioting throughout the country, which caused the fall of the Cabinet in Amman. Next, in a marvelous piece of knavery, Egyptian-trained Palestinian commandos in Jordan were influenced to escalate their raids into Israel from Jordanian territory, bringing down upon Jordan the severe reprisals for which the Israelis are noted. King Hussein, his country plunged into chaos, decided he had more to fear from Nasser than to gain from Great Britain, heretofore his aid and protector, so he renounced the Baghdad Pact, dismissed his British commander in chief, General Sir John Glubb, replaced him with the pro-Nasser General Ali Abu Nawar, dissolved the Jordanian Parliament, and called for a new election, which produced a landslide victory for the Nasserites.

The basis of Nasser's public policy, Florentine though it was in execution, was simple in theory. Egypt (and other Arab states without oil riches) could not make it economically without massive aid, nor could its masses be energized out of their age-old apathy without a succession of pride-stirring national achievements. So long as Egypt was safely within the sphere of the Western powers, those powers would never give it more than minimal aid, for they would see no necessity for doing so. Nor could an Egypt under Western tutelage expect any aid or arms from the Communist bloc. Nasser's solution to the dilemma of being unable to pursue meaningful independence without outside help was to break off from the West, to drive out all vestiges of its dominant presence, to stake out a foreign policy independent of the West's, to raise up a bloc of Arab states that followed his lead in their foreign policies. Only when Egypt and its sister states had established independence from both West and East could they make *both* bid against each other for their support. And in the successive steps by which the Western presence was driven out of Egypt, Nasser found the series of "victories" by which he could re-create pride in his people, a sense of progress, and a feeling of unity among Arabs as a whole.

The policy worked to a surprising degree. American CIA agents in Cairo computed that Nasser, as a man capable of making trouble for the capitalist world and an inveterate crusher of domestic Communist parties in Arab countries under his sway, had finagled several times the financial aid and credits and barter deals from the West *and* the Communist bloc that Egypt could have otherwise expected. "He grasped as no other Arab leader had before," wrote biographer Robert Stephens, "the bargaining power the Arabs could expect through their geographical position and oil resources, provided they were united, at least in their policies if not in their political institutions."

There were periods of breathtaking expectation when Nasser's triumphs came strung together in clusters, lending them the aspect of a rising crescendo that, given one final spark, could ignite the conflagration that would sweep the whole Arab world, with all its oil and strategic importance, into his camp.

From February to July 1958 Nasser and his Pan-Arab forces achieved a merger of Syria and Egypt in the United Arab Republic, an unsought confederation with the monarchy of Yemen, an uprising in Lebanon resulting in a civil war that ended in Nasser's playing a key role in the selection of the new government, a near coup in Jordan that further weakened King Hussein's position, and as the capstone the overthrow of Iraq's royal family and the replacement of its pro-Western government with a radical nationalist regime. "We have reached a stage in our struggle," Nasser told welcoming-home throngs in Cairo in the midst of these successes, "that places us on a sunlit summit."

A similar wave of coups, street triumphs, and Nasserite confederations billowed up in 1962–63, and another appeared to be rolling in again in early 1967; but in between Nasser had each time lost his gains. Why he had lost them contained as much instruction for Nixon policy makers as did the sources of his resilient strength.

The combination of Nasser's unique leadership capacities, his grip on the Arab imagination, and the questing temper of the times was such that if the Arab world could be united at all, he was the man to do it and this was the time it could be done. "Since Nasser," an Iraqi editor told an American official, "I feel as an Arab, not as an Iraqi." That his various unifications had thus far crumbled like pyramids of sand was due to two forces that could be as efficaciously employed in the Nixon era as before it: the centrifugal force of Arab divisiveness and the easy exploitability of that divisiveness by the "imperialists."

Nasser found that Syrians were as quick to pull out of a merger as they had been to come in, bearing out the salutation-warning given him by Syria's President Kuwatli at the time of the original merger: "You have acquired a nation of politicians; fifty percent believe themselves to be national leaders, twenty-five percent to be prophets and at least ten percent to be gods." Iraq's Kassem government broke with Nasser almost as soon as it got on its feet, and for a decade Egyptian-Iraqi relations would be "up" during one Iraqi regime and "down" during the next. Jordan would slip out of Nasser's grasp each time his grasping hand was about to close. The Saudis, long dazzled by Nasser and manipulatable by him because of their hereditary clan hatred for Jordan's ruling Hashemites, came at last to fear that their interests as oil-rich feudalists spoon-fed by Aramco were inimical to Nasser's designs to redistribute oil wealth Arab-wide and to drive out the Western oil companies.

All these divisions and more were, of course, widened and deepened by Oil-influenced Western stratagems. British Prime Minister Anthony Eden stated the basic Western objection to Nasserism and Arab unity in a September 1956 missive to President Eisenhower, at the time of Nasser's expropriation of the Suez Canal oil artery from a large consortium of nations led by Britain and France:

> . . . the seizure of the Suez Canal is, we are convinced, the opening gambit in a planned campaign designed by Nasser to expel all Western influence and interests from Arab countries. He believes that if he can get away with this, and if he can successfully defy eighteen nations, his prestige in Arabia will be so great that he will be able to mount revolutions of young officers in Saudi Arabia, Jordan, Syria and Iraq. . . . These new Governments will in effect be Egyptian satellites if not Russian ones. They will have to place their united oil resources under the control of a united Arabia led by Egypt and under Russian influence. When that moment comes

Nasser can deny oil to Western Europe and we shall be at his mercy.

The United States came only slowly to adopt the Eden view. The State Department was basically pro-Nasser during the early years of his revolution. Gradually, however, the American international oil companies were stimulated to active opposition to Nasser, first by the instabilities he created after 1957 by his topplings and near topplings of government after government that Oil was doing business with; then, after 1961, by his concerted campaigns of propaganda and subversion against the conservative Arab monarchies, which dominated in Arab oil country and which Nasser accused of being in league with imperialism (read "oil companies and their parent nations") to destroy Arab socialism. When in 1964 Nasser began to demand incessantly the evacuation of American and British military bases from Aden, the Persian Gulf, and Libya, Oil's case seemed vindicated.

According to Miles Copeland, a State Department-CIA insider in the formulation and implementation of American policy on Nasser, Oil and allied business interests effectively turned around both the executive branch and Congress. Copeland and other experts credit their campaign with emasculating the pro-Nasser influence in the State Department, weakening congressional support for aid to Egypt, influencing Eisenhower to use American naval and marine forces to contain the 1958 wave of Nasserite uprisings, and helping persuade the Kennedy-Johnson administrations to strengthen Saudi internal security against Nasserites and to aid the Saudis in staving off Nasser's 1962–67 60,000-man invasion of the southern Arabian Peninsula—the high-water mark of Nasser's drive to gain control of Middle Eastern oil. Meanwhile, as has been seen,* the oil companies effectively used their control of oil production and its attendant revenues in country after country to thwart domestic Nasserism and radicalism.

There was more, then, than met the eye to many of the snags that had repeatedly broken the Nasser momentum within sight of vast victories: the durable resistability of Hussein, propped up by Anglo-American gold and force; the overthrow in Saudi Arabia of Nasser dupe Saud and his replacement by Nasser enemy Faisal; the economic tailspin in Egypt that accompanied the gradual shutdown of American

*See pages 9 to 16.

aid; the draining away of Nasser's military strength and prestige in the Yemen war while a U.S.-financed and -supplied buildup of the Israeli military proceeded. One is driven to sympathize with Nasser by the size of the odds against him and because of the heroic scope of his ambition compared to the dearth of his physical resources, even as one admires the effectiveness with which Oil and Washington were concerting to safeguard the oil security of the West.

Despite all, in his minimal unity goal—the adherence by Arab nations to a common stance toward the foreign power blocs, under Egyptian tutelage—Nasser achieved considerable success. His combination of pervasive propaganda throughout the Arab world and active campaigns of subversion against regimes that ignored his baton "*did* eventually make it virtually unthinkable," Copeland points out, "for any Arab leader to make an agreement with a major power, Eastern or Western, without considering the 'union' and even Nasser's personal wish."

Indeed, Nasser seemed within reach of a historic success in early 1967. At least, U.S. policy planners thought so. A secret high-level policy study commissioned by the Johnson administration, directed by former Ambassador to Iran Julius Holmes, concluded that Nasser-inspired tremors in Arab politics, heightened by boosts from the Soviet Union, "threatened major American and allied interests in the region" and that "a continuation of the process, which would involve the Nasserization of Jordan, Lebanon, Libya, Tunisia, Morocco, Saudi Arabia and the Persian Gulf, would present . . . a security crisis of major and potentially catastrophic proportions."

The threat which this drift of affairs held for the oil industry was voiced in the spring of 1967 by the *Middle East Economic Survey,* a publication widely thought to reflect the posture of the Saudi Arabian government: "If the worst [sic] comes to the worst, the present structure of the oil industry in the Arab world might be irreparably shattered. . . ."

Nasser ran terrible risks, however, in trying to carry things off as the author of a great social revolution and the reviver and dominant figure of the far-flung, disjointed, unruly Arab world. With a treasury barely kept afloat by alms cadged from other nations, an army dependent for weapons and training on the credits he could wring from half a dozen governments whose cooperation waxed and waned, an indigent population growing by 175,000 mouths each month, he had to lead not from strength but by prestidigitation—by slogans, intrigues, alliances with irreconcilable interests, demagoguery to hold the many, terrorism and blackmail to intimidate the few, all of which involved bluffs that could be called at any time.

To maintain his ascendancy among a shifting motley of allies that at times included such wild men as the Ba'athist rulers of Syria and Iraq, he had to support their dangerous adventures with appropriate posturings and braggadocio. To keep on his team the more zealous religious and patriotic haters of Israel, he had to keep up a constant hit-and-run campaign of gestures, public blackguardings, and petty raids, hoping against history that the Israelis would prove as tolerant a punching bag as the British had been. And to stoke the fires in the bellies of his radical following, it behooved him to press his verbal attack on the oil companies and on the "quisling" kings and sheikhs who collaborated with the oil companies and who demurred from Nasser's panacea that the oil-rich Arabs share the common patrimony with their poorer Arab brothers in oilless states, thus making hard enemies of the oil companies and the ruling establishments of the "have" nations of the Arab world.

So Nasser—this figure of self-conjured strengths hobbled by vulnerabilities inherited from a bleak history—was juggling away on his high wire as the summer of 1967 approached, trying to stand by Syria in its provocations of Israel while heading off the consequences through quiet negotiations with Washington, when his string ran out.

As a boy Gamal had played the role of Shakespeare's Julius Caesar in a school production. Walking in Caesar's sandals, if only before the footlights, was an experience that had excited his youthful fancies. Over the passing years his fascination with the life of that peerless figure deepened, the more so as he discerned a lengthening number of parallels between it and his own. Like him, the great Caesar had inherited a society on the edge of ruin, had built his career on the multiplication of debts, and had passed through the various stages of street agitator, back-alley conspirator, urban mob master, stager of circuses and other diversions, organizer of cabals, spinner of intrigues, manipulator of religion, military dictator, yet a populist and the embodiment of the general will—all in the pursuit of the regeneration of his country. Though unsurpassed in his mastery of the practical arts of politics and warfare, Caesar held that at times fate was the final arbiter, that for one who would attempt such undertakings as reviving his country and remaking the world, the time comes when reason, prudence, and preparation have carried the cause as far as they can. To gain the final victory, one must knowingly leap into the unknown and entrust his enterprise to the mysterious workings of fate.

In June 1967 Nasser had arrived at such a moment. An escalation of raid and reprisal between bumptious Syria and remorseless Israel had trapped him in a dilemma. He must either be visibly active on Syria's side and risk a war with Israel he was unlikely to win or hang back while Israel deflated an Arab ally and in so doing showed Arab unity to be a façade and his own leadership of the Arab union an empty imposture. Nasser chose to rise to the occasion through gesture and bluff he hoped Israel would not call—verbal fusillades over Radio Cairo, the closing of the Gulf of Aqaba to Israeli shipping, a menacing-sounding demand that the United Nations withdraw its peace-keeping force that stood between the Egyptian and Israeli armies.

He had plunged into the unknown, hoping that behind-the-scenes diplomacy would provide a face-saving exit while trusting that if the die came up war, the goddess of battle would smile on him.

In three days of early June 1967 the grand coups and tenuous achievements of fifteen years were undone. Gamal Abdel Nasser, so long the vibrant giant of Arab hopes, was suddenly mute and haggard, an oracle without answers, trapped in the unwinding of a nightmare from which there could be no awakening.

Three *hours* had in fact sufficed. The first blips on the screens had appeared after breakfast, heralding a preemptive first strike by 300 Israeli planes that had swooped low over the desert to avoid radar detection and now streaked in from unlikely directions to obliterate Egyptian air bases. By lunchtime 80 percent of Nasser's air force—the repository of so much effort, pride, and hope—was a burning wreckage, never having gotten into the air. The rest, completely overmatched now, was destroyed within twenty-four hours. What had followed, though it piled in chaotically in day-by-day calamities, was but the playing out of a relentless script. The Egyptian army divisions in the Sinai, 100,000 men without eyes aloft to forewarn, without any protection from Israeli air strikes, and receiving contradictory orders from a command structure unstrung by the initial disaster, stumbled and fled to their encircling ruin. By the third day, June 7, the Israelis were at the Suez Canal, and nothing stood between them and Cairo but their own sense of restraint. With the Egyptian Army and Air Force, the principal Arab combat forces, knocked out of the war, Israel turned to the lightning annihilation of Nasser's allies, Jordan and Syria.

Even for one whose fortunes had from the beginning ridden on substances as chancy as dares; as unpredictable as the outcomes of palace intrigues and alley scuffles, as evanescent as cheers from a million throats, as revocable as foreign loans, as uncontrollable as the gambles of revolutionary comrades, the reverses of a few days in June, turning Nasser's world on its head, were more than could be assimilated.

The Suez Canal he had seized from great powers was now under the guns of unrecognized Israel. The Egypt he had freed of the restricted presence of unassertive British guard troops was now half occupied by rampaging warriors, the only counter to whom was to invite Russian troops into the other half. The Muslims he had summoned to an Islamic restoration had under his banner lost even their holy places in Jerusalem. The shining arms pacts he had in triumph brought home to his soldiers and propagandists survived only in the mocking litter of the Sinai. His high dam at Aswan, at last on the eve of completion, stood defenseless, surviving only through the forbearance of his declared mortal enemy. The confidence and spirit of unity he had inspired in Arabs must now undergo the disillusions and mutual recriminations that follow a monumentally abject defeat. And a thousand domestic ventures that were to breathe life into "Arab socialism" faced the poised gavel of national bankruptcy unless Egypt, casting aside the independence and pride he had brought it, were somehow to obtain massive loans by indenturing itself to the oil monarchies and the great powers.

Such enormities reduced Nasser, so resilient in past crises, to the edge of a nervous and physical collapse which he later likened to the condition of "a man walking in a desert surrounded by moving sands not knowing whether, if he moved, he would be swallowed up by the sands or would find the right path."

For three days, as the catastrophe mounted, he had made no public statement. On June 8 he accepted the inevitable and the responsibility for it. Publicly he announced a cease-fire; privately he decided to end his career and to announce this to the nation on the following evening. Hearing that the commander in chief of his army, General Abdel Hakim Amer, was beside himself with grief and preparing to commit suicide, Nasser went to Amer, an old and close friend, to dissuade him, telling him that the primary responsibility was his own, that he was going publicly to accept it and resign on the following day, appointing Vice President Zacharia Mohieddin as his successor.

On June 9, while Israeli forces were finishing off Jordan and Syria, Nasser spoke to his people over radio and television. The figure on the screen was not the buoyant, glowing Nasser familiar to Egyptians but a shadow, his face drawn and haggard, his manner hesitant, his voice at times choked. Egypt had been caught by surprise attack, he said, because he had listened to the requests of both the American and Soviet governments that he not open fire first. The enemy's attack had been stronger than expected because "there were other forces behind him which came to settle their account with the Arab nationalist movement."

Declaring that he was "ready to assume the entire responsibility," he announced that he was resigning from all his official and political roles and would return to the life of an ordinary citizen and that he had asked Mohieddin to take over the presidency.

While Nasser was speaking, there was unfolding an extraordinary scene, or at least one that must seem so to Americans, whose statesmen never resign except when in the clutches of the district attorney and who would not be mourned if they did. At his announcement early in his speech that he was stepping down, people began to come out of their houses to mill in the streets or to crowd the windows of their apartment buildings, gesticulating in anguish and venting hysterical pleas that Nasser not abandon them.

According to *Le Monde*'s on-the-scene correspondent in Cairo:

> In the twilight and the semi-blacked-out streets, hundreds of thousands, some of the men still in pajamas and the women in nightgowns, came out of their houses weeping and shouting, "Nasser, Nasser, don't leave us, we need you." The noise was like a rising storm. A whole people seemed to be in mourning. Tens of thousands gathered round the National Assembly shouting "Nasser, Nasser" and threatening to kill any deputies who did not vote for Nasser. Half a million people massed along the five miles from Nasser's home at Manshiet el Bakri to the centre of Cairo to watch over Nasser during the night and make sure he would go to the National Assembly the next day to withdraw his resignation. Millions more began to pour into Cairo from all over Egypt to make sure that Nasser stayed.

Such a spontaneous outpouring could not but have revived the flickering hope in the old gambler's bosom. And there was something more. In the few days since the onset of the Israeli blitzkrieg, there had been an impressive demonstration of Pan-Arab solidarity, not sufficient to have impact on a lightning war but significant enough to impress upon Nasser that the seeds he had sown had after all not been cast on barren ground. Jordan had entered the war against Israel, despite Israeli assurances that if it stayed out, it would be spared; more to the point, King Hussein's decision to intervene was in part forced by public opinion. "Such was the pressure of Arab public feeling," writes Robert Stephens, "that if he had tried to keep out of the war, he would have lost control of both his army and his throne." Iraq had sent a division, which was on its way to the Jordanian front. Algeria was sending troops, and a few of its jet fighters and advance army units had already arrived in Egypt. Morocco, Sudan, Kuwait, and Saudi Arabia had promised troops. While the awesome swiftness of Israel's conquests of Egypt, Jordan, and Syria had rendered moot most of these aids and promises, they indicated that had the Arab frontline combatants been able to stay in the field, Israel would have faced the kind of Arab solidarity that Nasser throughout his career had sought to call into being.

Moreover, several Arab governments had broken off diplomatic relations with the United States and Great Britain, the presumed allies of Israel, and a number of Arab oil states had stopped the loading of tankers servicing the West.

Nasser, in the hours of crushing defeat, was thus tempted back from the abyss of despair by the irresistible demand from his own people that he stay on and by signs from across the Arab world that his vision of Arab unity might yet be vindicated and might rescue on the diplomatic front what had been lost on the battlefield. On the day after his resignation he went before the National Assembly to declare he would continue on, an announcement that, so far as the world's press correspondents in Cairo could determine, was greeted with universal joy.

There was a precedent that encouraged Nasser to hope that an Arab oil cutback could create such intolerable economic problems for the Western industrial nations that to end it, they would force Israel to withdraw from its newly conquered territories, thus giving him the ultimate victory. This was precisely how he had emerged triumphant from the Suez War of 1956, though the military phase had ended with his forces shattered and Israeli, British, and French forces occupying roughly the same areas of Egypt the Israelis now held.

Then he had closed down the Suez Canal, through which more than half of Europe's oil passed. He simply had ships sunk in the canal to render it impassable, and he got the Syrians to sabotage the pipelines that carried Iraqi oil through their country on its way to Europe. Together these actions shut off two-thirds of the oil used in Western Europe—2,165,000 barrels a day. At bottom, it was the need of Britain and France for this oil—or rather, their reluctance to meet indefinitely the higher costs the oil companies charged for replacing it from elsewhere—that forced them to withdraw ignominiously from Egypt along with the Israelis, who snorted with anguish over having gotten themselves tied up with such weak sisters.*

Could Nasser do it again? The transportation stranglehold Egypt and Syria retained in 1967 did not affect as high a percentage of the West's oil needs as in 1956. On the other hand, oil now accounted for almost 90 percent of all world energy exports, and Europe's dependence on imported oil, most of it Arab oil, was approximately 95 percent. And Nasser had had an additional decade to perfect his techniques for both persuading and intimidating Arab governments.

Immediately on the heels of the first Israeli knockout blows, Nasser's far-ranging political machine saw to it that popular upheavals gave direction to the swell of Arab emotion against Israel and its assumed Western backers. In Saudi Arabia strikes and demonstrations disrupted the oil fields, resulting in 800 arrests and large-scale deportation of Palestinian laborers. Radio Baghdad warned of "spontaneous" sabotage of Western oil installations. In Libya Nasserite oil and dock workers battled the royal government and achieved a substantial oil shutdown. In Abu Dhabi the terrorist bombing of a British bank jolted the government into heeding popular demands for an oil shutdown.

Arab governments quickly got in front of their stampeding publics. During the first week of the war Lebanon banned the loading of oil at

*To avoid oversimplification: Great Britain and France, unless they could replace the oil being blocked by Nasser, faced economic and political difficulties. They could replace most of this oil only with the cooperation and financial aid of the United States, but the United States would not proffer it unless the invaders withdrew, for Washington respected Nasser's capacity to rouse the Arab world and the third world in general against the West over this affair and recognized the possibility that he might be able to involve Russia actively in their behalf. While Britain, France, and Israel could have carried it off by continuing their military action, by clearing the canal of obstructions, by reopening the Syrian pipelines, and by fending off guerrilla attacks indefinitely, it was a gamble they did not have the nerve to take.

its terminals. Saudi Arabia announced a prohibition against oil shipments to countries which supported Israel. Kuwait followed suit. Iraq's exports dropped by 72 percent, Libya's by 79 percent, Saudi Arabia's by 37 percent. By June 10 the amount of Arab oil production cut back had reached 10 million barrels a day! From Beirut, Abdullah Tariki, the rejected prophet of joint Arab action, exulted: "This is the chance of a lifetime to seize control of the world oil market."

Already Nasser, from the chasm of defeat, had achieved something of possibly momentous consequence. "This development of instituting an oil boycott against consuming countries because of political considerations," writes Middle East oil historian Benjamin Shwadran, "was a radical innovation in the history of the modern Arab movement." If a cutback of such magnitude could be maintained for a period of several months, it would so undermine the economies and political nervous systems of the West as to force a crackdown on Israel. This could force the Israelis to withdraw from the conquered lands—and usher in a mighty factor in the power politics of the world. More than a passing victory over Israel, it would signify a long-term victory in Arab relations with the West.

The defeat of this cutback/boycott—the turning of it from a landmark innovation into an object lesson not to be repeated—was the province of the great international oil companies. For more than a decade they had been Nasser's behind-the-scenes nemesis. Now they were to be frontline adversaries, determined to break the cutback/boycott of a dozen nations because it was a direct and, if successful, a mortal challenge to their control of world oil.

In part, Oil's defense was the product of the routine business precautions against insecurity of supply taken individually and spontaneously by its many companies so as always to be able to meet their contract commitments. And in part, the organizational and conceptual framework for the effort dated back to the 1956 creation of the Middle East Emergency Committee by Eisenhower's interior secretary, Fred Seaton, made up of senior Interior officials and executives of the fifteen largest American oil companies, who together evolved the operations that during Nasser's five-month Suez Canal crisis of 1956–57 furnished 3 million barrels a day to Europe—90 percent of its normal demand.

Thus, after 1956, to lessen dependence on oil that would be subject to interception by Nasser and his allies, the oil companies had undertaken large-scale oil development to the west of Suez—in Libya, Algeria, and Nigeria. By 1960 oil from North Africa had already attained a 4 percent share of the Western European market; by the late 1960's that share had risen to 30 percent. The share of European oil imports originating *east* of Suez had declined from about three-fourths in 1956 to one-half in 1967—a critical difference when one is putting together a bailout operation.

Moreover, even with regard to oil from the East, the companies had lessened their reliance on the Suez Canal and the Syrian pipelines. Conversion to larger tankers—tankers too big to use the canal—was under way. They were able to carry so much oil that they could haul it 3,500 miles around the Cape more cheaply per barrel than smaller ships taking the short route through the canal. The latest models, ordered in 1966, could carry 3.5 million barrels of oil. If Nasser's "oil action" were to be protracted, these supertankers would be coming on-line in time to play an important role.

Oil was ready for Nasser in another respect: There were large stockpiles of oil throughout Western Europe. Whereas in 1956 Great Britain had only a six-week supply, for example, in 1967 the average European inventory was good for sixteen weeks or more. And in the United States the oil companies, under government regulations, maintained in readiness the spare capacity to quickly increase production for export to Europe by 1 million barrels a day, 2 to 3 million if the affair dragged out.

By 1967 the strategy was so finely honed that the Middle East Emergency Committee did not even have to declare a formal emergency. Because Oil had at the ready the plans and resources with which to reassure Western consuming nations, there was no panic in Europe of the sort that might have given Nasser a quick victory. Assured of normal supplies for several months, the Western governments calmly settled back and let the oil behemoths "do their thing" —which was to pick apart Nasser's uneasy coalition of boycotters, playing upon their conditioned fears that the oil companies, the bearers of wealth, could all too easily do without them and might shut off their spigots for good.

Oil deployed its strategy of divide and conquer with a panache founded on three positions of strength: (1) Oil could replace a considerable amount of lost Arab production from its non-Arab oil provinces, so company tankers were ostentatiously rerouted to Iran, Venezuela, Indonesia, and the United States; (2) over the years Oil by prudent manipulation had kept its Arab hosts on limited rations, seeing to it that their oil income growth did not exceed their spending growth so that no big surpluses accumulated to make the hosts financially independent of the companies for any length of time; (3) Oil knew that most of the big Arab producers—Saudi Arabia, Kuwait, Libya, the Gulf sheikhdoms—had joined the cutback/boycott reluctantly, out of deference to a wave of popular emotion and fear of Nasserite subversion. The one was a transitory phenomenon; the other could be brought under police control in a matter of weeks, whereas Oil's stockpiles would last for months.

Skillfully Oil played these cards, letting the Saudis know that its long-term oil rival Iran was experiencing a 50 percent rise in production that could become permanent at the Saudis' expense, feeding the Algerians' suspicion that Kuwait was faking on its cutback claims, showing Libya that despite its sacrifice, Europe was getting all the oil it needed—and getting it from other Arab countries to boot.

So the lines were drawn: If the Arab states could exhibit from the outset firmness of purpose and a credible capacity to persevere, not all the resourcefulness and psychological warfare of the oil companies could prevail for long against a 10-million-barrel-a-day shutoff in production, but if parts of the Arab front flinched or shirked, its cause would quickly disintegrate and oil company supremacy would be reinforced.

Before June was far advanced, defections were chipping away the Arab ranks. The loss of oil revenues and the possibility of permanent shrinkage of market shares were riddling Arab resolve more quickly than even the oil company strategists had hoped. Oman, an oil sheikhdom on the Gulf, was *stepping up* its oil exports. The Saudis had moved quickly against the Nasserite oil field disrupters, arresting 300 and beginning deportation actions against suspect Palestinian workers. In mid-June, as security returned, Saudi Arabia ended its measurable participation—i.e., the absolute ban on the loading of any tankers—while continuing its more nebulous participation: an evadable stipulation that none of its oil could go to the United States or Great Britain. On June 14 Kuwait resumed oil exports on the same basis. In Libya it took almost a month to put down the Nasserite uprisings, but by July 5, with "movement" leaders in prison, the Libyan government resumed the loading of tankers. In July surreptitious defectors felt emboldened enough to go public, with Saudi Arabia as their chief spokesman.

The ball was now in the court of King Faisal, and he had no intention of playing Nasser's game. If Faisal was not large enough to envision the unlimited possibilities that sparkled in a successful oil action by a united Arab bloc—united temporarily, to be sure, by accident, but a unity perhaps preservable for a few months by statesmanship—neither was he small enough to permit himself to be dragooned, either by the Arab emotionalism of the moment or by the maneuvers of Nasser's apparatus, into an effort that could resurrect his fallen enemies.

He had gone through the motions of deferring to Pan-Arab pride and had taken the steps to defuse radical fury. Now, by gradations, he signaled that Nasser and the cutback/boycott were to be scuttled. The Saudi oil minister, Ahmed Zaki Yamani, let Aramco know that the Saudi government would not object to a resumption of full production so long as the production figures were concealed. Yamani sent out a public signal that disengagement was near by announcing that as a result of the boycott, his government's oil revenue was seriously reduced, and it was necessary to cancel a number of economic development projects and even to impose certain tax increases. Libya's oil minister emulated Yamani by giving the oil companies oral permission to ignore his written orders to cut production (he refused to put his rescission in writing). Soon Libya was demanding that the embargo be called off altogether on grounds of inefficacy, charging that Europe was getting its oil anyway. Even the radical states were caught up in the growing demoralization. Iraq, ascribing perfidy to the conservative oil monarchies, determined to go its own way and to restore its own oil exports unless other countries joined in an ironclad cutoff. Algeria accused Saudi Arabia of permitting its oil to reach countries which supplied Israel and expressed doubt that Kuwait was participating in *any* respect.

Nasser watched in dismay as within a month of the Arab disgrace in arms the spectacle of Arab inconstancy once more materialized to bedevil his cause. He attempted to stanch the informal bleeding away of oil state solidarity by calling for a meeting of Arab foreign ministers. In solemn official conclave, nations could not so easily abandon their recent rhetoric. The foreign ministers met at Khartoum, in the Sudan, in early August 1967. With the support of Egypt, Syria, and Algeria, Iraq proposed an "economic war plan" built around an airtight oil embargo. The Iraqi plan "for a total shutdown of oil production until Israel had withdrawn from the newly occupied territories" was resolutely opposed by the oil-producing states.

Delay is the great drainer of patriotic emotions and imperatives of "honor," and the majority of foreign ministers deferred action on this hot potato to a mid-August meeting of oil, finance, and economic ministers in Baghdad. There final action was again postponed to yet another foreign ministers' conference at the end of August in Khartoum, but the shape of things emerged when Saudi Arabia, Kuwait, Libya, Jordan, Morocco, and Tunisia tentatively opposed the Iraqi oil embargo plan. Saudi Minister Yamani pressed the view that the cutback/boycott was hurting the Arabs themselves more than anyone else and that the only ones to gain were the non-Arab oil producers of Iran, Venezuela, and America, which were moving into markets the Arabs were vacating.

Almost three months had passed since the Israeli conquests of early June and the undertaking by the oil companies to defeat the spreading Arab oil shutdown. Time had done its work for King Faisal. The patriotic fires of June now flickered desultorily. Nasser's fifth columns had been brought under the knout, and Egypt's economic straits had become so desperate that it was now more a supplicant than a menace. The oil-producing societies among the Arabs, Faisal's natural allies, had suffered enough losses even from a disintegrating embargo to be hurting and fearful of the future. The oil companies seemed to have proved, as Faisal had warned, that they could supply oil to the West indefinitely without the Arabs. Arab politics, once a stage from which Nasser emoted and a skein of conspiracies he directed, was now a web the strands of which were the economic realities. And at the center sat Faisal.

The Arab chiefs of state were to meet in Khartoum immediately after the foreign ministers' conclave, and it was Faisal's goal not only to put an end to the oil action but also to compel Nasser's endorsement of it so that it could be made to appear not as a pusillanimous abandonment of the "frontline" Arab states but as a united action aimed at the rebuilding of Arab strength for the battles yet to come. So at the foreign ministers' meeting Faisal's spokesmen dealt the deathblow to Nasser's hopes for victory through oil pressure by declaring flatly that Saudi Arabia would not shut down its oil industry. Libya and Kuwait followed suit. All three contended that their economies depended on the oil revenue and that without it their economic structures might collapse, leaving them unable to bail out the war-ravaged states. The Egyptian newspaper *Al-Ahram* later summed up the common fear of the oil monarchies when it declared that if the boycott had continued, the Arab oil producers would have lost their oil markets *forever.*

Nasser knew now that as the agent for uniting and regenerating the Arab world, he was at the end of the line. His final coalition, like all the others, had disintegrated. Its coerced members—Saudi Arabia, Kuwait, Libya—waited at Khartoum to put a leash around his neck as the price of a financial bailout; its sympathetic members despaired of collective action and were going their separate ways (Syria and Algeria had refused to go to Khartoum and associate with "reactionaries"; Iraq, unable to get its oil war plan accepted, would not cooperate with anyone else's). So he left Cairo for Khartoum, knowing that this time there would be no repeat of the public triumphs that had so many times apotheosized him, nor had he any hope of returning to Cairo bearing aloft a glittering diplomatic achievement as he had returned from Bandung, Moscow, Addis Ababa.

At Khartoum he was to meet with the newly dominant figure in Arab councils, King Faisal, at the residence of the Sudanese prime minister, Muhammad Mahgoub. The prospect must have brought home most acutely to Nasser the unfairness of life. He, Gamal Abdel Nasser, was widely accepted as the greatest Arab leader not just of his generation but for eight centuries, since Saladin, the victor over the Crusaders. He had achieved this stature by what *he* had called into being, not through any family inheritance or accident of national resources. But now he had to bend the knee to Faisal, that thin, cheerless man who beyond the range of his tribal vassals could not inspire a full-throated cheer in all of Araby, whose one puny brigade had typically arrived too late to fight the Israelis, who had no visions for the future of the impoverished millions of his race, who represented in this age of cataclysms no cause except the protection of his family inheritance and the perpetuation of a benighted religious sect, whose power and influence stemmed solely from the money paid him by the Western oil companies, who, for all his weeping about Jerusalem, was armed by the same Western overlords that armed the Israelis, who had never staked his fortunes on the defense of Arab states or Arab causes, yet who, by picking up the pieces of the Arab defeats he had sat out, had now emerged as the dominant influence.

Whatever his resentments, Nasser, the adaptable politician, parleyed with Faisal, as his allies would not, accepting the penalties of defeat in return for the alms that would stave off the collapse of Egypt and buy it the chance for another day.

What emerged was called the spectacular rapprochement. Nasser was to give his blessing to the formal calling off of the oil boycott and would publicly endorse it. He also agreed to withdraw his troops from Yemen and, in effect, to cease further to contest Saudi dominance in that area. Henceforth Nasser was to play the role not of the leader of the radical Arabs and the aspiring leader of *all* Arabs but of moderator of disputes between the conservative kingdoms and the radical dictatorships. Faisal, speaking for his fellow oil kings, agreed to an annual subsidy to Egypt and Jordan, "the war-ravaged states," of $378 million a year, two-thirds of which was for Egypt, one-third for Jordan. Syria got nothing, the price of its anti-Saudi invective and its refusal to come to Khartoum. These subventions were to be paid by Saudi Arabia, Kuwait, and Libya and were to continue for several years. This was a hefty sum for those days and those economies, amounting to about 20 percent of the annual revenues of the bankrolling states.

Nasser paid the first installment of his new subservience in his speech to the Khartoum Conference. Its subdued departure from the robust ebullience of fifteen years brought poignant moments even for those to whom the abdication of the "Hope of the Arabs" brought profound relief.

> I think that the matter has now been clarified. I do not want to repeat what I said yesterday, and none of us wants to go around in circles. We want everyone to do what he can. We do not ask anyone to give more than he is able to, or to do anything which, in his opinion, conflicts with the interests of his country. Yesterday we agreed with you that the oil embargo should be discontinued since you were unable to go on with it; of course, I am referring to our brothers in Saudi Arabia, Libya and Kuwait. And now I say we are not asking you to withdraw your deposits [from Western banks] since you are unable to do so or see no point in it.

Though outwardly Nasser seemed to roll with the punches with his customary aplomb and to bargain with his old-time chutzpah (despite the poor hand he held, he emerged with a huge annual subsidy for Egypt, and the mutual disengagement from Yemen had at least the façade of evenhandedness), his inner assessment that his visions were never to be and that defeat had irreparably enveloped him was to be written in that progressive breakdown of health that often attends the onset of despair.

Though he had ruled Egypt for fifteen years, he was only forty-nine years old at the time of the June war and its Khartoum aftermath. Within several months he would suffer a collapse attributed to a combination of strain, exhaustion, and a "diabetic complication" in one of his legs. Thereafter, according to his intimate Mohammed Heikal, he lived and labored on in an "agonizing pain" which for the most part he managed to conceal. In September 1969 Nasser was felled by a heart attack, officially disguised as influenza, and was incapacitated for six weeks. He had but a year to live until, at the end of a day of hectic negotiations that brought about a truce between King Hussein of Jordan and Yasir Arafat of the PLO, another heart attack took his life.

The justifications put out by the Khartoum participants (both conservatives and radicals) were true ones—they couldn't get along any longer without full-scale oil revenues, and they risked losing their market shares to those who had come forward to fill them during the cutback—and glossed-over ones—that by resuming oil production in full, they would strengthen their economies for the battle to come.

Radio Cairo, now sounding like Sheikh Yamani as an earnest of Nasser's future good behavior, justified the decision on the ground that a total embargo would have harmed the Arabs more than the West. Customers would have turned to Venezuela and Iran, explained the Egyptian Information Ministry.

President Abdul Salem Aref of Iraq explained to his countrymen that the Arabs agreed to end the embargo because they needed the money, needed it "to strengthen their armies and renew their military efforts against their enemies."

Had these explanations been fashioned by the psychowarriors of the oil companies themselves, they could not have been more convincing testaments to the supremacy, nay, the indispensability of Big Oil, or furnished more encouraging proof that the desired object lesson had been taught.

Faisal had thus seized upon the Arab debacle of June 1967 to end the ascendancy of the radicals in Arab affairs and to establish his own. The oil monarchies, because they were being made rich by the oil companies, because they had been preserved from the aggressions and subversions of their radical rivals through the help of the oil companies' parent nations, and because they were spared by geography from military confrontation with Israel, had come to Khartoum in a position to call the tune in Arab affairs. At Khartoum, through the tethering of Nasser, the ostracizing of Syria, and the isolation of Iraq, the Faisal bloc had made operative its supremacy in Arab affairs. And before leaving Khartoum, it was to institutionalize this paramountcy in a way designed to safeguard the status quo in oil and to insulate its oil revenues and its good relations with the West from being swept up in the anti-Israel jihads and anti-Western spasms that periodically blew up typhoonlike in the have-not Arab states. In other words, as Arab potentates—but potentates subject to radical insurgencies and transitory runnings amok—it was incumbent on them to make obligatory gestures of solidarity whenever their poor relations got involved in brawls and ruckuses which twanged the chords of Arab emotion. But though they were willing to cheer from the fringes of the brawling and to spring for the bail money and the hospital bills, they wanted to make sure that their properties and family jewels were in no way liable or endangered.

So before leaving Khartoum, they pushed through a resolution that outlawed "spontaneous" boycotts: "Nothing should be done to impair the financial capability of the Arab oil-producing states to back the unified Arab effort . . . the responsibility for deciding on appropriate measures should be left to the producing countries themselves. . . ." And Faisal laid the base for an exclusive organization that would supersede such pretenders as OPEC and the Arab League as the forum for inter-Arab oil policy. It was to be called the Organization of Arab Petroleum Exporting Countries (OAPEC), and its charter members were the Big Three of Arab oil—Saudi Arabia, Kuwait, and Libya—conservative monarchies all. Its membership rules were designed to exclude regimes which favored Arab-wide confrontations with the great oil companies or which saw oil as a foreign policy weapon at the service of *all Arabs,* to be used particularly for the retrieval of the Israeli conquests.

The exclusion of obstreperous voices from this high council of oil was achieved through bylaws promulgated at Beirut in January 1968: A member must be Arab (which excluded Venezuela, the premier remaining advocate of collective action against the oil companies); its major export must be oil (excluding the radicals, Egypt, Syria, and Algeria); and new memberships were subject to the blackball of the founding members (excluding oil-endowed but radical Iraq). Since no effective joint action on oil production or prices—either as an economic strategy or as a foreign policy protest—was thinkable without the participation of the Big Three that constituted OAPEC, no real challenge was foreseeable to the oil status quo—a system dominated by the great oil companies but subject to enough competition and consumer-nation influence to have long been identified with the policy of all-the-oil-the-world-wanted, available at under $2 a barrel.

El Rais had been terminally defeated. But as in the Phoenician legend, from the teeth of the slain dragon there would spring up new enemies shaped in Nasser's image: Muammar el Qaddafi of Libya; Saddam Hussein of Iraq; the offspring of Nasserite revolutions in Syria and of guerrilla-comrades in Algeria. In themselves, even in unison, they were a lesser threat than their progenitor. But the strength of the aggressor is only half the equation. Would the dragon's teeth be resisted in the future as skillfully and perseveringly as Nasser had been by the combination of oil companies, Western governments, and client states that, on the eve of the Nixon presidency, so totally dominated the oil world?

Chapter Seven

Thunderbolt from Tripoli

All the concern in 1967 about European dependence on Persian Gulf supplies had been resolved at the cost of a much heavier dependence on this single country. . . . The consequences of rapid and intense exploitation of Libyan oil resources included a strategic vulnerability the industry had not experienced in any other country at any other time."

—Joe Stark, 1975
Middle East Research and Information Project

Tripoli on its rocky promontory, tan and white against the Mediterranean blue, had been accommodating itself to invaders for twenty-six centuries by the time the oilmen came in the 1950's. A marble triumphal arch implanted by the Romans, Arabian palaces with golden domes, a Spanish cathedral that still projected a medieval stolidity, Turkish mosques of stucco and decorated tile, Catholic crosses atop Italian churches alike looked down on a harbor so much more ancient than they as to preserve no trace save whispers of its first users, Phoenicians, who would stop at this place they named Oea to buy precious stones brought from across the Sahara and set off from here to cross unknown seas to the west in search of rumored lands where tin, silver, or gold might be had, back in that venturesome dawn of things when history emerges from legend.

The faces of Tripoli—and there were now half a million of them—were in their varied hues and structures testaments both to a history of conquest, colonization, and conversion and to a fierce resistance against that history. Most of these faces reflected a mixed Berber-Arab descent that bespoke the gradual yielding by indigenous Caucasian to marauding Semite, begun twelve centuries before. But there flourished here large minorities that epochs had not digested, nor cataclysms dispersed: Greeks of the small entrepreneur class, whose forebears had vied with Carthaginian and Roman for the city's trade and who had even ruled here for a 100-year hour between the leaving of the Vandals and the coming of Mohammed's swordsmen; Jews—artisans and shopkeepers—who could trace their Tripolitanian identity back for 1,000 years to colonies of refuge from that era's persecutions; black laborers, descended from those hardiest of slaves who had survived the herding from Central Africa across the Sahara to Arab flesh auctions; Italians of but a generation or two in Libya, who had come to till the lands and fill the offices that Italy seized from the disintegrating Ottomans in 1911 and who wondered what the future held for them in a Tripoli whose street signs were now lettered in English and in a Libya embarked by the Anglo-Americans on its first experiment in self-rule under a native constitutional monarch, King Idris I.

Upon this scene of picturesque but threadbare stratification there burst in 1956 the shock of new life and the promise of undreamed-of prosperity, for oil had been found in Libya. In its wake came a new wave of adventurers in business suits, most of them Americans, which reactivated Tripoli's power to adapt. Quickly the rectangles of oilmen's office towers were hoisted on the skyline, to take their jarring place alongside dome, curve, cross, and minaret, and in certain parts of town the blare and hustle of nightclubs, bars, and casinos usurped the evening air that before had floated the melancholy wail of Muslims at prayer.

The oilmen were in Libya in unprecedented numbers for a whole complex of reasons; one of them was the key place assigned to Libya in Big Oil's strategy for diluting and scattering the potential power of host governments.

The directors of the major companies never lost sight of the fact that their dominance in the world of oil—hedged about, to be sure, by political-consumer interests of the Western governments and by revenue-and-pride pressures from the host governments, but dominance nevertheless—would last only as long as they avoided becoming really dependent for oil on any cohesive group of Islamic states. From the time Nasser first exhibited the potential for putting together just such a bloc, they had pursued a many-faceted program for lessening their dependence on oil that had to be shipped or piped through the spheres of Nasser and his Syrian confederates or pumped in countries subject to Nasserite inflammations and Israel-centered dislocations. All the elements of that program—stockpiling, supertankers, global diversity in exploration, keeping up U.S. spare—were successfully accomplished during the decade or so after 1956, but the most spectacular success of all was the transformation of barren Libya into a major oil producer.

The first sustained wave of oil activity in Libya followed upon

Nasser's Suez embroilment in 1956–57 and his successes in Syria, Lebanon, and Iraq in 1958. Expenditures for exploration surged. Geological and geophysical crew-months soared from 70 in 1955 to about 100 in 1959. In 1955 there was only 1 exploratory well in Libya; in 1959 there were 343.

Exxon, still called Standard Oil of New Jersey in those days, was the first to make a big strike in Libya, in 1959; others quickly followed suit. Whereas in 1961 Libya was exporting 20,000 barrels of oil a day, by 1966 the figure was 1.5 million, a level it had taken the Saudis more than twenty years to reach. What followed was truly astonishing. In 1968 oil exports hit 2.6 million barrels a day; in 1969, 3.1 million; 3.5 million were scheduled for 1970, making Libya the production equal of the established Middle East giants—Iran and Saudi Arabia*—and the far and away superior of onetime giants Iraq and Kuwait.

The phenomenal character of this upward leap—a 49 percent increase in 1968, a 23 percent increase in 1969—could be properly appreciated only by oilmen. But its impact could be easily grasped by Western statesmen. In 1960 Western Europe had depended on the Nasser-clouded Middle East for 72 percent of its oil; by 1965, only 62 percent; by the end of the 1960's, 48 percent. Conversely, Libya and the rest of oil-producing North Africa, which had provided 4.4 percent of Western Europe's oil in 1960, were furnishing 24 percent by 1965 and 41 percent at the end of 1969.

And the Libyan oil explosion was scarce begun. Future production schedules called for Libyan oil output to rise to 3.8 million barrels a day in 1971 and to 5 million barrels a day in 1973! This leap of 2 million barrels a day in only four years, from only one country, meant that all the consumer-favoring conditions of surplus—excessive supply, declining prices, a deteriorating bargaining position for host governments—must continue far into the 1970's. Between 1961, when Libyan oil first appeared on the market in modest amounts, and 1965, its infusion caused the wholesale price of gasoline in Rotterdam to fall 22 percent. In the crude oil market the doubling of Libyan exports between 1966 and 1969 was a key reason why, as 1969 progressed, the world price of crude had dropped down close to $1 a barrel. With a near doubling of Libyan production in the works for the early 1970's and the even more spectacular reserve additions just recorded in Saudi Arabia and Iran, only a miraculous intervention, a major upheaval in the way the

*In one quarter of 1969 Libya actually surpassed Saudi Arabia in oil production.

world oil industry functioned, could prevent the 1970's from being a decade of gradually declining oil prices and corresponding economic growth throughout the Free World.

The intervention that was to derail this train of happy probabilities occurred on September 1, 1969, and though it was far from miraculous in itself, one of a hundred third world coups, it was to be raised *ex post facto* to the stature of a pivotal event by the failure of perception and supineness of response by Richard Nixon and his Western peers.

In area Libya was a large country, one-fourth the size of the conterminous United States, but it contained fewer than 2 million inhabitants. It was a country of a few population centers, isolated from one another by vast stretches of desert and primitive transportation facilities. In fact, 99 percent of Libya consisted of the desert lands of the Saharan plateau. So vast a space with so few people was difficult to maintain centralized rule over, especially when the few people had scant sense of nationhood, a tradition of tribal patriarchalism, and a history of division that was symbolized by the fact that Libya had two capitals, a western capital—Tripoli—and an eastern capital—Benghazi. Moreover, its ruler was seventy-nine and ailing, the fading hero of bygone escapades, with a tiny army of dubious loyalty and a lax security apparatus that took its lead from an overrelaxed government the officials of which were chronically distracted with lining their own pockets.

That the sway of the regime of King Idris I was thin and tenuous was of little concern before oil came, when Libya was so poor and unprepossessing a place that no one had designs on it and when its government, with its minimal services, subsisted on aid from the United States and Great Britain, much of it in the form of rent for military bases. But by 1969, when Libya was producing 3 million barrels of oil a day and had become the centerpiece of an oil strategy formed to free the West of oil dependence on the Middle East, the stability of its pro-Western constitutional monarchy was the type of concern that State Departments and National Security Councils presumably exist to wrestle with.

Though the tempests loosed by Nasser were spent, and his stricken days numbered, he was to reap a last harvest from the seed he had so long cast upon the winds, some of it wafted across desert reaches by Radio Cairo in nighttime broadcasts heard by boys who lolled outside Bedouin tents under the stars, stirring to visions of glory and revenge:

"Dear Arab brothers, raise your heads from the imperialist boots, for the era of tyranny is past! Raise the heads that are bowed in Iraq, in Jordan, and on the frontiers of Palestine. Raise your head, my brother in North Africa. The sun of freedom is rising over Egypt and the whole of the Nile Valley will soon be flooded by its rays. Raise your heads to the skies."

Here and there the seed had taken root and silently pushed up stalks that were to bloom late in Nasser's life. In July 1968 militant Ba'athists seized control of Iraq; the destined leader among them was young Saddam Hussein, who had for a time literally sat at the knee of *El Rais.* In the spring of 1969 a young Nasserite officer, Jafaar al-Numeiry, seized power in the giant Sudan. And in Libya a group of twelve army lieutenants and captains, all of them in their twenties and modeling themselves after Nasser's Free Officers, were plotting a Nasserite revolution. Twice in the past year they had postponed their uprising, but as September 1969 opened, the conditions seemed optimal and the code word "Palestine is ours" was flashed across the 700 miles that separated the key units in Tripoli and Benghazi.

King Idris was out of the country, in Turkey undergoing medical treatment, and was scheduled to go from Turkey to Greece for a three-week rest at an exclusive spa. In the king's absence the security of the regime was in the hands of his praetorian guard, the senior officers of the national police force, who were loyal enough but, it appears, none too bright. When they were invited en masse to a party being given in their honor by some young army officers, none sniffed anything amiss, and all attended. As a festive evening was drawing to a close, the king's men were suddenly surrounded by soldiers, at gunpoint, arrested, and carried off to prison—all so neatly done that not a shot was fired. With the back of the potential resistance broken, the rebels crept through the darkened streets in the wee hours to arrest the army's chief of staff and to surround and overwhelm key government and military installations.

In the eastern capital, Benghazi, a rebel unit commanded by a communications lieutenant, Muammar el-Qaddafi, occupied the radio station and, after scattered fighting, immobilized the Cyrenaica Defense Force by capturing its armored vehicles.

The victorious young officers, who had named themselves the Revolutionary Command Council after Nasser's example, immediately proclaimed a Libyan Arab Republic, adopted the Arab nationalist slogan of "Socialism, unity and freedom," dissolved the Libyan parliament, established a military dictatorship, and sent a public message to President Nasser pledging their support of his "anti-Zionist principles."

In its first days as claimant to power, while bracing for a counterrevolutionary riposte from within or without Libya, the regime issued its pronouncements as a group—as Nasser's RCC had done—but by week's end Qaddafi had emerged as strongman and felt able to announce that "supreme authority has been vested in me."

Though only twenty-eight, Qaddafi cut a forceful figure, and his manner conveyed authority. Tall, athletic, swarthy, he was, when animated, handsome in a hard way, with curly black hair under his high-crowned military cap, blazing black eyes, and a pleasing array of white, even teeth; in contemplation, however, his face darkened into a hawkish aspect, with the classic Bedouin scowl. He was one who, as events would show, could routinely give orders for assassinations and issue decrees that would uproot from their homes tens of thousands of Jews, Italians, and Greeks. He spoke as one who carried the double burden, albeit confidently, of leading the Arab race to a glorious destiny heretofore denied it and of shaping up that race's miserable contemporary specimens for the journey. In his first public speech in Benghazi after the takeover, he digressed from painting a vision of the coming Pan-Arab unity to admonish his hearers: "From now on you will have to work and sweat to produce." On another occasion he revealed his fear that Allah had given him poor, frivolous clay with which to erect His monuments: "What the Arabs need is someone to make them weep, not someone to make them laugh."

He had been, in the not-so-long-ago, the bright boy of his desert tribe who was selected to be sent to town for schooling and who suffered humiliations there because even to small-towners in desolate central Libya, a Bedouin apparently seems a lout. He told me in an interview, "We were lions in the desert, jackals in the city." The interview took place in a tent he had set up on the manicured lawn of his headquarters building in Tripoli which he claimed filled a need to be outdoors. "I think better out in the sun," he explained. This may have been an affectation for image-building purposes, just as Nasser, city-bred, loved to dramatize his tenuous link with peasant Egypt, but in any event Qaddafi had accumulated a hoard of abrasive convictions that indeed smacked of aeons of solitary brooding among the dunes.

He appears to have been the product of three basic molding influences. As a nomad of Libya's Sirtic Desert he had grown up in a region long influenced by the Sanusi, a proselytizing sect that taught a puritanical Islam and inculcated a fierce hatred of things foreign. As a Bedouin, nourished on campfire-bred tribal memories and antipathies, he bore the psychic marks of the Italian onslaught on the Bedouins of Cyrenaica and the Fezzan, which reached its peak in the early 1930's, the decade before Qaddafi's birth, when many a desert tribe was imprisoned in concentration camps for its refusal to embrace the grand new world of cities, harbors, roads, and farms that *il Duce* was creating. And as a 1950's Arab adolescent of pride and ability he found his hostilities, frustrations, and hopes given gripping expression and purposeful synthesis in the broadcasts and deeds of Gamal Abdel Nasser, whose mythic image he zealously worshiped and romanticized as only youth can and who became his model in both goals and tactics, in the most methodical sense.

And so Muammar el-Qaddafi, advertising himself as the "Sword of Islam," the disciple of Nasser, determined to succeed where the prophet had failed, had seized power. The first question was: Would he be allowed to keep it?

Here was a situation, once the fun and adventure of it were set aside, full of grave implications both for those charged with upholding the norms of legitimacy and for those who bore the burdens of protecting Western advantage and preserving Western security. A legitimate Libyan government, an authentic *native* government, succeeding long ages of foreign rule and imposed puppets, a government called into being by the United Nations General Assembly, its constitution drawn up and ratified by a Libyan National Assembly which also selected its king, who was a Sanusi grand master and a hero of the fight for national liberation, a government aligned with the United States and Great Britain in such vital undertakings as military bases and defense arrangements, an ally which had in general followed a pro-Western policy in international affairs, an ally, moreover, which had become the centerpiece of the post-Suez Western oil strategy and was now the preeminent oil supplier of several NATO countries—all this overthrown at gunpoint by a corporal's guard of young conspirators representing no constituency in particular, whose first act was to abolish the parliament and who sought to steal a realm of vast area, incalculable wealth, and great strategic significance, a realm they could not possibly hold except by the aid of old Libya's enemies and the sufferance of its friends.

And even if American and British statesmen were not especially moved, their routine proclamations notwithstanding, by legitimacy violated in a murky world or by an ally pulled down, they were presumably concerned to see the usurping pretender displaying all the symptoms of a deep hostility.

On the morrow of its takeover the Revolutionary Command Council (RCC) sent, and brandished in public, a message to President Nasser pledging the new regime's support of policies that boiled down to the destruction of Israel and reprisals against Israel's Western friends, notably the United States. The coup attempt itself had used as its code name Operation Palestine—all in all an emphasis which, as one's motivating symbol and first thoughts after victory, revealed a bellicose hunger for rekindling the Mideast tinderbox, an aim to use its snatched control of Libya to shift the post-1967 balance of power in the Arab world away from conservative, pro-Western, commerce-as-usual quiescence to radical, anti-Western, commerce-be-damned agitation. And in its initial policy statement, issued on the same day, the RCC proclaimed its alignment with Arab nationalism and socialism—that is, with Algeria, Iraq, Syria, and revolutionary Egypt and against the Arab oil monarchies that were seen in radical-state rhetoric as stooges of the United States and the oil companies. To be sure, the RCC added an assurance that existing agreements (with oil companies) would be observed, but this assurance ran the risk of reminding people that the same pledge had been given by its predecessors—the radical military officer revolutions led by Nasser and Kassem—in order to dissuade the Western powers from counterrevolutionary action and that once Nasser and Kassem felt entrenched enough, they had seized the properties they had pledged to honor.

But what of acts, as opposed to words, slogans, bravado? From its first day the Qaddafi revolution acted out a progressively widening hostility to the Western presence, an animus so across the board as to make the new regime's survival a distinct frontal challenge to Western interests.

Diplomats, intelligence operatives, and their superiors in Washington and London were wont to view the goings-on through rose-tinted lenses which softened data that would otherwise add up to the need for hard decisions. One could, if so inclined, attribute to laudable Islamic piety the closing down of all bars, nightclubs, and casinos, those ubiquitous adjuncts of Western civilization. One could lay to unfortunate but pristine scriptural zeal the lopping off of hands for theft, the public flogging of fornicators, and other repudiations of civilized law and spirit. One could applaud as healthy the racial pride that caused the

Roman lettering on street signs to be replaced with Arabic. One could, in the same spirit, understand the outlawing of the teaching of English in the schools and the harassment of American Peace Corpsmen—the least offensive and "exploitive" of Westerners—into pulling out of Libya. One could regret but dismiss as "not yet our affair" the closing of Christian churches and synagogues and the turning of them into mosques* and the decree that all businesses (except oil businesses) must be owned by "Arab Libyans," and related omens which meant the expropriation of the livelihoods and the ultimate expulsion of tens of thousands of Libyans of Italian, Jewish, and Greek descent from this population-starved, talent-starved country.

But when the RCC abruptly canceled Libya's $1.2 billion air defense pact with Great Britain amid signals it was shopping for arms in Moscow; when, after a week in power, Qaddafi chose as prime minister the most conspicuous enemy of the oil companies, Suleiman Maghrebi, who had been in prison since 1967 for organizing the strikes and demonstrations which forced the Libyan government to join the 1967 oil embargo against the West; when, after two weeks, the RCC announced that it was in effect undertaking to expel the Americans from their great air base at Wheelus and the British from their bases at el-Adem and Tobruk, the broad picture ought to have come into focus, making it clear that what was going on was very much our affair after all.

As early as September 5, during the first week of the revolution, the *Middle East Economic Survey* published this estimate of the turn of events in Libya:

> [The coup] could well signify a momentous realignment of the balance of forces in the Arab world. Up to now the inner dynamics of inter-Arab politics have been regulated by a precarious equilibrium between the so-called "progressive" or "revolutionary" states on the one hand and the traditional monarchies, which include most of the major oil producing states, on the other, with the political leverage of the former being to a certain extent offset by the financial strength of the latter. Now, with the abrupt transfer of Libya into the "revolutionary" camp . . . the previous balance between the two groups has been dramatically upset. And

*For example, the magnificent Italian Cathedral of the Sacred Heart of Jesus was "deconsecrated" and reopened as the Mosque of Gamal Abdel Nasser.

> this is likely to have important long-term implications on a number of fronts—not the least for the remaining oil-producing monarchies, for the oil industry, and for the Arab struggle against Israel together with the related strategic interests of the big powers.

The early assessment of the oil company executives in Libya is reflected in an internal memo which the Oasis Consortium—a combine of Marathon, Continental, Amerada, Hess, and Shell which accounted for 31 percent of all Libyan oil production—circulated among its officials during September. "Once the regime is stable it will launch a frontal attack on the oil industry," Oasis predicted. Though believing it likely the RCC would stop short of the total confrontation with world oil that outright expropriation would provoke, Oasis warned its affiliates to brace for everything else: "Driven by missionary zeal of secure absolute economic sovereignty, the regime will use every possible means of 'persuasion.' . . ."

A 1974 inquiry by the Subcommittee on Multinational Corporations of the Senate Foreign Relations Committee, headed by Senator Frank Church (D-Ida.), found that from the start the Qaddafi regime was blatantly hostile: "Their first order of business was the prompt removal of British and American bases from Libyan soil. . . . Colonel Qaddafi made it known from the outset that his government sought to extract better terms from the companies and thereby to discredit the Idris regime [and other pro-Western oil state governments]."

Henry Kissinger, then the dominant foreign policy adviser to the President, concedes in his memoirs that there was from the start ample evidence that Qaddafi was "avowedly hostile," was espousing "implacable anti-Western doctrines," and had "set out to extirpate Western influence" from a country previously pro-Western and vital to NATO's oil security.

These assessments and the digest of deeds, words, and associations that prompted them are cited in perhaps laborious detail to emphasize the discontinuity between provocation and response, between what was known and what was done.

If the Qaddafi grab for power were to be thwarted, the impetus would have to come from the United States. Great Britain, the other Western power with a large responsibility and stake, had no stomach to resist yet another third world rousting. This was established at the start, on September 3, when Whitehall turned away an appeal from King Idris. The tepidity of the British in the matter is illustrated by the English statesman-historian J. B. Kelly: "The British Foreign Office, running true to the form it has increasingly shown of late years, also tried to ingratiate itself with [Qaddafi] by counselling British nurses in a hospital in Benghazi, when Libyan soldiers went on an anti-European rampage, to submit to rape rather than provoke the Libyans' ill will."

Amid the complexities which the Qaddafi coup presented to American officials charged with maintaining the oil security of the Western world, three realities emerged, as concrete and unambiguous as they were unpleasant: (1) Having reduced the West's oil dependence on the several oil states of the Middle East by concentrating more and more of that dependence on one reliable, proximate country, thereby violating Winston Churchill's basic axiom of oil statecraft—"Safety and security in oil lie in variety alone"—it lay the more heavily on Washington to make certain that this one country—Libya—remained securely in its orbit; (2) the shift of a great oil producer such as Libya from traditional oil monarchy to revolutionary regime risked, if countenanced, an upset in the balance of power within the Arab world, swinging it from accommodationist to confrontationist so far as the Western powers and oil companies were concerned; (3) one reason for the accommodationist oil monarchies to adhere to the oil status quo and avoid confrontation with the Anglo-American oil companies was their assumption that under the oil status quo, they could rely on the Anglo-Americans to help safeguard them from radical overthrow, from without or *within;* Western acquiescence in Qaddafi's coup would explode this assumption and remove a key reason for their deference.

This last point was elementary in the view of Henry Kissinger, though at the time he did not see fit to uphold it forcefully: "Accepting Qaddafi's revolution," Kissinger later wrote, "was bound to affect also the West's political relations with the conservative oil producers. Libya taught these rulers a fateful lesson: The industrial democracies would not protect friendly governments [such as the Idris government] so long as their radical, avowedly hostile successors did not [openly] challenge the democracies' access to oil. Hence, there was no point in seeking to buy Western good will by restraint on oil prices or anything else."

As a determined revolutionary and religious zealot driven at once by idealism and hatred, as a usurper at gunpoint who must justify himself as the avenger of terrible wrongs, as a mere boss of a tiny gang who must quickly stir up a mass following at home and who craved one throughout the Arab world, Muammar el-Qaddafi was obliged to take steps that were dramatic, shocking, and punitive. But for the world's puniest regime to do so at the expense of the world's greatest combination of power had its hazards—on the record.

Only eleven years had passed since President Eisenhower, with no greater provocation, had put ashore 14,000 marines in Lebanon from the same Sixth Fleet which still patrolled the Mediterranean, while the British were landing a companion force in Jordan, to forestall the establishment of RCC-like regimes, a chore they had five years earlier accomplished less visibly to save the shah from Mossadegh. Eisenhower's successors, Kennedy and Johnson, had upheld this tradition by giving the Saudis the military aid they needed to put down another RCC type of revolution—in Yemen, not to speak of the U.S. provisioning of Israel for its recent wipeout of "frontline" Arab armies, for which Qaddafi held the United States responsible. Would the new American President, Richard M. Nixon, allow what his predecessors would not? (Five years after these events, during which Qaddafi had enjoyed ample time, oil wealth, and arms sales from the Soviet bloc with which to make his regime incomparably stronger than it was in 1969, it was still so vulnerable, according to Pentagon assessments I obtained access to in November 1974, that an invasion by two marine divisions was considered sufficient to topple Qaddafi and secure the oil fields from sabotage, and CIA sources were insisting that it would still be relatively simple to engineer a coup and replace Qaddafi with a leader more friendly to the West. To take such steps in 1974 would have raised far greater practical and moral problems than in 1969, when we would have been acting not as a conqueror but as the defender of the lawful government, interceding at its request.)

Though inclination and circumstance prompted Qaddafi to be bold, history recommended to him a pattern of selective prudence. His assaults on the U.S. presence in Libya, and on the oil companies, proceeded by measured steps, as if to test Nixon's resistance, thereby to learn where to back off. As preliminaries to *acting* against the Americans, the RCC would pursue the safer course of menacing gesture and ugly complaint, always getting results that relieved its anxiety about retribution and usually getting results that obviated the *need* for action.

In the first days after the coup, when its fate hung on the success of King Idris's envoy, Omar al-Shelhi, in his trip to London and then to the United States to seek great power intervention, RCC grumbles were loosed in the direction of the American Embassy in Tripoli, alleging that the United States was conspiring for an Idris restoration, while at the same time there came official reassurances that the oil pacts would be abided by and the 10,000 American civilians in Libya kept safe. The State Department promptly responded, on September 5, instructing its chargé d'affaires in Tripoli, James Blake, to assure the RCC that Shelhi had not been invited to the United States but had come on his own initiative and to add somewhat unctuously that he had been received not by the State Department in Washington but by "a junior official" of the U.S. Mission to the United Nations in New York. The news, needing no embellishment from the RCC propagandists, was broadcast over the controlled media as a proof of the triumph of the coup and the futility of resistance. The next day the United States extended official recognition, a greater triumph, even though the argument on whether or not the coup would be ultimately tolerated had scarcely begun in Washington.

The ousting of the American official presence, cultural as well as military, was a revolutionary imperative, and the latter a prerequisite to safely laying hands on the oil companies, but it was approached gingerly, the rhetoric notwithstanding, with a pause every step of the way to take the pulse of American resistance.

A quick and easy anti-imperialist trophy was the American Peace Corps, which had a 143-person contingent in Libya, teaching English in the public schools. The Peace Corps was an organization known to come only when invited and to leave when disinvited, but even here the ferocious Qaddafi approached its ouster indirectly and cautiously. A hullaballoo was whipped up about the cultural imperialism at work in the teaching of English to Libyan schoolchildren. (Qaddafi himself well knew the value of having learned English as a schoolchild and exploited his facility in it at every opportunity.) If the RCC suspended the classes —its unchallengeable right—the Peace Corps mission might take the hint and voluntarily self-destruct. And so it did. On October 3 the United States announced that its Peace Corps detachment was voluntarily pulling out of Libya, rather than wait for the unpleasantness of expulsion.

The identical process was under way, for infinitely higher stakes, with regard to the great Wheelus Air Force Base east of Tripoli, home of thousands of American airmen, training center for NATO bombing forces, and bastion of Western security in the Mediterranean. So long as American might bristled at Wheelus and British forces menaced from their bases at el-Adem and Tobruk, the RCC was inhibited from its platform goals of pogroms, oil control seizure, and an anti-Western shift in the region's alignment. Throughout the first month of the revolution the junta continually created incidents and asserted claims designed to build up an ascendancy over Wheelus and to test American resolve on eventual ouster. The initial test came on the second day after the coup, when the RCC asked Wheelus authorities temporarily to suspend flight operations, thus to block off any arrivals or departures not under junta control. The United States immediately complied, as did the British, thus according sovereignty to what was merely a splinter group of pretenders.

Next, Ambassador Joseph Palmer II surrendered to a barrage of complaints about "smuggling" by accepting the principle of junta controls at Wheelus and the presence there of RCC customs and immigration troops. Accused correctly of rescuing a Libyan Jew from junta clutches by hiding him in a crate supposedly carrying musical instruments and flying him to safety in Malta aboard a U.S. Air Force plane, the United States, which in a more robust age would have gloried in the deed, officially apologized, promised to do no such thing in the future, and agreed to yet more RCC troops and controls at Wheelus.

Qaddafi was moving toward a longtime favorite cause of Nasser: the ouster of American and British troops from their Libyan bases. But gradually. Libyan Premier Maghrebi had begun talking about "not renewing" the lease at Wheelus when it expired a couple of years hence, the least confrontational manner in which an ouster could be approached and one which left plenty of room for graceful retreat if the United States signaled resistance to this unprovoked breach of faith and relations. No such resistance was signaled. Instead, Qaddafi could read in the American press the reaction of unidentified U.S. officials to the effect that "U.S. Air Base Looks Doomed." So he pressed ahead. On October 28 the United States announced it would soon begin evacuating all military personnel from Wheelus without even contesting the matter. The British were to follow suit. The Nixon administration thus withdrew its military strength from our oil fortress of the Mediterranean just as obligingly as it had removed the Peace Corps. So far each probe by Qaddafi for the pulsebeat of U.S. resistance had failed; there just wasn't any to be found. In December, at an Arab summit where he broke bread on terms of equality with his lifetime idol, Nasser, Qaddafi publicly pledged "to put all my country's resources at the disposal of the confrontation states."

The West's armed forces having commenced their pullout in December, Qaddafi could begin his crackdown on the oil companies in January—and he did, demanding on January 20 price increases unprecedented in the history of oil negotiations and implying that if he didn't get what he wanted, he would tear up the existing oil agreements by stopping production unilaterally, an act which would constitute a jugular attack on the control system through which the West dominated international oil.

These events took place, it must be remembered, during an era in which the United States was still ready and willing to assert both its influence and its power to protect its vital interests in this part of the world. Two years before the Qaddafi revolution the United States and its oil companies had automatically mobilized to defeat the Arab oil embargo. A year *after* the Qaddafi coup the United States would join actively and openly with Israel to thwart the toppling of King Hussein of Jordan by Syria and the Palestinian fedayeen. *Four* years after the Qaddafi coup Nixon would launch the unprecedented air resupply of the Israeli armed forces in the midst of the Yom Kippur War and mount a worldwide atomic air alert aimed at intimidating the Soviet Union from intervening on the Arab side. So the sun had not yet set on the day of an Eisenhower-style Lebanon response or a Johnson-style Dominican Republic response or a covert response of the sort that had toppled Mossadegh.* And we have it from Henry Kissinger that the U.S. government debated within itself whether to accept or reject the Qaddafi seizure of power and that Kissinger had analyses made of various actions that could be taken against it.

What then, with the redoubtable Richard Nixon at bat, was the rationale behind what Kissinger would later condemn as his administration's "passivity" in Libya?

The State Department explanation, given to a Senate inquiry by David Newsom, U.S. ambassador to Libya until just before the coup and deputy assistant secretary of state for African affairs thereafter, was that there was a local intelligence failure, which made misjudgments unavoidable. According to the subcommittee's summary of Ambassador Newsom's testimony: ". . . the U.S. Embassy in Tripoli had not anticipated the Libya coup. The young military officers who made up the Revolutionary Command Council (RCC) were not known to U.S.

*In fact, in 1973 Nixon would publicly raise the specter of the Mossadegh precedent when Qaddafi expropriated the Bunker Hunt oil company as part of his third-anniversary celebration of his takeover of the Wheelus Air Base.

Government officials. The U.S. Government therefore did not anticipate the radical changes which were to follow in the Libyan Government's pressure against the oil companies."

Moreover, U.S. officials thought that the Revolutionary Command Council's "religious fervor and strong nationalist bent were compatible with the U.S. objective of maintaining a secure investment presence in Libya."

On its face, this has the whiff of either disingenuity or of an impenetrable naïveté, raising rather than answering questions. The brand of religionationalist fervor prominent in Libya, and evident in the rhetoric of the Qaddafi-ites' earliest broadcasts, that of the Sanusi, had for almost seventy years been known in diplomatic literature as incurably antiforeign and especially anti-Christian. As far back as 1902 it had been analyzed by the French colonial officer Captain Gentil LaMotte: "As distinguished from other Moslem brotherhoods, this one has pan-Islamic dreams. It wishes and pursues above all else the expulsion of the infidel from the Dar al-Islam (the Land of Faith) and for a long time it has been preparing for this goal."

Considering that King Idris was aged and infirm, that Libya had been racked by Nasserite upheavals in 1964 and 1967, and that Idris had had to install six different governments in six years, how could the regime's shakiness and the question of the successor to Idris not be the constant preoccupation of American policy makers responsible for Libya? Why did not departmental ignorance of the usurpers, whose battle cries were that King Idris was a puppet of the United States and a stooge for the oil companies and whose very broadcast proclaiming the coup described the period of the American-Idris cooperation as "the era of reaction, bribery, intercession, favoritism, treason, and treachery" excite American caution rather than collaboration? Why, given the unbroken anti-Western record of radical officers' revolutions in the Middle East, given the Nasserite identity proclaimed by the RCC in all its visible trappings, first liaisons, and stated goals, did not Qaddafi's RCC strike Washington not as a blank piece of paper but rather as a carbon copy of all the troubles it had had with Nasser's RCC, Yemen's RCC, and the radical officers running Iraq and Syria? Why did not the coup immediately call to mind the Johnson administration's Holmes report of two years before, which cited Libya as ripe for "Nasserization" and which unequivocally analyzed Nasserite revolutions not as unknown quantities but as threats to "major American and allied interests in the region," the spread of which would "present . . . a security crisis of major, and potentially catastrophic proportions"? How, given the anti-Western, anti-oil-company, anti-Christian, anti-Jewish character which a "strong nationalist bent" joined to Islamic "religious fervor" had historically implied, could these qualities reassure Washington as being "compatible with . . . U.S. objectives"?

Henry Kissinger has come forward, out of retirement, with a partial answer. Carefully explaining that he himself was not to blame and was only on the periphery of the decisions made about Qaddafi ("I did not in Nixon's first term take an initiating role in Middle Eastern policy"), Dr. Kissinger contends that the lower echelons of the U.S. foreign policy apparatus were soft-headed about America's true interests and had a schoolboy's crush on left-wing revolutionaries. Or, as Dr. Kissinger puts it:

> The working level of our government, especially in the State Department, operated on the romantic view that Third World radicalism was really frustrated Western liberalism. Third World leaders, they believed, had become extremist because the West had backed conservative regimes, because we did not understand their reformist aspirations, because their societies were backward and eager for change—for every reason, in fact, other than the most likely: ideological commitment to the implacable anti-Western doctrines they were espousing.

Kissinger thus supports the explanation given by British historian J. B. Kelly:

> . . . there was a lingering belief in both capitals [Washington and London] . . . in the cleansing and therapeutic powers of revolutions conducted by ardent young officers of austere habits, stern demeanor and patent rectitude. If change meant progress, so the accepted theory went, then arbitrary change meant even greater progress. Colonel Qaddafi's ascetic ways, his conspicuous piety and his relentless vigor greatly impressed the British and American governments, so that when he peremptorily demanded the immediate removal of their military and air bases from Libya they hastened to accommodate him.

Not only were our "working-level" statesmen hot for the anonymous guerrillas from the dunes, but they were cold on the government of

King Idris. That over its eighteen years the Idris government had left its neighbors alone, had oppressed its own people less than most regimes in the region, had shunned the depravities of koranic criminal justice, had cooperated with American defense policies, welcomed American enterprises, tolerated American creeds and diversions, and supported American foreign policy were all matters that apparently earned it no points with the American State Department.

Our people were moved by less practical, more aesthetic considerations—the bribe taking of Libyan officials, in which the Idris monarchy aped most governments in the region, especially where oil was gushing. In the opinion of the State Department's resident oil expert, James Akins, "The Idris regime was certainly one of the most corrupt in the world. Concessions were given, contracts were given on the basis of payments to members of the royal family."

Throughout the apologias spawned by the Libyan debacle sounds the fell phrase "the corrupt Idris regime," sometimes rendered as "the feeble and corrupt Idris regime," as if American interests were better served by a vigorous, uncompromising enemy than by a lethargic, purse-lining friend or as if the traditional crimes of five-percenter politics were more repellent to the descendants of Boss Tweed and John D. Rockefeller than the traditional crimes of radical totalitarianism. Since millions of people are today unemployed in America and Europe, and tens of millions suffer more acutely in the less developed world, because of the progression of events that snowballed in part from this selective and irrelevant moralizing, the matter is worth a digression.

So far as our studies reach, the custom of bribery was widely and venerably institutionalized throughout the Eastern Hemisphere, and the practice of paying off authorities for oil favors has been coterminous with the presence of Oil in all hemispheres. We cannot really know whether the Libyan government was more corrupt than the Saudi or Iranian or Nigerian or other fabulous contenders for the title, or, for that matter, if the king's officials in Tripoli and Benghazi were any more corrupt than the cross section of U.S. congressmen tested by the Abscam unit or the cross section of New York City police officers caught in the searchlight of the Knapp Commission, or whether the indictable percentage of the Idris palace guard was any higher than would prove to be the case with the Nixon palace guard.

What is knowable, however, is that the Idris government stood head and shoulders above its oil state peers in larger tests of integrity—the tests of how it structured its oil industry toward Libyan goals, not oil industry goals, how it forced the pace of oil development that *Libya* wanted, not the one the Seven Sisters wanted, how it methodically saw to it that Libya controlled its gigantic Oil guests rather than being controlled by them. By these standards the Idris government, far from being a bought-up tool of foreign capitalists, was unique in its successful assertion of Libyan mastery in its own house, a mastery that, as we shall see, was now to make it possible for the other oil states to challenge the international companies for dominance.

The discovery of oil in Libya came decades later than in the other big oil states of the region, and the Idris government was determined to profit from their experiences. "I did not want Libya to begin as Iraq or as Saudi Arabia or as Kuwait," said the Idris petroleum minister, Fuad Kabazi. "I didn't want my country to be in the hands of one oil company." Not only were many companies to be invited to compete, but Idris wanted a lot of them to be smaller companies, preferably American independents, that were anxious to produce and were not tied into the worldwide network of the Seven Sisters, which would naturally subordinate Libyan production to that of its older claimants. "We wanted to discover oil quickly," recalls Kabazi. "This was why we preferred independents in the first stage, because they had very little interest in the Eastern Hemisphere outside Libya." Companies that were dependent on Libyan oil would be amenable to Libyan wishes.

Idris wanted the great companies, too, for they had the greater developmental resources and held the keys to much of the world's market for oil, but the great companies were brought in on Libya's terms. They did not receive the rights to vast contiguous areas, in virtual perpetuity, which they could develop or hold out of development as suited their overall needs, as was the case in other concessionary states. If their bids were right, they were awarded small strips for limited periods and under strict requirements for exploration and development effort, strips that were interspersed between plots awarded to independents. In the first round of bidding, seventeen winning companies were awarded eighty-four separate concession areas. The performance of Exxon or Mobil was thus to be tested against that of its hungry neighbors. Kabazi: ". . . if in an acreage I found an independent who had spent more energy and discovered oil, this would permit me to talk to his neighbor and tell him, 'Look, here's your neighbor who has discovered oil. You are almost on the same structure. Come on, now; try to drill.' "

The Idris system of ready access to all, favorable tax policies for oil companies, a positive political atmosphere, the requirement of earnest effort, plus the natural advantages—the high quality and low sulfur content of Libyan oil and its proximity to consumer markets in Europe—accounted for Libya's 200-fold surge in output and revenues over eight years.

This was done, however, by punching a hole in the control system through which the international oil companies regulated the world industry. The heart of that system was this: No oil company was as dependent on any concessionary state as that state was upon the company; no concessionary state could work its will on any resident oil major in regard to production level or price because the major, working in tandem with other majors, could get along without either the production or the profit from any one country, by making it up elsewhere, and could, to protect the system, cease operations in any intractable state, whereas no state could hope to prosper from oil except through the facilities of the majors. In turn, the oil majors, though they could hold the host states in check, were not a law unto themselves, whatever gaudy visions their founders may have dreamed in the long ago. Ultimately the authority of their parent governments, voted in by consumers, hovered over them, benignly as long as supplies were ample and prices cheap, but capable of turning meddlesome, even hostile, in case of shortage or inflation at the pump. So the great companies, strong enough to dominate the concessionary states but not strong enough to abuse the Western public unduly, had pursued the bland, safe, automatic profits of stability. The result had been, for most of the non-Communist world, a happy one: all the oil it could make use of at a price that for many years had been slowly declining from $2 a barrel toward $1. Upon the twin foundations of certain energy supply and cheap price had been built the greatest and most widespread prosperity the world has ever known, a boom that in 1970 was well along into its third decade and seemingly stretching toward infinity.

Fortunately the hole punched in that system by the Idris regime was sealable, for Idris sought not to challenge the system but rather to get more of its benefits for Libya. His demand for ever-increasing production would within four or five years be mooted, the majors predicted, as Libya reached its natural maximum production and leveled off. For that long, the majors could continue to make room for Libyan oil in the world market by holding down production growth in their Middle East provinces, though not without continued price slippage and mounting protests from these provinces. On their charts the Libyan hole would be plugged by 1973, when Libyan production reached its upper limit of 5 million barrels a day. After that the steady 6 or 7 percent annual rise in world demand would allow them to permit the kind of production jumps in the Middle East that would ease aggrieved tempers—just in time, they calculated.

If Idris's Libya was a mixed blessing to the oil majors, it was an unalloyed one on the higher plane of the Western statesman (except for those politicians, as in Great Britain, who had coal miner constituencies and who saw cheap oil as a monopolist plot to drive out coal), to whom the Libyan oil bulge meant not only security of supply but also the looser market that made the price cheaper and the oil companies less haughty. The Libyan oil surge thus served twice, by making the Middle Eastern oil states more amenable to the oil companies, which managed oil security for the West, while making the oil companies more amenable to consumers and their governments. By the same token, however, the toppling of King Idris by a clique that Kissinger termed "avowedly hostile" could change Libya from the reserve weapon of the West to a cannon pointed at its very structure of oil control and at the mystique of invincibility that undergirded that control.

With so much clearly at stake (clearly enough for American reporters to speculate about and for journalists at the *Middle East Economic Survey* to spell out in the first week of the revolution) Washington's unconcern about the survival of Idris and its hasty recognition and acceptance of the RCC can be explained not by fastidiousness over bribe taking by Idris ministers or by a predilection at the State Department's "working level" for Arab radicals but only by a fundamental, from-the-top-down attitude, a failure to be concerned about the paramount thing our government was supposed above all to be worried about and watching for in that region, which was as visible as a sandstorm whirling on a desert horizon.

Henry Kissinger claims to have belatedly caught faint glimpses of that sandstorm but to have lost sight of it in the dust raised by foot-dragging bureaucracies:

> There were desultory discussions in the Washington Special Action Group (WSAG) on what attitude to take toward the new

> Libyan regime. In a meeting of November 24, 1969,* I raised the question whether to have the 40 Committee [interagency committee that supervises intelligence activities] canvass the possibility of covert action. A study was prepared of economic and political pressure points on Libya; but the agencies did not have their heart in it. All options involving action were rejected, causing me to exclaim that I was averse to submitting to the President a paper that left us with the proposition that we could do nothing. My reluctance did not change a consensus precisely along those lines.

Dr. Kissinger wallows inextricably in the political subjunctive here—"desultory discussions" on "attitude," a "study" of unidentified "options" left unconsummated because unidentified agencies "did not have their heart in it," "I raised the question whether . . . to canvass the possibility," "my [unavailing] reluctance"—suggesting that had he released the raw WSAG minutes rather than provided us with his own blindfolded tour of veiled monstrosities, we might have found him even less effectual than he appears here. We are indebted to him, however, for such glimpses as he does give us into the classified realm of top-level decision making.

Kissinger has provided, from the interagency analysis made three months after the Qaddafi coup, excerpts which describe the rationale behind a Nixon administration policy obviously in effect from day one of the coup on through the many months of propitiating responses to test probes by Qaddafi down to the ultimate disasters of late 1970 and early 1971: "The return to our balance of payments and the security of U.S. investments in oil are considered our primary interests. We seek to retain our military facilities but not at the expense of threatening our economic return."

The stakes are trivialized here—described in terms of monetary balance sheets instead of strategic necessities—either because of an ignorance difficult to comprehend, given the oil diplomacy of five successive presidents before Nixon, or because if one's thesis is to end up by rejecting resistance of any kind and by entrusting the nation's cause entirely to propitiation, it behooves one to trivialize that cause.

The "primary interests" of the United States in Libya were nothing so routine as preserving investment returns and trade balances. Our vital concerns were the continuance of Libya's key roles in: (1) the

*Quite a leisurely response, considering that the coup took place on September 1.

maintenance of the favorable oil status quo and of the free world prosperity and unity based in part on it; (2) the maintenance of the pro-Western balance of power in the Arab oil world; and (3) the maintenance of basic oil security for the NATO allies and Japan. To these interests, the Idris government was committed and aligned by a network of defense agreements, aid programs, oil pacts, and other joint undertakings. Moreover, as Dr. Kissinger points out in his somewhat anomalous critique of the Nixon foreign policy, it was a primary interest of the United States to maintain its own credibility, both as savior of threatened allies and as retributor when provoked, among the other conservative oil monarchs with pro-Western alignments, the peers and colleagues of Idris in OPEC. We of the West could not let Idris be pulled down by insignificant forces on a Sunday night, while watching from our bristling military bases and offshore fleets, and then be seen trying to cuddle up to his overthrower, without causing them to write us off as their deliverer in time of need and therefore to begin to cogitate new alignments and to conspire over ways to gouge us they had hitherto forborne.

Having deleted our larger interests from its analysis, the government's position paper finds no threat to us except that which we might create ourselves by resistance: "We see no immediate threat to these interests, although such could result if the regime is threatened, or becomes increasingly unstable, or if there were a real confrontation over Wheelus, or in the event of renewed hostilities in the Middle East."

Even aside from the historic underestimate of the threat, the statement is notable for its upending of past policies that had been so successful. Heretofore, America's military presence, such as at Wheelus, and its other capacities to exert pressure and excite foreboding had been seen not as impediments or embarrassments to our oil security but as safeguards of it, as not so subtle reminders that the world's strongest power was prepared to preserve by all lawful means, and sometimes by extralawful ones, the economic pacts and regional alignments it deemed vital to its position. Heretofore, when faced with a new, rhetorically hostile regime that seemed tempted to tear up the old agreements under which a now-flourishing oil industry had been originally created, Western oil strategy had not met the challenge by making itself scarce and by avoiding the assertion of Oil's rights but rather had relied on the quiet but visible possession of the capacity indeed to threaten and indeed to destabilize that regime by cutting its economic umbilical chord to the rest of the world—the very sort of thing that Nixon and Kissinger were being advised to avoid even a whisper of.

Kissinger's instinctive request for a study of the kinds of "covert action" possible in Libya, for an evaluation of the "economic and political pressure points," was in that tradition; Qaddafi's cautious, bologna-slicing tactics described above, designed to give Washington successive opportunities to reveal its line of resistance, his postponement of any substantive moves against oil interests until the Western military presence was being dismantled, showed respect for that tradition, memory of the isolating of Iraq, the three-year boycott of Iranian oil, the covert overthrow of Mossadegh, the occupations of Lebanon and Jordan to forestall coups, the breaking of the Arab oil embargoes of 1956 and 1967, the cumulative ruining of Nasser.

Had Qaddafi known, in those early months, about the counterrevolutionary coup being hatched, he undoubtedly would have assumed it to be but the latest chapter in a very long book. A group of Libyan exiles had organized an expedition to Libya of mercenaries, whose battle plan was to break into the fortress holding Qaddafi's political opponents, many of them military men, liberate them, and with their help seize Qaddafi and attempt a general rallying of forces to overthrow his regime.

The American Embassy in Tripoli was apprised at an early stage of this attempt at restoration of the lawful government of Libya. But the U.S. government, instead of helping out or watchfully waiting or preparing to wrest advantage at the propitious moment, informed Qaddafi of the plot. The State Department would later explain that it "did not wish the Libyan Government to think that the United States was in any way associated with the anti-Qaddafi movement." Here was a scent of American weakness that doubtless did not escape the nostrils of Muammar el-Qaddafi, that hardened plotter of coups and bankroller of international terrorist groups: The State Department has never acknowledged, however, that—as I learned from CIA sources—the CIA actively took part in thwarting the operation, in cooperation with British and Italian intelligence, interfering on no less than three occasions to prevent the counterrevolutionary forces from ever reaching Libya.

The reasoning behind this sort of response by U.S. policy makers has been explained by Dr. Kissinger: "According to the dominant view, the real danger of radicalization resided in our *opposition* to Qaddafi [italics Kissinger's]. . . . [E]nergy supplies were in jeopardy *only if* we did something to antagonize the new Libyan revolutionary regime [italics added]."

Kissinger, though he perceived this analysis as quackery, did not choose to mount the effort necessary to overrule it. Qaddafi, more and more satisfied that he had nothing to fear from Washington, now moved to head-on confrontation with the oil companies.

Chapter Eight

The Dragon's Teeth

One man that has a mind and knows it, can always beat ten men who haven't and don't.

—George Bernard Shaw, 1929

United States energy policymakers are responsible for assuring the availability of huge quantities of energy at lowest cost.

—Professor Richard B. Mancke, 1974
Fletcher School of Law and Diplomacy,
Tufts University

On January 20, 1970, Muammar el-Qaddafi, now prime minister as well as RCC leader, convened Libya's twenty-one foreign oil producers for a meeting that was to inaugurate the junta's challenge to Oil and has since come to mark, better than any other formal occasion, the kickoff of the process that was to unravel the economy of the Western world.

By the time Qaddafi strode into the conference room at the Oil Ministry, his minister, Ezzedin Mabruk, had jerked the oilmen to attention with intimations of a new order: "What we wish to emphasize, with absolute clarity and frankness, is that the new revolutionary regime in Libya will not be content with the previous passive methods of solving problems."

Until Mabruk spoke, the oilmen present, representatives of six of the seven international majors—Exxon, Mobil, Socal, Gulf, Shell, and British Petroleum—and of the leading independents, had assumed the matter on the table to be the 10-cent-a-barrel transportation adjustment in prices that King Idris had reasonably, even belatedly, requested in his last month and that had remained the RCC's request during its almost five months of rule, an adjustment to compensate Libya for the lower cost of transporting its oil only 700 miles to Europe as against the 3,500-mile trip around the African coast that competing crudes from the Persian Gulf had to traverse since the closing of the Suez Canal in 1967. The companies recognized that unless the canal opened soon, they would have to do something along the lines Idris had proposed, but they intended for as long as possible to pursue the oilmen's tactic of simple foot dragging since each month's postponement saved 10 cents a barrel—on 90 million barrels.

Perhaps the term "foot dragging" is unjust. Negotiations "demanded" by an oil state were not akin to the bargaining that periodically takes place between management and labor or between a supplier and a producer when an existing contract expires. With Oil and its hosts, no expiration date was involved. Concession agreements typically ran for a period of between forty-five and seventy-five years; the productive life of an oil field can exceed forty years, and it frequently has taken many previous years to find it. What was involved were attempts by a host government to change in its favor the terms of agreements that lawfully had many years to run, and the oil companies felt justified in approaching this backsliding as gingerly and deliberately as possible.

The surprise from Mabruk—"impertinence" is perhaps the better word—was that the junta was no longer interested in the small change of transportation adjustments but was demanding an out-and-out increase in the price paid per barrel (which had remained unchanged since 1960, when it was *reduced*) and a change in the 50-50 profit sharing, the keystone of the equilibrium that had prevailed in the Eastern Hemisphere for twenty years. The precise details of Libya's demands would be made known to each individual company when its turn came, for the RCC intended to approach them each separately so as to chip away their traditional unity, but the general thrust was a demand for a dramatic revenue hike—about 40 percent—at a time when world oil prices were in the midst of a long decline the end of which could not be safely predicted.

If Mabruk's role was to reveal a new and shocking intent by the RCC, Qaddafi's was to demonstrate its inflexible resolve, even unto the use of force. His surprise appearance gave the oilmen their first close-up look at him, and it was to be the last. His message, conveyed as much by a stern, peremptory demeanor as by his words, was that this time things were to be different. This time the corporations were confronted not by soft princes wheedling for higher production quotas but by lean revolutionaries who threatened to stop all production. Qaddafi was willing to risk all, he said, and would confidently pit the Arab people's capacity to endure the loss of necessities against the oilmen's stomach for forgoing luxuries. There was a cryptic, scriptural roll to his lingo that was impressive and forbidding in a young man of the twentieth century and that delivered the punch without entangling him in specifics: "People who have lived for five thousand years without petroleum are able to live without it even for scores of years [longer] to reach their legitimate right."

It is interesting—and significant, for it was to be a consistent stratagem—that whereas Qaddafi came on like the heavy, as if to confirm the rumors of nationalization, expropriation, confiscation, and other such resorts to force that had been floating around Tripoli since the revolution, Oil Minister Mabruk adopted a stance of lawful negotiation, of arguing things out on their merits, as if the junta sought at once to frighten them just enough to put them in a mood for concession while reassuring them just enough to forestall their mobilization for battle.

Mabruk maintained that Libyan oil was worth more than other oil, far more than the oil companies were paying for it, for four reasons: (1) Lower shipping costs were involved; (2) the oil had a lower sulfur content which made it sought after, an argument that Libyan students in the United States had picked up from environmentalist agitations; (3) Libyan oil yielded, when refined, a particularly high proportion of high-value products (gasoline and heating oils); and (4) the oil companies had been underpaying the collusive, complacent Idris government for years and now must make amends for it. "The just demands we seek here," Mabruk perorated, "are not intended to bring about any basic changes in the existing structure of the world oil industry, not specifically in the price system. . . . What we ask of you gentlemen is to recognize the changed circumstances in our country and, accordingly, let your actions be guided by flexibility and reasonableness."

In the days that followed, the junta launched its divide-and-conquer strategy by culling two companies out of the pack for separate negotiations—Exxon and Occidental, the two largest single producers of Libyan oil. The terms being demanded now became precise—a rise of 44 cents per barrel (four and a half times the proposal of Idris), a hike in Libya's share of profits from 50 to 58 percent, and a written acknowledgment by the companies that they had, in effect, been shortchanging Libya all along and were thus to pay reparations.

To the oil companies, the RCC demands represented exactly and uncannily the threats which Mabruk had disclaimed—a reversal of world price trends, a torpedoing of the existing industry structure, and a vitiation of relationships between government and companies. Qaddafi's strutting notwithstanding, the oilmen were confident enough of their strength to dismiss the proposals and to decline serious negotiations on them.

With a twinge of suspicion it could be asked, were the oil companies really *against* a government-mandated (and therefore guilt-free) price increase? If so, why? Why not gladly accept it, explain it as the doings of government, and pass it on to the consumers while adding to it their own exaggerated markup, in the manner with which we have latterly become familiar?

The two fundamental reasons were: (1) The companies did not believe they *could* pass the added costs along to their consumers, and (2) they had developed over the generations an ingrained craving for stability as the industry desideratum, growing out of all too frequent clashes with the reality that sharp price increases were historically followed by sharp price decreases, disrupting operations all along the line both when they boomed and when they busted. Behind the first assumption was the educated fear that because of the world oil glut, which continued in 1970, and the possible imposition of price controls by irate consumer governments, a Libyan oil price rise would *not* be passed on; that would mean the companies themselves must absorb it and be left permanently with a smaller share of a shrinking profit. Behind the second was a history of withering precedents going back to the very origin of the petroleum business. Not only did the first successful driller of oil, "Colonel" Edwin Drake, die a pauper, but all the chaotic propensities of the industry were on display, as if microscoped, in its first two years. During the year after the discovery of oil, at Titusville, Pennsylvania, in 1859, the price reached $20 a barrel; within a year overproduction caused by that alluring price had created the first glut, which sank the price to 10 cents a barrel, ruining most of the people involved. The great companies were nowadays headed by executives who had an ingrown fear of too quick profits and who had been reared in the traditions of those who had transformed recurring panic into lasting order. "What the major corporations most wanted," wrote Professor Mira Wilkins, "was predictability, stability and security of supply."

The oil companies were therefore markedly cool to the sweeping changes being demanded by the RCC—to the extent they took them seriously at all. They were willing to consider a 5- or 6-cent hike to reconcile the differential in transportation costs, but even here they hesitated, regarding the Suez Canal closure as a political ploy that could be lifted at any time, leaving them holding the financial bag for a transportation differential that no longer existed. Beyond this they saw no common ground. In their eyes, a 44-cent-a-barrel hike would set off a clamor for the same increase across Africa and the Middle East and would upset the pricing system throughout the world. A 58 percent profit share for Libya would sound the death knell for the 50-50 principle that had for so long served as an ethically defendable line. To plead guilty to past rookery, and to compensate for it, would invite a plague of attempts to reopen settled arrangements. Above all, to make such unprecedented surrenders to so insignificant a regime, and to so ostentatiously hostile a one at that, would be to destroy the mystique of Oil's invincibility and would both embolden and shame all other concessionary states into insisting on changes no less costly and revolutionary than those won by the upstart Libyans. George Piercy, one of Exxon's top officers, later testified, with the oil executive's functional understatement, that at the time of the Libyan challenge it was clear to the oil companies that the RCC was asking "something that would establish a new mark and would have spillover possibilities into the Gulf."

Besides Oil's apprehension about the future impact of Qaddafi's proposals, all the present market conditions and projections, all the traditional factors in oil bargaining—current trends in prices and production costs, the state of supply and demand, the imbalance of power between the Oil Colossus and *any* host government—argued to Oil that the junta's demands ran perversely against the grain of economic reality.

At a time when Qaddafi was demanding an unprecedented 40 percent rise in the overall "take" he received, world crude prices were declining, had been declining for years, and, barring some miraculous intervention or grotesque blunder, would continue to decline for years to come (a trend the cost of which had for eleven years been borne by the companies, not the hosts, which had since 1960 been shielded by company adherence to a fixed posted price). Persian Gulf crudes—the world price standard—which in 1965 brought $1.60 a barrel could now be had for $1.20 to $1.30 a barrel. In fact, if one discounted for inflation, in early 1970, when Qaddafi's demands were being bruited over, the true price had dropped to 92 cents a barrel, in constant 1963 dollars. Between 1960 and 1970 the wholesale price of gasoline on the Rotterdam market fell from 7 cents a gallon to 5 cents. Moreover, long-term oil delivery contracts being negotiated in 1970 featured lower prices than short-term contracts, showing the market expectations for the 1970's.

Though it has been widely proclaimed that the oil price explosions of late 1970 and early 1971 were caused by a "surge in demand," an explanation that coincides with the official need to explain away those precursory catastrophes in the nonjudgmental terms of impersonal and inexorable forces, the fact that actually faced the oil companies was a *slackening* of demand. By mid-1970 it was clear that the year's growth in world demand over 1969 was below the average year-to-year increase of the 1960–70 period (and continuing this trend, the 1971 consumption growth in Western Europe and Japan, the relevant markets for North African and Middle Eastern oil, would fall to about *half* the decade average).* So the oil company negotiators were looking not only at declining prices but also at a slowdown in growth of demand and an alarming increase in unused production capacity—a trio of indicators portending lower prices for a long time to come.

Another sobering reality of the 1970 oil scene into which Premier Qaddafi was protruding was falling production costs, which should have militated toward falling prices. Between 1960 and 1970 the investment needed per unit of crude oil capacity had fallen by more than half. Efficiency innovations had reduced labor needs, for both constructing oil installations and operating them, even more drastically. "Supply has not only not tightened, it has been getting easier," wrote Professor Adelman in 1972.

So persuasively did these economic facts of life argue against attempts to bargain up the price of oil that the oil-producing states themselves had long acquiesced in both the fixed price and the right of the oil companies to fix it. Up until the appearance of Qaddafi no state in the entire region had sought to increase the price of oil,† nor had the

*The survival of this misconception is the more curious in that it was authoritatively exploded as early as 1972 by the renowned scholar of oil markets Professor M. A. Adelman of MIT, in an article written expressly for students and practitioners of foreign affairs (*Foreign Policy,* Winter 72–73): "The unanimous opinion issuing from companies and governments in the capitalist, Communist and Third Worlds is that the price reversal of 1970 and 1971 resulted from a surge in demand, or change from surplus to scarcity, from a buyers' to a sellers' market. The story has no resemblance to the facts. The 1970 increase in consumption over 1969 was somewhat below the 1960–1970 average in all areas. The increase in 1971 over 1970, in Western Europe and Japan, was about half the decade average. In the first quarter of 1972, Western European consumption was only 1.5 percent above the previous year. [The average 1960's increase was 11 percent.] By mid-1972 excess producing capacity, a rarity in world oil (i.e. outside North America) was almost universal and had led to drastic government action, especially in Venezuela and Iraq. The industry was 'suffering from having provided the facilities for an increase in trade which did not materialize.' A drastic unforeseen slowdown in growth and unused capacity would make prices fall, not rise, in any competitive market."

†The contemporary activities of Algeria are excepted here, first because Algeria's oil strivings related essentially to France and grew out of their colonial union and struggle and therefore fell

states ever taken action collectively through OPEC to do so. For example, in June 1970, five months after Qaddafi had made his oil demands, the shah of Iran, eager though he was at all times for higher revenues, publicly and specifically disclaimed any intention of seeking an increase in the price, for it would mean, even if granted, less Iranian oil sold and hence lower revenues. Thus the posted price paid to host governments by the oil companies had remained frozen for twenty years, excepting the *reductions* of 1959 and 1960.

So convinced, in fact, were the host governments of the unwisdom or futility of Qaddafi's price increase approach that in their ceaseless efforts to obtain more revenues from the oil companies they had employed every device *but* price increases, mainly the pressure for increased production quotas.

Oil-producing countries saw higher revenue as a function of higher production. Throughout the 1960's and into the new decade the consistent thrust of the host governments, amounting to a single-minded preoccupation, was their drive as individual states to increase their oil exports at one another's expense. They were indifferent to three connected truisms: (1) Overall oil production could rise only as much as world consumption; (2) beyond that level, increased production weakened prices; and (3) in this circumstance, the production rise of one OPEC country must cause cuts in the production schedule of another.

This universal agitation for higher production quotas was at once the bane of the companies' relations with their hosts (for the companies' inability to pump more oil than they could sell was seen by the hosts as a refusal to do so and was their main grievance against the companies) and a key to oil company supremacy over the hosts. While Qaddafi was demanding a price increase and issuing vague threats of a production cutoff, other host governments were clamoring for the privilege of filling any gap that might open up in the world oil pipeline—at existing prices. Iraq, for example, was demanding a long-term doubling of its permitted output. Saudi Arabia was pressuring Aramco for a doubling of its production in the coming five years. Iran was pressing for a 20 percent increase in 1970 and was to announce in 1971 that its "principal concern" in oil was for an ongoing 20 percent annual gain in production —i.e., a doubling in four years. The concessionary states were therefore competing against each other for a boon that only the international oil

---

outside the main currents of the world oil market, and secondly because after Qaddafi's coup Algeria acted in collaboration with Libya, and therefore, its oil stances will be treated here in that light rather than as a separate phenomenon.

companies could grant, a formula that, though it strained the oil companies' capacity for the diplomatic balancing of conflicting interests, left them in the catbird seat, able to play off one producing country against another. How this was viewed from the White House has been described by Henry Kissinger: "Richard Nixon came into office in a world economy that treated cheap oil as natural and excess production capacity as the main economic problem. If we worried at all about the political dimension, it was how to satisfy oil-producing countries who competed for the favor of access to our market and offered foreign policy benefits in exchange."

Qaddafi, with trumpets blaring, was thus launching his challenge in the face of inauspicious world market conditions and in the teeth of an international oil industry that was still acting vigorously and calculatingly to perpetuate its power.

In particular, the industry was moving steadily to reduce inordinate bulges of dependence on sources of supply that otherwise might become emboldened by their importance to rebel against Oil's control. Despite years of oversupply, weakening prices, and now tepid demand growth, Oil was searching and finding in West Africa, Indonesia, South America, Alaska, and the North Sea. The industry was in the midst of what *Business Week,* in October 1969, called "the wildest—and most widespread—oil rush in history . . . [notwithstanding] an oversupply of crude so massive that if not one additional barrel of oil were found the world could maintain its current consumption for more than 30 years [an estimate which the oil discoveries of the early 1970's would at least double].

As Oil had cut down the reliance of its prime import area, Western Europe,* on Middle Eastern oil—from 73 percent of all oil used in 1960 to only 45 percent in 1970—so it had also been moving to shrink the resultant "west of Suez" primacy of Libya by massive oil development of Nigeria.

Non-Communist Asia, a negligible producer in 1960, would in the early seventies be pumping more oil than prodigious Kuwait. Newly found Alaskan oil could soon provide the United States with four times its 1970 imports from the Middle East, if the U.S. government granted the pipeline permit.

*Remember that U.S. imports from the Middle East in the 1969–70 period were negligible.

In April 1970 years of costly failures in the North Sea were rewarded by the historic East Ekofisk find, the first of a series that by 1972 would confirm that the North Sea held larger proved reserves than the United States—at least 42 billion barrels—and would by the 1980's be producing 4 million barrels a day. This would mean oil and gas self-sufficiency for Great Britain, Holland, and Norway and a subsiding of that level of European imports which gave to Saudi Arabia, Iran, Libya, and the rest their hopes of power.

Moreover, many of the oil-consuming nations, as the 1970's began, were purposefully developing substitutes for Middle East/African oil, especially through the development of native oil and natural gas and by the acceleration of nuclear power. "Given time," wrote oil historian Peter Odell, "the expansion of these energy powers would have diminished the rates of increase in oil output in the major producing and exporting countries."

How weak a single country like Libya was vis-à-vis the Oil Colossus can be seen from the subordination to Big Oil of the incomparably stronger Persian Gulf *group* of oil states. While the above developments were steadily chipping away at the potential of these several states, if they could ever unite, to become indispensable and therefore capable of forming a monopoly, the situation in the Gulf itself—a fantastically mounting superfluity of proved reserves matched by an intensified jockeying by its various regimes for more revenues through more oil pumping—not only made any *individual* oil exporter, such as Libya, dispensable but also tended toward making the Persian Gulf its own neutralizer, inasmuch as its fundamental condition of vast oil surplus capacity held in several hands was the natural nemesis of monopoly. In a crunch, as long as any considerable portion of the Gulf adhered to the Western companies (and most Gulf states considered the companies, in Yamani's word, "indispensable"), the rest of the Gulf could be gotten along without.

Consider this cornucopia through the anguished eyes of oil executives who were trying to keep oil production from bursting upward to levels that, should chaos ever return, could plunge the price back down to 10 cents a barrel.

Saudi Arabia was in the midst of exploration finds by Aramco that within a couple of years would expand its proved reserves by 100 billion barrels, so that its *increased* reserves alone were equal to three times the *total* reserves of the world's leading producer, the United States, causing the Saudi sheikhs to press Aramco to raise production from 3.5

million barrels a day in 1970 to 20 million barrels a day by the end of the seventies, a level it could physically maintain for many decades.* Next door the Iranian Consortium was discovering eighteen major fields in the 1964–73 period, which it had to hide for the most part because those fields made achievable the shah's constant exhortation to hike Iranian production from the 3 million barrels a day of 1970 to 10 million. Nearby Iraq, according to a Senate report, had larger reserves than Iran, which had been only minimally developed because of Iraq's history of hostile behavior toward the companies.

To assure their ability to deal with any emergency, whether it arose from a misestimate of future demand in any given year or a political upheaval or from the need to discipline one or more recalcitrant concessionary states, the various oil companies maintained, beyond their stockpiled inventories in the consuming nations, spare capacity (i.e., extra, ready-to-pump wells operable on short notice) that in 1970 amounted to 5 million barrels a day. Almost 2 million of this spare was in the United States;† more than 2 million was in the Middle East. The shut-in reserve exceeded the output of Libya or any other concessionary state. This was for short-term emergencies; for longer-term problems the proved reserves of the Persian Gulf were so large that production there, now 50 percent of the world's export trade, according to the calculations of Dr. Richard Mancke, "could be easily doubled or even tripled within just a few years and with no appreciable rise in unit production costs."

The seeming immensity of the odds facing Qaddafi as he maneuvered to penetrate the defenses of Oil, the unbeatable hand which mounting world surpluses dealt the United States as orchestrator of the free world's response to any host's attempt to derail the plentiful flow of cheap oil, are illustrated by two happenings of that period.

At the time of the Qaddafi revolution there was pending before President Nixon an offer, indeed an importuning, from the shah of Iran to enter into a pact whereby the United States would purchase 1 million barrels a day of Iranian oil for ten years at the fixed price of $1 a barrel —about 40 percent of the price Qaddafi was contemplating. Two years after Qaddafi's great battle with the oil companies Saudi Arabia would approach the White House with an offer to meet all of America's

*As things turned out, of course, lack of demand and monopolistic restrictions kept Saudi production from ever rising above 11 million barrels a day. At mid-1982 the entire output of the thirteen OPEC nations was only 16.5 million barrels a day.

†Down by one-third from the late 1960's.

long-term supply needs at a stable price, in return for admitting the Saudis to a commercial partnership in the refining and marketing of its oil.

So it was that Sir David Barran, the chairman of Royal Dutch Shell, observed in 1969 that "the future as I see it is on the whole a reassuring one," and that George Piercy of Exxon saw the oil industry, as 1970 opened, as more than able to hold its own in negotiations with host governments, and that Professor Adelman saw the forces advocating higher oil prices as "slowly retreating" and doomed to continue retreating if the West's political tactics were equal to its oil strength, and that oil historian Joe Stork could declare (unhappily, for his sympathies lay with the radical host states), "Western control of Middle East oil in 1969–1970 seemed more secure than could have been predicted in June 1967."

Faced with Qaddafi's demands, the oil companies deployed in their traditional defensive formations. They rejected the RCC's proposal, either making no offer at all or, like Occidental, making a token response of a 1.2 percent a year increase for five years. To put the junta in mind of the slow strangulation of Iraq in the 1960's and perhaps of the quick and total ostracism of Iran in the 1950's, the oil companies slashed their development activities in Libya. The number of operating rigs fell from fifty-two to twenty-six in early 1970 and then to thirteen as the year wore acrimoniously on. Occidental cut back and then closed down altogether its programs for developing a natural gas industry, a favored goal of the Libyans. Most of all, the companies just ignored the RCC's ultimatums month after month while producing and selling record volumes of oil under the existing terms, as was their contractual right. Winter gave way to spring, and Qaddafi made no response.

In the fight which David had provoked with Goliath, however, David was not without some hurtful rocks for his sling, nor was Goliath without a gaping chink in his armor—of his own devising.

The oil industry in Libya deviated from its configuration elsewhere in that it did not consist of a few major companies in partnership but was rather an uneasy condominium of majors and medium-sized "independents." In that peculiarity lay a defensive weakness which Qaddafi was alert to and planning to exploit.

That there were independents sophisticated enough and well heeled enough to operate in the complex and costly international arena, despite the prescient efforts of the majors to lock them out, was due in part to a persistent antimonopolist undercurrent in American political thought which distrusted the collusive power of the powerful few and feared that one great consortium dominating a host country would be the inevitable target of fierce nationalist hatred, whereas many small and competing companies would present no such target, would pump more oil for the host at lower prices for the consumer, and would be easier for the host country to control for its own purposes. This philosophy scored a breakthrough in 1954 at the time the U.S. government rescued the Iranian oil industry after the Mossadegh crisis, when as a minor experiment it brought about the injection of several nonmajors into the new Iranian Consortium—Continental, Arco, Getty, Sohio, and Charter. Though their shares were tiny, they got their feet wet and their curious, adventurous noses inside the main tent of Oil.

That fifteen years later grown-up independents—Arco, Occidental, Continental, Marathon, Amerada Hess, Grace Petroleum, Bunker Hunt, Gelsenberg—held half of Libyan oil production, as opposed to under 10 percent elsewhere in the concessionary world, was due to the deliberate design of the Idris regime, which, as we have seen, wanted Libya developed by a lot of small companies hungry for oil instead of a few sated giants inclined to sit cautiously on new oil so as to protect the price of their old oil in other places.

Not only did Idris invite the independents in, but he assured they could compete successfully by letting them help write the oil tax laws. The result was that the tax laws abounded in loopholes and write-offs of a type the Libyan-centered independents could claim more aggressively than the world-oriented majors, whose Libyan operations must not disrupt an international, multilayered tax, price, and production structure. Thus in 1964 independents such as Continental, Marathon, and Amerada Hess were paying 30 cents a barrel in taxes to Libya while Exxon was paying 90 cents.

The Idris regime was on the same wavelength as the independents: Both were happy to sacrifice per barrel revenues and to rely on greater volume to give them a larger return. Paying lower taxes than the majors (the principal element in production costs), willing to take smaller profit margins in order to muscle in on the majors' markets, uncaring that their booming low-price production was threatening the price and production levels of oil in Saudi Arabia, Kuwait, or Iraq, where, unlike the majors, they had no holdings, the independents cut a swath into the hitherto exclusive preserves of their betters.

It was a small swath, to be sure—for instance, between 1962 and 1965 the majors' share of the European market fell by about 4 percent, and between early 1965 and late 1968 the price of Saudi and Iranian oil marketed in Japan fell by 10 to 12 percent—but large enough to cause resentment in those who had long disciplined themselves to achieve the goal of a stability based on self-sufficient strength and moderation in its use. Resentment flared into animosity when the majors counterattacked on the tax front and maneuvered the Idris regime into large tax hikes for the independents by welcoming a small one for themselves, thus narrowing the tax differential which had given the independents a competitive edge. Animosity deepened into feud when, after the government of Peru had expropriated Exxon facilities in that country and Exxon had begun a by-the-book boycott of Peru, the independent Occidental violated the code of Oil by offering to take over Exxon's seized installations and run them on Peru's terms.

With their institutional memory of 100 years, the majors looked upon the independents with contempt and alarm, contempt because they were rushing headlong into all the mistakes the majors' forebears had outgrown long ago, alarm because they were ignoring the iron rules dictating that mere corporations could keep the upper hand only by controlling the international market in a world in which sovereign states had a monopoly on local force.

Escalating production madly in Libya for today's profits, heedless of tomorrow's penalty of glutted markets, the independents aroused horror in the boardrooms of the majors. George Piercy of Exxon explained: ". . . if some capacity was brought on anywhere else in the world . . . it is like a balloon and if you bring it on in one place, you punch it in one place, something has to give somewhere else, because the fact that oil was brought on here or there does not in any way mean that there is more consumption. The consumption is fixed by what customers want to buy."

By the same token, lowering one's prices so as to invade the other fellow's market, or joining with a host government against another oil company so as to gobble up the leavings, had effects far beyond their local context. To the majors, cutting prices anywhere weakened prices everywhere. A breakdown in Oildom's control system anywhere, even Peru, threatened discipline everywhere.

Most grievous of all, in the majors' eyes, the Libyan independents had neglected the defenses that were vital to maintaining Oil's position versus host governments. An oil company operating in a foreign country—invariably pumping less oil than the government presses it to; paying lower prices than the government would like; awarding a smaller share of profits than the government, with the singular vanity of the nonproductive, thinks it deserves; conspicuously able to resist the government, which is sovereign over all else within its borders—always faced the day of eventual confrontation. After the company had lived up to its part of the original bargain by converting barren desert into wealth-giving oasis, by building up a mere gamble into a vast and profitable concern, with the oil found and charted, the installations and pipelines and harbors gleaming and efficient, the markets created and buying, the host government would be tempted to welsh on its end by seizing the going concern so as to take *all* the now-assured profits for itself or by rewriting the original covenant unilaterally so as to grab *almost* all.

When that day came, the company's only defense, since courts were ineffectual against governments and gunboat diplomacy was now out of vogue, was to have created in advance the circumstance that the host country needed the oil company more than the company needed the country. To achieve this was a complex work of many years, even decades. Against the day when government violated the covenant and the company responded by shutdown and boycott, the company must have oil reserves in other countries to draw upon, must maintain spare producing capacity in these countries that is immediately pumpable, must be part of an alliance of peers it can call on for oil and transport and market switches, must have developed the connections and the standing so that worldwide commerce will honor its boycott of the offending country, and fear *not* to do so, must have stashed away financial reserves that will enable it to weather the costs of protracted economic warfare, must have seen to it over the years that it rationed its revenue allotments to government X so that X had no real financial reserves. If it had developed all these nonviolent recourses against an expropriating junta—as each of the Seven Sisters had—the oil company, though it commanded not a single platoon, was more than a match for its sovereign antagonist.

But the independent parvenu in Libya, such as Occidental, which was each year boosting oil production by unheard-of leaps and shaving prices so as to be able to move it, had built up none of these defenses. Typically the independent had no source of replacement oil if the Libyan supply—upon which it depended to meet all its penalty-strewn contracts—was interrupted. Typically the independent, because it operated as a loner at the expense of the oil establishment, could not call on that establishment to respect its interests and to bail it out if its key source were cut off. Typically it was all offense and no defense, a "hot stock" operation overcommitted financially to new refineries which it must keep going at all costs and therefore unable to afford the expense of a long impasse. And by heedlessly boosting its production each year, oblivious to the excess revenues that were piling up in the Libyan treasury, the independent was giving to the potential adversary the means of economic warfare even as it stripped itself helpless. By its very mode of operation, then, it broke the cardinal rule of Oil defense, not just by making itself dependent on the host country but by making itself the total captive of that country.

The glass jaw of the aggressive Libyan independents had been revealed a few years before Qaddafi came to power. When the Idris regime ended the preferential tax breaks of Marathon, Continental, Occidental, and the others by hiking their per barrel assessments, they resisted payment on the grounds that the host had no right to change unilaterally the terms of a concession. Whatever its merits as theory, this was a defense that ought not to have been undertaken except by oilmen determined to prevail, for defeat or backdown would enable a dangerous precedent to get its foot in the door. In November 1965 the Idris government issued a royal decree threatening to halt the exports of companies that did not immediately comply with the tax hike. The ultimatum was fortified by legislation empowering the Council of Ministers to terminate the production and expropriate the assets of any delinquent company. Their bluff having been called, the independents found themselves in no position to raise, for they had no comparable source of low-cost oil and minimal financial staying power; by January 1966 all of them had bowed to the king's order.

The question that dogged Qaddafi was how to be forceful enough to wring epic concessions from the oil companies, dashing enough to claim revolutionary leadership of the Arab world, without crossing the invisible tripwire that would rouse Oil and its parent nations into the kind of mobilization which only two years before had routed the combined Arab states. Qaddafi's answer, audacious in its totality, can be seen upon dissection as a succession of testing probes, of retractable bluffs, of muted confrontations, of small penetrations of selected weak points followed by halts to assess reaction, resuming only when no counterattack surfaced, becoming truly bold only in the eighth month of the campaign after a dozen tests had shown that there *never would be* a counterattack.

Qaddafi's early days in power were filled with reassurances for Oil; his early months, a permissive carnival of laissez-faire. The U.S. Embassy and the companies were assured that the oil contracts would be honored and that Oil was exempt from the bans on alien proprietorship. Qaddafi did not even pursue the Idris-scheduled negotiations for the 10-cent price increase. From September until June he prudently reined in his passion for "conservation" and allowed oil production to soar to a new record each month until in May 1970 it reached 3.7 million barrels a day, placing Libya on a plateau where, among the Arabs, only Saudi Arabia resided. Though he had proclaimed to the oilmen on January 20, at the start of his oil offensive, that his people were prepared to give up oil revenues for aeons in order to obtain justice, he did not for a moment contemplate acting on that bombast. Before he risked the loss of a barrel, he would see to it that his first nine months of rule had gathered in two to three years' worth of average revenues by past standards, so that his great oil rebellion would be bankrolled in advance by the oil companies themselves.

The disciplined wraps which this often rash figure put on himself where Oil was concerned, the hesitance and mutability of his drive, were perceivable in things many and small. For months the normally propaganda-minded junta refrained from making public the scope of the demands it pressed so stridently behind closed doors, leaving open the path for a nonhumiliating retreat. For months the junta sat still for the refusal of the oil companies to entertain seriously *any* of its demands or even to suggest realistic settlement of the old 10-cent-a-barrel proposal of King Idris. Qaddafi responded only by moderating his demands and changing his negotiators.

Even the verbal bravado had telltale overtones of caution. When in March an RCC spokesman threatened "unilateral action" if the oil companies did not come to terms, the spokesman was only a lesser, disavowable figure—the director of technical affairs in the Oil Ministry—and in April the chief negotiator, Ezzedin Mabruk, stepped back, saying that such steps as nationalization were not even being considered. When, in response to the severe cutback in exploration activity by the companies after January 20, Qaddafi thundered, "Drill up or give up," a threat to revoke concessions, he was apparently surprised when three of the smaller operators in Libya—Gulf, Arco, and Grace

Petroleum—decided indeed to pull out rather than sink more money into a politically uncertain climate, for no leases were revoked, though the exploration slowdown deepened.

And when Occidental cut back on construction of a giant liquefied petroleum gas installation at Zuetina and then closed it down altogether, after having spent $60 million on it, there was no RCC response. The dichotomy between words and action caused it to be accepted in oil circles that the more bombastic pronouncements of the council were mere revolutionary fustian directed at the Libyan people and the Arab world, not at the oil companies. An especially delicious example was Qaddafi's climax to an attack on the oil companies for hiding Libya's oil potential: "There will be no more dry wells in revolutionary Libya!" One need not slight the potency of revolutionary élan to doubt its capacity to alter the industry norm that 9 out of 10 wells drilled will be dry holes, of which there were already 700,000 in the United States alone by 1970. So it was that at various stages Exxon's chairman, Monroe Briscoe, and its president, J. K. Jamieson, put out public statements dismissing as not-to-be-concerned-about the RCC rhetoric about "expropriation" that had reached the ears of Exxon stockholders.

In early June came a mild escalation. The RCC imposed higher port dues on export operations. This was a safe way for Qaddafi to show his impatience and resolve, for the device had been used by Iraq a few years before at the port of Basra with impunity. (Well, not quite impunity. The oil companies paid the increased fees all right, but they slashed exports from Basra in half, claiming the higher costs per barrel they caused made Basra oil uncompetitive.) In the same month the RCC blocked the commencement by Exxon of shipments of liquefied natural gas from the plant it had erected at Marsa al-Burayqah at a cost of $350 million, pending an agreement on prices; this was an annoyance, another step-up of psychological pressure, but hardly a cause of confrontation in a world where such impasses were commonplace. In early July Libya nationalized the distribution facilities for internal marketing of oil within Libya belonging to Exxon, Shell, and ENI, the Italian company. An escalation of pressure? An earnest of Qaddafi's resolve? Yes, but again, gestures that were eminently safe. Any form of "nationalization" was a charged act, but this form was the least objectionable that could be found; internal marketing within Libya was nickel-and-dime stuff and of its nature within the government's acknowledged province; besides, full compensation was offered.

When the RCC ordered Occidental to cut back its June oil production by 128,000 barrels a day, here at last was a potential casus belli, but it was lawfully and pacifically packaged. The ostensible ground for the cutback was "conservation." Occidental was said to be pumping too fast for the proper management of the reservoir. Since Occidental was detested by the majors as a pushy cutter of corners, it made an ingenious target, hardly a subject for a call to arms, unless there had been statesmen in the ranks of Oil. As for possible reaction by the United States, the ground that Qaddafi had invoked was the same as that by which the Texas Railroad Commission regularly cut back oil production in Texas. President Nixon himself was blocking *America's* most promising oil projects—in Alaska and offshore—on the ground of conservation. In comparison, Qaddafi's cutback embraced only 3 percent of Libyan output.

The rash of actions by which Muammar el-Qaddafi had begun to move out of his half-year sufferance of oil company stonewalling was, as Dr. R. C. Weisberg points out, "widely interpreted as an indication of the government's determination to take further steps if its demands for revenues were not met." But though these actions succeeded in creating an ambience of apprehension within Oil, especially in Washington, what ought to have stood out about them was their cautiousness, their attempt to evince determination on the cheap. Qaddafi had painstakingly searched out the least provocative, most plausibly justifiable acts he could take and still avoid giving the defeatist signal of doing nothing in the face of total rebuff. He was in the posture of one who, fearful of the enemy's overwhelming but unmobilized strength, minces forward as far as he can while prepared to retreat at the first sign of counterattack.

All the while Qaddafi had been working energetically to diminish, or at least to appear to diminish, the junta's dangerous isolation, which might give international Oil the idea that revolutionary Libya was ideally cast for the role of object lesson. Conscious that his Libya was unloved even by most of the Arab oil world, was far removed from the Soviet sphere, the proximity of which had fortified Iraq in its nose thumbing at the Western powers, and was geographically a piece of cake for boarding parties from the U.S. Sixth Fleet, Qaddafi sought out allies both for practical help in the oil struggle and as demonstrations that any manhandling of Libya would create a troublesome stink from one end of the Mediterranean to the other of the sort that the Anglo-

Americans, world-opinion-pecked as they were, should shrink from.

In January 1970, before his appearance before the Western oil executives in Tripoli, Qaddafi's oil attachés met in Baghdad with delegations from Iraq, Algeria, and Egypt (lately become a bona fide, if modest, oil-producing state) to form a radical Arab oil bloc. The four nations pledged themselves to share information and coordinate their marketing of oil. In late January, just before Qaddafi singled out Exxon and Occidental for the real start of bargaining, Libya and Algeria issued a joint declaration of unity regarding their efforts to raise oil prices. In May, before Qaddafi opened the brass-knuckles phase of his oil negotiations, the oil ministers of Libya, Algeria, and Iraq met and produced an accord for a unified strategy for waging their separate confrontations with the oil companies. The agreement called for: (1) ending the company tactic of "lengthy and fruitless negotiations" by setting a time limit; (2) raising oil prices *unilaterally* if oil company negotiations were not forthcoming; and (3) pledging financial aid to any of the three who were singled out for oil company cutoff of production and hence revenue. The first two principles (if they could be taken seriously—which had always been the Achilles' heel of Arab proclamations)—asserted the right and intent to force the oil companies to accept sweeping changes in their long-term concession agreement and, if the oil companies insisted on adhering to their lawful rights, to impose these changes by fiat. The third principle proposed a joint defense against the traditional oil company reprisal of selective ostracism.

Moreover, Qaddafi had sent an oil mission to Moscow and had received a Soviet delegation in Tripoli, there to disjoint the noses of Western oil companies by conferring on such matters as Soviet oil technicians for Libya and the future marketing of Libyan oil in Eastern Europe. In fact, the RCC had cast its welcoming net for every foreign firm or interest from all three "worlds" that was willing to talk about Libya's oil (oil that was already tied up under existing contracts to the Western oil companies). Not only Russians were getting the red-carpet treatment, but so were delegations from Japan, Germany, France, and Hungary, along with a flock of curious OPEC oil ministers.

Most of this was more show than consequence, but amid the hot air and the untried paper pacts stood a solid dividend of Qaddafi's campaign for outside collaboration: In mid-May, Syria, the radical Arab state without oil but with pipelines that carried lots of oil to the Mediterranean, had cut off the trans-Arabian pipeline (the Tapline) by refusing to allow a one-day repair job to be done on a rupture the Syrians alleged had been caused by a bulldozer "accident." The loss of transit revenue to Syria was reportedly being subsidized by Qaddafi. Half a million barrels a day were thereby kept from European markets, increasing the demand for Libyan oil.

Qaddafi was now poised on the brink of a truly revolutionary attempt to turn the tables on the oil companies. Before him, when host producing states had rebelled, it was not to get their per barrel tax rates raised but to get production levels raised, the only thinkable approach toward higher revenues, and when the oil companies felt constrained to nudge a host state back into line, they did so by cutting its production level. Qaddafi, who began his campaign with a demand for a sweeping increase in prices—a 40 percent increase in Libya's per barrel take—was now preparing to prosecute his case by *cutting* the amount of oil that the oil companies would be allowed to produce—a claim of imperviousness to Oil's own weapon and a turning of it against Oil itself.

Here, with his first cutback of Occidental, he stood at a perilous juncture. *Control of production level was a fundamental contractual right of the oil company, the heart of its concession pact, the key to its power.* Qaddafi, as he prepared to expand the Occidental cutback and to extend the device to the other companies, was challenging them, and the nations behind them, for basic control of oil in the western Mediterranean. Would they sit still? Or would they strike back by refusing to produce *any* oil unless they retained full control of production levels?

Qaddafi had tried to confuse the issue by invoking conservation and to lessen the likelihood of reprisal by picking a target that could not long endure the shutting down of its pumps because the oil was fully committed to refiners, because the target had no backup source elsewhere, and because Occidental was probably too unpopular with the majors for them to initiate an industry-wide shutdown in its behalf.

That Qaddafi proved right encouraged him to inch forward another step, a further cutback order against Occidental, this time a truly crippling blow of almost 200,000 barrels a day, reducing Occidental's output from the 800,000 level of mid-May to 500,000 in mid-June. Occidental had to inform its customers in Europe that it could not fulfill its contracts and must begin rationing. Having drawn blood, Qaddafi took another step, on June 15, by ordering a production cutback of 120,000 barrels a day by the Amoseas Consortium (Texaco and Socal), a small cut compared to Occidental's, but a cut imposed on two of the Seven Sisters.

Besides the failure of the oil community and the diplomats to rise to *any* response to Qaddafi's first cutback orders, there had been other exhibitions of noncombativeness in Washington which had allayed his fears and fortified his hunch that he was now rolling along a track without obstructions.

In the first place, during the six months of slow escalation of oil hostilities since January there had been no sign of Nixon administration intervention to unite the many American oil companies into a common front, not even the official relaxation of the antitrust laws that would have to precede the oil companies' forming a united front by themselves.

The signs were, in fact, in the opposite direction: that Washington didn't want its oil companies making waves in Libya or anywhere else.

In the previous year Colonel Qaddafi's counterpart in Algeria, Colonel Houari Boumedienne, had nationalized several foreign oil companies including the American firm Phillips. These were modest enterprises by Big Oil's standards, and Algeria was outside the mainstream of American-British oil currents, but even so, this was the forbidden act. U.S. diplomatic pressure on Algeria was to be expected, and a commensurate interference in Algeria's international commerce. But though other countries protested, there had been no U.S. remonstrance worthy of the name; on the contrary, Washington had pursued a closer relationship with the Boumedienne regime. The Nixon administration thus became the first to preside over a no-reprisal-against-oil-nationalization policy, and this was not lost on Qaddafi.

The most visible explanation was the American effort to obtain large-scale imports of Algerian liquefied natural gas, apparently so important to gas-short America that Washington did not want to rock the boat. Well, *Libya* had lots of natural gas to export, too, and the United States also needed lots of Libyan low-sulfur oil if its power plants were to abide by the clean air bill Nixon was about to sign into law.

June had seen the final stage of Anglo-American military evacuation from Libya. In July, perhaps in the fullness of its exuberance at being relieved of the restraining shadow of Western troops, the RCC expropriated the property of all Jews and Italians living in Libya, and Qaddafi advanced to the decisive stage of his confrontation with the oil companies. Oasis, comprised of Marathon, Continental, Amerada Hess, and Shell, largest of the consortium producers at 1 million barrels a day, was cut back by 125,000; Mobil, along with its junior partner, Gelsenberg of West Germany, was cut by 55,000; and forlorn Occidental, by another 60,000, bringing its total production loss to 360,000 barrels a day, almost half the production on which it relied to sustain its precarious world empire—one of the great gambles of oil history. By the third week of August, as he looked toward the as yet unconstrained Exxon, the lion of the jungle, Qaddafi had slashed overall oil output by 800,000 barrels a day. With the companion cutoff of the pipeline by Syria counted in, 1.3 million barrels had been subtracted from the daily stream flowing to Europe.

Given the oversupply of oil in the world, and the precautionary entrenchments of the majors against the unforeseeable by virtue of which stockpiles and excess producing capacity were maintained in costly readiness, and the nimble mobility the majors had habitually shown in moving spare oil to where it was needed, a shortfall of 1.3 million barrels a day, even though fabricated with little warning, should have been a routinely manageable affair. Oil had but to step up the pumping in Nigeria and the Persian Gulf, reroute portions of its tanker fleets, draw down somewhat its European stockpiles and, presto, the loss of Qaddafi's oil is repaired with scarcely a ripple in the stream! And with Qaddafi the only substantial loser.

But this time the practiced maneuvers of Oil were being thwarted by a remarkable combination of coincidental events, some of them brilliantly exploited by Qaddafi, others interventions of that gratuitous luck which so often in great affairs decides victory and defeat: Nigeria was caught up in a civil war which paralyzed its oil production until just after the moment of need; an unusually cold winter in Europe had drawn down its stockpiles of oil, while an otherwise above-normal surge in European demand for oil, in perverse contrast with the behavior of the world at large and with its own graphs the following year, was straining those reserves further; the sudden and untimely appearance of a new major importer of low-sulfur Libyan oil, the United States, was diverting Libyan oil from its usual destinations and thus puffing up its importance; and the actions of Qaddafi and his allies had overloaded the circuitry of tanker deployment.

On the whole, the charter tanker business has proved marvelously adaptable to the rise and fall of demand and is acknowledged as being as close to a paragon of pure competition as can be found among larger enterprises. But there were limits even to *its* adaptability. In the late 1960's two to three years were required between the order for a new tanker and its launching, a bit too long to keep up with an extraordinary convergence of contradictory events: The shutdown of the Suez Canal since mid-1967 had created a need for long-haul tankers for the 3,500-

mile trip from the Persian Gulf around Africa to Europe, but it was a chancy need, for the canal might reopen at any time; the Persian Gulf's falling share of European oil consumption caused by the Libyan export surge had heightened the uncertainty; and now the cutback of Libyan and Syrian pipeline oil was causing a fourfold increase in the demand for long-haul tanker tonnage. Per barrel rates for a single tanker voyage from the Persian Gulf to Rotterdam rose from $1.10 in late May to $3 in September, an increase four times larger than the one Qaddafi was demanding for his oil. In time, Oil's transportation system would respond so well that by early 1972 that rate had fallen to 72 cents, but during the critical months of mid-1970 Qaddafi enjoyed a charmed period in which, though his oil itself could be easily replaced, the transportation to haul the replacement oil was temporarily short, and its lack was creating the exact effect Qaddafi sought: an oil shortage in the Mediterranean.

It was a Last Hurrah for Gamal Abdel Nasser. The protracted closure of the Suez Canal was what now necessitated the long-haul voyages that were the base cause of the tanker shortage. The shutdown of the Tapline in Syria by Nasser's old confederates would have been unimportant without the Suez closure; in conjunction with it, however, it was weighty. The production cutbacks by Nasser's disciple Qaddafi had an immensely magnified impact because of the blockages of replacement oil by Egypt and Syria. The airtight character of the cutback was guaranteed by the coordination with Libya of Algeria, whose revolution owed much to Nasser, and of Iraq, long an off horse in such ventures but now pulling together under the leadership of another Nasser protégé, Saddam Hussein, the strongman of Iraq, whom years before as a hunted man Nasser had brought to Cairo, protected, and molded in his own image. The Nasserite school tie had proved more serpentine and resilient than suspected.

"The oil companies will negotiate only if *they* are losing money," said Omar Muntassar of the Libyan Oil Ministry. This was the guiding insight of Qaddafi's RCC, and it was now able and willing to lose $1 million a day itself to test its theory. Would the oil companies with their fat treasuries so long accreting endure far smaller losses in the present to preserve the future? Would they come to one another's aid, as had the Nasserites?

The test case, singled out for that role months before by Qaddafi, was

Occidental. With its oil supplies cut almost in half, the company faced not only the mass flight of its customers and the costly shutdown of its still-mortgaged network of refineries but the contractual obligation to make good financially on what it failed to deliver.

So far as leadership was concerned, Occidental was a one-man show. The man was seventy-one-year-old Armand Hammer, who had begun his adult life as a medical doctor and had forsaken medicine for the life of an international entrepreneur.

From long experience in dealing with leftist dictators that began in the early 1920's, when he cut his first deal with Lenin, Hammer well understood the vulnerability of his company and had early recognized the danger posed by Qaddafi. When the junta first made its demands on Occidental and Exxon, Hammer's negotiator, George Williamson, went to Exxon's Libyan chief, Hugh Wynne, to enlist Exxon's participation in a joint defense. Exxon refused, contending that unless Washington came forward with a dispensation from the antitrust laws, it was prohibited from collaborating with Occidental. Hammer, feeling that this was as much pretext as scrupulousness, bided his time as Qaddafi's successive cutbacks eroded his position; then, on July 10, Hammer called personally on Exxon's chief of chiefs, J. K. Jamieson, at Exxon's international headquarters in New York City.

For Hammer to do this required a great deal of brass and self-abasement, tactical capacities men of affairs seldom lack. For years Hammer had been bargaining his way into Exxon markets with cut-rate deals that Exxon contended damaged not just it but the entire industry. For years Hammer had been maintaining a political lobbying campaign for the U.S. government to give him the import quotas now going to Exxon and others so he could build a 300,000-barrel-a-day refinery at Machiasport, Maine, from which he would sell his Libyan oil at 10 percent under the going price—a plan which to his orthodox rivals contained the worst aspects of demagoguery and disloyalty. And just a year ago he had volunteered to play the scab by taking over the running of Exxon installations shut down in a dispute with the Peruvian government, an act seen by Exxon as a perfidious blow against all oil companies' capacity to defend themselves from unlawful seizures by governments.

If Hammer was unchastened by these offenses as he arrived at their target's headquarters seeking a bailout, he must at least have been somewhat awed by the evidences of scope and grandeur on display. According to Anthony Sampson:

> Beside a bubbling fountain and pool on Sixth Avenue, the fluted stone ribs soar up sheer for fifty-three stories, and inside the entrance hall is hung with moons and stars. On the twenty-fourth floor is the mechanical brain of the company, where the movements of its vast cargoes are recorded. A row of television screens are linked with two giant computers and with other terminals in Houston, London and Tokyo. . . . They record the movement of five hundred Exxon ships from one hundred and fifteen loading ports to two hundred and seventy destinations, carrying one hundred and sixty different kinds of Exxon oil between sixty-five countries. . . . From the peace of the twenty-fourth floor, it seems like playing God—a perfectly rational and omniscient God, surveying the world as a single market.

Hammer's destination was the fifty-first floor, from which Exxon's directors ruled their empire. There he entered a two-story lobby, where a high balcony looked down on priceless tapestries, and moved along wide corridors, past palms and Middle Eastern art and Peruvian artifacts, through a padded silence broken only by the faint buzz of air conditioning. Jamieson, tall, his face characteristically expressionless, a Canadian by birth, a Texan by molding, coolly received this distrusted embodiment of the industry's opposite pole in an office that was more like a drawing room.

Dr. Hammer got right to the point. He could not hold out much longer with almost half his oil cut off. If he caved in to Qaddafi's demands, which he was struggling to avoid, all the oil companies would be damaged. To stay in the fight, he needed oil from Exxon to replace the cut-off oil, and he needed it not at Exxon's normal charge to outsiders but at a price just above cost. Otherwise (under the cut-rate contracts he had signed in order to invade Exxon's markets) he would be reselling Exxon's oil at a loss.

Gratifying though it would be to spurn the cheeky importunings of this mischief-maker Hammer and turn him away empty-handed to face his just desserts, advantageous though it would be for the majors if the improvident independents that had been "made" by the Libyan peculiarities should now be broken by them, Exxon's responsibility for oil leadership called on it to provide Hammer with the wherewithal he needed to continue resisting Qaddafi, the greater threat. For if Hammer, isolated and despairing, surrendered to the junta instead of being drawn into alignment with the majors, not only would Oil's invincibility be called into question around the globe, but a quarter of Libya's oil production would be in Qaddafi's hands—half when the other independents tumbled—giving him the revenue base he needed to wage protracted war on the majors and, if it came to that, making it legally and practically difficult for Oil's tactic of freeze-out to succeed.

But Jamieson could not rise above the low visibility of narrow considerations; he sent Hammer away without a decision, and two weeks later Hammer was informed that Exxon had rejected his appeal: It was willing to sell oil to Occidental, but only at the going price to buyers outside the Exxon family—a condition the overextended Hammer regarded as ruinous. During the last part of July and throughout August Armand Hammer wrestled with the decision of whether to capitulate or continue resisting, while Qaddafi tightened the screws by a third cutback and threats of outright nationalization. If nothing else, Occidental's prolonged foundering gave the parties responsible for oil security both the spectacle of a clear and present danger and the luxury of plenty of time to do something about it.

In Washington it would seem that alarm bells must have been ringing all over the White House, the State Department, and the Pentagon. By the late summer of 1970 the hopeful assumptions about the Qaddafi revolution that had dominated White House councils in the fall of 1969 had all been exploded. Qaddafi not only had expelled U.S. and British forces from Libya but had purchased Soviet tanks, engaged East German security forces, and was trying to get hold of a nuclear bomb from Communist China. U.S. oil policy and the oil responsibilities proclaimed and shouldered for a generation were under direct challenge. The instrument long relied on to carry out that policy and meet those responsibilities—the American-dominated international oil industry—was in Libya shown to be wobbling and divided against itself.

Qaddafi had by now shown that he was up to far more than commercial haggling over oil prices. He was striking at all four of the control devices by which the oil majors up to now had maintained Western oil security: (1) control of production levels, which was the power to prevent a world shortage; (2) control of oil transportation, the power to prevent a local shortage; (3) control of international marketing, the power to prevent expropriated oil from being sold and therefore to render expropriation profitless; and (4) the control of prices, which, combined with control over production levels, constituted the power to limit host state oil revenues to levels that prevented treasury surpluses, so that the host states were always as anxious to sell their oil as the West was to buy it.*

Qaddafi was in the midst of dismantling this control system. He had already seized the contractual right of the companies to set production levels and was using it to create a shortage; he had collaborated with Syria (and built on Nasser's Suez closure) to block the means of oil transit, canceling Oil's capacity quickly to move its excess stores of oil to where the artificial shortage was; he was pursuing negotiations with the Communist bloc that would give him markets for oil if he decided to cross the personal Rubicon of expropriation; and if he achieved his price demand, which would then be demanded by the other oil states, the extraordinary jump in government takes would give the hosts rapidly rising cash surpluses—a strike fund, as it were—that would one day enable them to get along without new revenues for longer than the industrial states could get along without new oil.

Moreover, Qaddafi had from the start linked his oil offensive to an avowed effort to undermine the foreign policy and security arrangements of the United States, thus giving it dimensions beyond the scope of commercial bargaining and placing it in the province of U.S. policy makers.

Qaddafi was using the oil issue, the Multinationals Subcommittee of the Senate Foreign Relations Committee later reported, "expressly to punish the United States for its support of Israel." To him, as he had told a great public rally in April, the oil companies were but the allies of "world Zionism" and of the "forces of local reaction" in the Arab world. Therefore, the whole Libyan people must "mobilize for the coming fight with the oil companies." He sought to demonstrate to the rapt Arab world that his revolutionary aggressiveness could achieve more in one year than the conservative oil state monarchies had achieved in decades. And as the RCC's strangulation of Armand Hammer's Occidental approached its climax, U.S. intelligence had detected a sharp increase in the traffic in coded messages between the Kremlin and Qaddafi. Their content could not be decoded, but the size of the traffic itself, at a time U.S. oil equilibrium was threatened with destabilization by Qaddafi, would seem to have been enough at least to perk up

*"The payments to exporters per unit of production were low enough to maintain their interest in increased output for increased revenue. . . . Most of them absorbed the revenues as quickly as they received the money . . . the low government takes were in fact determined by the oil industry which maintained firm control of the market."—Uzi Arad, Hudson Institute.

the interest of the White House and State Department in what was going on in Tripoli.

Why, then, had Washington not intervened to shore up its disarrayed forces? Why had the oil companies not trooped to Washington to entreat that intervention? The first of these questions points to one of the fateful enigmas of our time. The second is easily answered; shortly the oil companies *would* troop to Washington; they had not done so thus far (August 1970) because of the consistently unnerving signals coming from the State Department.

The signals were coming principally from James Akins, the State Department expert on international oil questions, soon to move over to the White House as its resident oil counselor. With the backing of his superiors and the assistance of his State Department colleagues, who apparently were not in the least humbled by the outcome of their advice to Kissinger nine months before, Akins had for months been taking the initiative to press upon the many oil companies in Libya the judgment that the RCC's 44-cent demand was "reasonable" and "fair" and that the companies' resistance to it was unwarranted and insensitive and would cause the Libyans to feel "they have been cheated," guaranteeing "a breakdown in relations with the companies and all sorts of subsequent problems." Beyond urging this view on all the American oil companies in Libya, the Akins group had attempted to line up the British government behind it in talks with the British Embassy in Washington.

The oil companies, accustomed to thinking of the U.S. government as its partner and last-resort backup, were bewildered by Washington's precipitate espousal of their adversary's position.

In supporting Qaddafi's demand for a price that was roughly one-third higher than the going spot price and one-fourth higher than the posted price for Saudi light—discrepancies so great, against a ten-year background of declining real prices and unchanging posted prices, as to revolutionize oil pricing—the State Department lost its credibility with the oil negotiators (and with many an oil scholar). First, Oil's private estimate of the Libyan value differential was 6 cents, not 44; even the public claim of the Libyan government, a month before the Qaddafi coup, was only 10 cents. Secondly, the oilmen well understood that rapid changes in tanker costs, refining techniques, and market preferences were eroding the Libyan advantage even as the negotiators sat, making dubious a long-term price hike for the Libyans of any substantial size. (Over the coming decade Libya would periodically either suffer staggering losses in exports for trying to maintain its price above the Saudis or have to cut its price *below* the Saudi standard [where it is at this writing] in order to move its oil as the technology of Oil made Libyan oil less and less preferable.) Moreover, the actual production cost of Libyan oil—including every factor from exploration to return on investment—was 7.5 cents a barrel. With Libya's take, pre-Qaddafi, already at 90 cents a barrel—twelve times the cost of production—how airy seemed Washington's preachment that equity demanded a price eighteen times the cost. Oil would love to pocket its share of that kind of profit, of course, were it not for the feared consequences, but fairness was an argument better never raised, lest it be examined too closely.

To Oil, the most confidence-shattering aspect of all this was Washington's apparent ignorance of the worldwide consequences of a 44-cent price increase for Libya. All the other host governments were bound to demand the same or lose their place to radical revolutions inflamed by the example of the easy victory Washington now wanted to give Qaddafi. And once they all had pulled even with Libya's price, Qaddafi was bound to demand yet another increase to restore the Libyan differential. The State Department policy was thus a sure formula for the continuous destabilization of world oil prices. So the oil companies kept away from Washington and drifted, rudderless, toward the rapids.

Though it was late, time had not yet run out. Had the Nixon administration, during those weeks when Armand Hammer was scrounging about for an alternative to surrender, intervened in the manner that had been routine for its predecessors, all was still rescuable, in the opinion of numerous oil scholars. Professor Adelman:

> When the first Libyan cutbacks were decreed in May, 1970, the United States could have easily convened the oil companies to work out an insurance scheme whereby any single company forced to shut down would have crude oil supplied by the others at tax-plus-cost from another source. . . . Had that been done, all companies might have been shut down and the Libyan government would have lost all production income. It would have been helpful but not necessary to freeze its deposits abroad. The OPEC nations were unprepared for conflict. Their unity would have been severely tested and probably destroyed. The revenue losses of

> Libya would have been gains to all other producing nations, and all would have realized the danger of trying to pressure the consuming countries.

It is amusing, in the genre of gallows humor, that after a dozen years of Qaddafi's entrenching himself in power with the arms and alliances purchased by oil wealth, Henry Kissinger, private citizen, took to lecturing the Carter and Reagan administrations, via television discussion shows, on the ease and necessity of dispatching Qaddafi. If the United States can't "take care of" a weak little country like Libya with only 2 million people, Kissinger would muse pregnantly, then how can our foreign policy have credibility anywhere in the world? But at the time when Qaddafi's dictatorship was in its swaddling clothes and Dr. Kissinger was in power and extreme provocation had been given, his administration decided to do nothing, not even to work out an insurance policy.

With no rescue posse forming anywhere, Dr. Hammer, who maintained enough aplomb to negotiate by day in Tripoli and fly home to Paris each night in his company jet for dinner and respite, by mid-August found the field of maneuver narrowed down to one bleak chance.

If he did not submit to Qaddafi and get his oil pumping again, Occidental was finished. If he did submit, and got his oil pumping, but nobody would buy it because, at Qaddafi's new tax costs, it must be priced higher than anyone else's oil, Occidental would still be finished. Only if he submitted, got his oil moving, and other companies followed his example and had to raise their prices, *too,* could Occidental survive. Qaddafi could thus expect total cooperation during the rest of his struggle from any company that surrendered, for that company would then have a vested interest in the success of his oil offensive.

On September 2 the isolated Hammer bit the one bullet of promise and signed Qaddafi's terms, in return getting the dictator to drop his hocus-pocus about conservation and permit Occidental to raise its production to 700,000 barrels a day. Occidental would pay 40 cents a barrel more—30 cents now and 2 cents more a year for five years. On top of this, its tax rate would be raised from 50 to 58 percent, retroactive compensation for the years of shortchanging Libya under Idris that the RCC was alleging. Occidental had to make a signed admission of this offense, with which Qaddafi was to make great propaganda sport.

The agreement was wrapped up in time for Qaddafi to announce it during the celebration attending the first anniversary week of his takeover. And he had much to crow about. According to the Multinationals Subcommittee, the Occidental agreement "represented the most radical increase in revenues that any oil producing country had won since the 1950 Aramco tax agreements."

The total capitulation of Occidental announced on September 4 stirred the polyglot oil companies in Libya to attempt a belated rallying of their forces. On September 7 John J. McCloy, spokesman to government for all the American Sisters, led a delegation of oil chiefs to Washington, where they conferred at length with Secretary of State William Rogers, Undersecretary of State U. Alexis Johnson, and the department's director of oil and fuels, James Akins. No decisions were reached, as this was intended as a briefing session, but the grave potential of the Libyan situation had been laid before the highest members of the State Department.

On September 10 representatives of the companies—majors and independents—met in New York to try to work out a common front of defense and a formula by which the "crude-long" majors would bail out the "crude-short" independents if they kept up their resistance. The meeting failed. The most conspicuous reason for that failure was the posture of the U.S. government. It was not just that Washington failed to come forward in its traditional role of knocker together of heads into a united defense; it was the actively demoralizing consideration that Washington had been all along the principal advocate of surrender. Its consistent advice to the oil companies had been to do what Occidental had now done. In the absence of crisis leadership from the government, or even moral support, the oil company representatives in New York could not rise above the deficiencies in the makeup and outlook of the modern corporate executive plunged into an essentially political confrontation, inadequacies that have been pinpointed by the political philosopher Irving Kristol: "Businessmen understand competition but not confrontation and conflict and in any case tend to be risk-aversive." And, more specifically, by Anthony Sampson: "The directors themselves [of Exxon] are something of an anticlimax. They are clearly not diplomats, or strategists, or statesmen; they are chemical engineers from Texas, preoccupied with what they call the Exxon incentive."

Engineers, of course, was what they were supposed to be. It was Washington that was supposed to furnish the strategists and statesmen.

Dr. Hammer played upon this default by the government and indulged in the sardonic pleasures of tit for tat with Exxon when he refused to provide the New York conference with the details of his settlement, contending that to do so might run afoul of the U.S. antitrust policies. Looming larger in this refusal, no doubt, was Occidental's reluctance to offend its new master, Qaddafi, by giving cooperation to those still holding out against him. This could have been a providential lesson to the Nixon people had it been heeded: If the U.S. government left, undefended and adrift, the oil companies which had for two generations been its buffer against the oil states and the acknowledged instrument of America's oil foreign policy, they would, being by nature organizations of profit and survival, go over to the other side.

With "Oxy" in the bag, Qaddafi turned to a quick rolling up of the other independents, the vulnerabilities of which were similar, giving them two weeks to "take it or leave it." And he fired his first big guns across the bows of the two great flagships of the majors, Exxon and Shell. On September 5 Exxon was required to cut back its production by 110,000 barrels a day; immediately thereafter an ultimatum was issued to Shell, as a partner in the Oasis consortium, to come to terms on the Occidental model or shut off its pumps. That Qaddafi now took on the most powerful of the oil dreadnoughts was an illustration of his boldness; that he had waited so long to do so, until oil had shown its disarray and Washington its lack of support, showed the wariness he still retained of the power of Oil.

On September 21 Marathon, Continental, and Amerada Hess, the independents in the Oasis consortium—the largest producer in Libya with a precutback output of 1 million barrels a day—bowed to Qaddafi's ultimatum and settled on terms almost identical with Occidental's. (Oasis would have a lower tax rate than Occidental—54 percent instead of 58 percent—because in the past it was deemed to have underpaid Libya less than Occidental.) Should total oil war now break out over his next move, Qaddafi could count on production from Occidental and Oasis of almost 2 million barrels a day at the enhanced take, more than enough to sustain him for an indefinite period.

Shell did not succumb, taking the position that to permit this precedent would be to endanger the entire worldwide structure of Oil by inviting repetitions of the same tactics and demands everywhere. On September 22 Qaddafi made two brass-knuckle responses. First, he ordered a complete cessation of Shell's 150,000-barrel-a-day production. Secondly, he presented all six of the international majors in Libya and the five independents still resisting with commands either to accept his terms by September 27 or to halt all production. There was no longer the disguise of "conservation" to mark the rupture of contractual rights, nor could there be since now that the money was right, Qaddafi had restored the production levels of Occidental and the Oasis independents to seven-eighths of what they had been before the RCC had begun to campaign against the rape of Allah's gift. There was just the assertion of naked force against undefended international law.

In the five days remaining to them before the expiration of Qaddafi's ultimatum, some of the majors agreed on a plan of resistance and turned to their governments for moral and diplomatic support.

A conference was scheduled in Washington for September 25, 1970, between all the involved oil companies and Undersecretary of State Johnson, who had now had more than two weeks to digest the warnings of the September 7 meeting, which in the meantime had been borne out by the fall of the Oasis independents and the Qaddafi ultimatum to the majors. The stated purpose of the conference was to discuss the implications of a possible massive shutdown of oil production in Libya. On their way to Washington the chairmen of the British-based Sisters—Sir David Barran of Royal Dutch Shell and Sir Eric Drake of British Petroleum—broached to their government their plan of resistance, in the person of the British foreign secretary, Sir Alec Douglas-Home, who was in New York attending a session of the United Nations.

The leaders of Shell, unlike the run of oil chiefs, had long been noted for their world view, for Shell's origins lay in trading and importing oil from around the globe; for decades it had been a company axiom that nothing could happen anywhere in the world without its affecting Shell. Sir David Barran was an exemplar of that tradition, and he dominated the presentation to Foreign Secretary Douglas-Home:

If the companies gave in to Qaddafi, the defeat would be repeated around the world. Where the damage would stop escalating, no one could know. The oil companies must therefore "defy the Libyans, even at the risk of losing their valuable concessions." To cave in before such tactics as the RCC's would "undermine the intricate nexus of relationships between producing government, oil company and consumer." Therefore, if Libya carried out its stated intentions to stop production altogether and then should seize the installations and try to operate them and market the oil, the companies, as in the past, should block that "hot oil" from commercial channels.

Barran and Drake estimated that with a shutoff of oil from Libya, the combined oil companies could nonetheless supply Europe with 85 to 90 percent of its normal consumption for at least six months. The shortfall was well within the range in which rationing could limit economic dislocation to a minimum. Libya would probably desist in a few months; if not, by six months' time the oil companies would have come up with the combination of alternate sources and shipping necessary to restore European supply to its normal rate.

The oil companies were asking very little of the British and American parent governments. No naval aid for blockading, as in Iran in 1951. No imposition of a new monarch or restoration of a fled one, as in Iran in 1944 and 1953. No troop landings, as at Suez and in Jordan and Lebanon in the latter half of the 1950's. No mobilization of shipping and resources, as was required for overcoming the cutbacks and transit blockages of 1956–57 and 1967. No cutoffs in government aid, as in Venezuela in 1948 and Peru in 1964. All the British and American governments would be asked to do was to put behind their oil "instruments" the weight of their influence; to help persuade the European governments that, as during previous contrived shortages, wisdom lay on the side of putting up with temporary inconvenience in order to prevent permanent financial extortion and insecurity of supply; and perhaps to enforce those commercial sanctions routine to such disputes.

The British foreign secretary was sympathetic and said he would sound out his European counterparts. Barran and Drake flew on to Washington.

Undersecretary Johnson received the oilmen, accompanied by James Akins. No transcript of the colloquy is available, but its parry and thrust can be pieced together from the testimony, depositions, and interviews subsequently given by the participants.

During the course of a sometimes contentious conference the position given earlier to Foreign Secretary Douglas-Home by Barran and Drake was presented, but in the teeth of an across-the-board resistance by Akins which at times approached a filibuster. Near the outset Akins took the floor and delivered himself of an hourlong lecture, remembered by some as a "harangue," mainly about the plight of the Palestinian guerrillas currently being run out of Jordan by King Hussein after having been run out of Palestine by the Israelis, the gist of which was that the oil companies should direct their attention to a just settlement of the Palestinian problem, which would in turn lead to a resolution of their own problem.

This was to the oilmen a curveball. It was a keystone of U.S. diplomacy to resist all attempts by the Arabs to use their "oil weapon" to suborn American foreign policy. The West's friends among the oil states were identifiable by their stated determination to keep oil policy divorced from politics—particularly the politics of restoring the Palestinians. Ironically, the oilmen had long felt disadvantaged by the difficulties which U.S. partiality for Israel over the Arabs had caused them with the oil states. Did the State Department really want suddenly to link oil peace to a pro-Arab foreign policy—one of Qaddafi's favorite threats—and if so, why did not Alexis Johnson and Akins pursue this policy in their own house instead of lugging into this meeting an intractable, open-ended monstrosity like the Palestinian problem when an oil security crisis with a forty-eight-hour fuse was the real issue?

Bewilderment was succeeded by dismay when Akins restated his position that there was equity in Qaddafi's oil package and that the course of practical wisdom was to accept it. The fact that Johnson would permit Akins to be the Nixon administration's chief spokesman here, and to repeat the arguments the oilmen had been hearing from him for six months, confirmed their fear that the Akins position was not the eccentricity of a minor official, as had been hoped, but represented the settled policy of the administration.

It was when Akins expounded his familiar conclusions that Qaddafi's package—now a 40-cent hike in per barrel price, a 58 percent tax rate, acknowledged compensation for past shortchanging—was "reasonable" and his apprehension that the Libyans would feel "cheated" and become obstreperous if denied a fair response that Sir David could no longer restrain himself.

With his monocle, his aristocratic carriage, the precisioned elegance of his language, and the authority bestowed on him by recognized expertise and obviously wide horizons, Sir David made an arresting, even intimidating advocate. He rejected Akins's characterization of the Libyan demands as "reasonable" and said that he resented the insinuation that the Libyans were being "cheated." Having just had his company's rights, under its Libyan pact and under international law, stripped away by the pen stroke of a dictator, he claimed to be able to speak with some relevance. "The dangers to our interests, and to the consumers' interests, lie much more in yielding than in resisting the demands being made upon us," he said. He restated with cool articulateness the arguments he had made to Foreign Secretary Douglas-Home and then capped it all with a conclusion both stirring and prophetic: "Sooner or later we, both oil company and consumer, will have to face an avalanche of escalating demands from the producer governments. We should at least *try* to stem the avalanche."

A number of the oil barons rose to take their stand with Barran—Drake of British Petroleum, Rawleigh Warner of Mobil, William Tavoulareas of Mobil, and others. Jamieson of Exxon managed to avoid expressing a clear opinion.*

But it was the administration which, at this watershed moment, held the power and bore the responsibility, and everyone in the room knew it. If it refused antitrust immunity, it placed at hazard all American companies which cooperated together on a price response to Qaddafi. If it sat on its leadership leverage, no one else had the power to force the internal divisions within Oil to resolution. If it merely withheld support, it would thereby send a signal to Tripoli and to the capitals of Europe and Islam alike that would dissolve that *force majeure* which was the necessary backdrop if corporations were to prevail at defeating juntas while soothing the fears of parliaments.

More than withhold support, the tenacious Akins, supported by the undersecretary, gave vigorous opposition. "Every suggestion put forward at the State Department meeting of 25 September of ways in which the companies might fight back," concluded J. B. Kelly, "was rejected as impractical, dangerous or counter-productive."

Akins's vision of the future was crowded with bogeymen, real or imagined. He insisted that the Europeans would "under no circumstances do without Libyan oil . . . if the companies try to block the sale of Libyan oil, as they say they would, through controlling their tankers or people that bought the Libyan oil . . . they would find themselves nationalized in Europe as well [as in Libya]." Sir David, whose interpretation of recent history was that there was little the European countries would *not* put up with, saw this as farfetched. He was willing to take his chances on hypothetical nationalizations by Europeans; when it came to it, they would see that they could get not an extra barrel of oil by nationalizing anything.† What needed dealing with now was the laying of hostile hands on the oil companies that had *already* occurred

*Decision making at Exxon, according to one of its former managers, Robert Stebbins, "is somewhat analogous to grasping Jell-O. Authority is usually vested in committees rather than in individuals who can and want to be held accountable. A bureaucracy does things collectively so nobody can single out an individual for making the wrong decision."

†This proved correct. During the supply crisis of three years later various European governments tried to issue orders to oil companies, found themselves totally ignored, and put up with it, leaving the job to the companies.

in Libya, while all Islam looked on, transfixed, waiting for the Western response.

Barran suggested that a strong intervention by the U.S. government against these half seizures would either sober the Libyans into backing off or inflame them to a total expropriation that would wake up the consumer governments in Europe to what was at issue here. But Undersecretary Johnson poured cold water on hopes of U.S. intervention of *any* kind. The United States had little or no influence with the Libyans, he said limply, and its intervention would be "ineffective at best." The "nothing can be done" syndrome, so often hung on President Jimmy Carter, was, so far as oil is concerned, at least a long-established legacy from the Nixon years by the time Carter was sworn in.

It was when Akins dismissed an oilman's estimate that the Qaddafi regime could not endure for very long a total shutoff of revenues that Barran and his group became resigned to defeat. Akins elaborated that the RCC had four years' worth of normal government expenditures in its financial reserves. But surely these reserves—most of them in American and European banks—would be automatically frozen, in the time-honored way of such disputes? Akins's demurrer, in the presence of Johnson, brought home the depth of U.S. aversion to *any* resistance.

Akins expressed confidence that the "avalanche" of Sir David was illusory, for the Saudis would never join in a price rise stampede. "You must be joking," one of the oilmen said, and, as he later related, was on the verge of walking out.

But he did not walk out. Even the most aggressive oilmen now knew that there was no recourse, symbolic or practical, to the refusal of the United States to support resistance to Qaddafi.

Douglas-Home's consultations with European foreign ministers could not possibly generate a positive response in the face of the negative position of the United States—and did not. Two days later, on September 27—ultimatum day—Texaco and Standard Oil of California capitulated, accepting terms very similar to Occidental's. On September 28 Exxon and British Petroleum succumbed, presaging quick surrenders by the remaining holdouts, even the one headed by Sir David Barran, which issued a statement that "continued resistance and consequent isolation had become pointless."

On the day that Exxon and BP followed Texaco and Socal in yielding, Gamal Abdel Nasser was stricken with a massive heart attack in Cairo. Within a few hours he was dead. That he knew of the formal capitulations of his career-long adversaries is certain, for at the time Qaddafi was with him in Cairo.

Scarcely had the State Department's celebration over its avoidance of danger subsided when the avalanche Sir David had warned against began to thunder down all the passes of the oil world. Iraq immediately demanded and received increases in prices and taxes essentially equal to those won by its partner. Algeria demanded and obtained from France parity of price with its Libyan comrades. The shah of Iran, suddenly a lion, threatened to expropriate the operating assets of the oil companies unless they raised his oil income, and they did. Kuwait forced the opening of price negotiations. In November the beleaguered companies offered to raise the tax rate from 50 to 55 percent for Iran, Saudi Arabia, Kuwait, Iraq, Nigeria, and the Gulf sheikhdoms, an offer the host states, instead of jumping at, coolly decided to let OPEC adjudicate collectively. In early December Venezuela unilaterally raised its tax rate to 60 percent. To the suddenly swaggering oil states these were to be regarded only as interim gains. In the second week of December OPEC met at Caracas and issued a demand to the oil companies that negotiations begin at Teheran within thirty-one days, to be conducted under a time ultimatum, with the goal of OPEC-wide price and tax increases that would equal Qaddafi's.

In making their "leapfrog" argument to the State Department, back in the old world that had still existed in September, oilmen had warned that the other oil states, as a result of the nature of things and the political pressures on them, could not sit still for Qaddafi to outdo them, would pay no attention to those differentials in oil quality and transit costs that Akins laid such exaggerated store by, and would insist on the same rates that Qaddafi got and that Qaddafi would then up the ante to restore his edge. It took but three months for the second shoe to drop. On January 2, 1971, Qaddafi repudiated the pacts of September as inadequate and demanded 50 cents a barrel more, plus a further tax increase, plus a larger freight premium, plus more retroactive claims for past cheating, and he threw in an extra 25 cents a barrel "reinvestment requirement."

Chapter Nine

Capitulation at Teheran

The control of oil, the lifeblood of an advanced industrial state, by potentates who have no other instrument of power and who are accountable to nobody, morally, politically or legally, is in itself a perversity. It is a perversity in the sense that it defies all rational principles by which the affairs of states and the affairs of humanity ought to be regulated to put into a few irresponsible hands power over life and death of a whole civilization."

—Hans Morgenthau, 1975

It is customary in the memoirs of Nixon administration figures to place the crisis point of the oil catastrophe in late 1973, when the eruption of the Yom Kippur War and the whirlpool of Watergate furnished both an outside impetus to explain the price explosion and an inside distraction to excuse the administration's unpreparedness. But in fact, the critical confrontation over oil took place almost three years earlier, during the first weeks of 1971.

In terms of oil economics this was a time of stagnant demand growth in the world and of burgeoning productive capacity that was reaching new peaks each month, a time when the basic objective of most Middle Eastern states was to attain higher and higher levels of production and when the oil companies still saw it to their advantage to resist demands for higher prices. In other words, it was an opportune bargaining time for the consuming West. Nonetheless, the still-divided OPEC countries, half-emboldened, half-mystified by the West's exhibition of powerlessness before Qaddafi, were mounting a precise challenge to all the international oil companies and consuming nations, the response to which would determine the direction the free world's economy would take for decades to come.

The oil-producing states, claiming the right to tear up their contracts and concession agreements on the grounds of "changed conditions," were demanding revenue increases of historic size. They were requiring them to be met within two months. They had prescribed a bargaining format by which successive blocs of hosts could succeed one another at the bargaining table, building on the previous group's gains. And they were threatening "concerted and simultaneous action" if their demands and conditions were not met—an OPEC-wide cutoff of oil production, but a cutoff directed against the oil companies, not the consumer countries, which would be permitted to buy OPEC oil if they paid the higher prices. The hosts, having closely observed U.S. behavior in the Qaddafi go-round, had thus picked out what they thought was the weak link in their adversary's camp and were inviting nervous Western governments to scab against their own companies—that is, if those companies had the temerity to resist.

Sir David Barran was determined that this time the oil companies should be organized among themselves and systematically lay the groundwork for the steadfast support of Washington. He perceived that the OPEC design for at least two separate negotiations with the oil companies—one limited to the Persian Gulf bloc of conservative nations, the other to the Mediterranean radicals—had not only the offensive purpose of dividing and leapfrogging the oil companies but the defensive one of avoiding the intra-Arab divisions which the presence of Libya, Iraq, and the other revolutionaries around the same table with the hated Arab conservatives—with all their bitter-end differences of ideology and approach—would inevitably catalyze. Rivalries among the Arabs had always been the hole card of the oil companies, and Barran wanted a format that would give them full play.

What could be done with the independents? Barran was in contact with Armand Hammer and heard him out on the predicament facing Occidental and the other oil-vulnerable independents. When Hammer asked him if, when Qaddafi cut back production during the next showdown, Shell would consider taking part in a "safety net" arrangement, Barran did not turn him away, as Jamieson of Exxon had done six months before, but assured him that if the legal obstacles could be cleared away and other majors brought in, Shell would join in an oil bailout of the independents. With hope of unity between independents and majors in sight, Barran turned to the problem of strategy and U.S. cooperation. He circulated among all the involved oil companies his view that "the avalanche has begun" and that "our best hope of withstanding the pressure being exerted by the members of OPEC would lie in the companies refusing to be picked off one by one in any country and by declining to deal with the producers except on a total global basis." Sir David proposed that the companies meet to discuss a joint strategy. The meeting was held in New York on January 11, 1971, at the law offices of John J. McCloy, who had arranged for representatives of the U.S. departments of State and Justice to be on hand.

McCloy's luxurious offices, atop the Chase Manhattan skyscraper in mute testimony to his long representation of Rockefeller interests, his anteroom adorned with personally inscribed photographs of six U.S. presidents he had advised, formed an auspicious backdrop for the conference, full of symbolism of the linkage that until now had bound together the great oil companies and the U.S. government whenever oil security was at stake.

For three days the oil barons conferred. The results were equal to the expectation of Barran and to the hopes he had conveyed to Hammer. The oilmen reached an agreement by which the worldwide majors would share crude with the Libyan independents at close to "tax-paid cost" if any company's production were cut back—the safety net that Hammer had once sought in vain from Jamieson. In return, each company pledged itself not to come to terms without the assent of the others. The oil companies drafted a joint message to OPEC which was to become known as the Message to OPEC in Western accounts but as "the poisoned letter" and the "dirty trick" in OPEC parlance, calling for one simultaneous negotiation on a global basis. "We have concluded," said the message, "that we cannot further negotiate the development of claims by member countries of OPEC on any other basis than one which reaches a settlement simultaneously with all producing governments." They were joined in this declaration by Compagnie Française des Pétroles, Petrofina of Belgium, Hispanoil of Spain, and the Arabian Oil Company of Japan—the more welcome because they represented critical consumer nation sentiment. The companies agreed to establish a unified command post, the London Policy Group, which was to coordinate the negotiating teams and authorize final terms. The U.S. State Department approved the joint approach, and the Justice Department gave the companies what amounted to a waiver of prosecution for any antitrust violations the joint approach might otherwise entail.

McCloy, having demonstrated himself worthy of the huge retainers he pocketed annually from all the great oil companies by the take-charge manner in which he had facilitated the agreements of the State and Justice departments to support the united stance of the companies, now served up his *pièce de résistance.* On January 15 he took the heads of many of the oil companies, major and minor alike, to Washington for a conference with Secretary of State William P. Rogers, who was accompanied by his new undersecretary, John Irwin II.

The oilmen reviewed with a genial Rogers what they had just achieved in New York and what they hoped to accomplish in Teheran. McCloy requested that a "high-ranking" representative of the U.S. government be sent to the leaders of the major Middle Eastern governments to stress the administration's support of the oil company stance. McCloy was trying to undo the impression created by the Libyan fiasco that a wedge could be driven by the oil states between the oil companies and their parent governments, and he suggested to the receptive Rogers three stances such a spokesman should take: He should make it clear to the heads of state that the U.S. government supported the global "collective bargaining" approach of the companies, should urge them "at least to engage in fair bargaining practices" (i.e., ease up on the threats to seize oil company installations and stop all oil production if the companies did not submit by a given day), and should ask them in general to "moderate their demands."

Rogers agreed immediately, said he would that day go to the President to get his input, that the President's representative would leave for Teheran tomorrow, and on the spot it was agreed that the mission would be undertaken by Undersecretary Irwin.

It is evocative of President Kennedy's comment "Success has a hundred parents but failure is an orphan" that to this day it is unclear who first suggested Irwin. But one thing is clear, according to the report of the Multinationals Subcommittee of the Senate Foreign Relations Committee: ". . . the Irwin mission was designed to place the power and prestige of the U.S. government behind the company negotiations position, and to convey that the multinational oil companies and the U.S. government . . . would insist on compliance with the explicit terms of the message to OPEC, calling for one global negotiation to avert the Libyan 'leap frog.' "

The joint company position communicated to OPEC struck a balance between liberality in the present and firmness in safeguarding stability for the future. It acknowledged that revenue gains would be forthcoming for all. Moreover, its letter to OPEC offered automatic adjustment of revenues for inflation and for fluctuations in tanker rates. Its insistence on one global negotiation, binding on all hosts and companies, and on a long-term settlement, was aimed at protecting the weakest link in Oil's chain from being ganged up on and becoming the trend setter, at preventing *this* settlement from being merely the precipitator of new demands, and at breaking an out-of-control spiral whereby what Qaddafi wins in September, the Caracas Conference demands in December, and what Caracas demands in December, Qaddafi enlarges on in January.

Thus it was not so much a commercial bargaining over prices that would take place at Teheran as an attempt to reinforce the balance of power between the oil companies and the oil states on which the Western consuming world had so long depended for the oil stability that was the key to economic growth.

The Libyan junta immediately confirmed the true issues at stake. Learning of the safety net agreement, the RCC moved to break up the unity pact before the ink on it was dry. Various independents were haled in by the oil minister and threatened. Occidental and the Bunker Hunt Company, selected for their weakness as the test cases, were given an ultimatum: Dissociate yourselves from the alliance by January 24, or suffer "appropriate action." But the two companies, no longer isolated and now assured of backup oil, refused to leave the London Policy Group and rejected the latest demand for higher per barrel payments, sticking instead to their just-signed contracts with Libya that had almost five years to run. The junta swallowed its threats and took no action. The Barran strategy had already begun to stabilize the defensive line.

But while the oil companies were fitting out their London negotiating headquarters at Britannic House, the home of British Petroleum, with all the trappings of an international conclave of nations, the familiar crabbed motion of backsliding was already under way in Washington. On January 15, while in Secretary of State Rogers's office the agreement was being struck to send Irwin to carry the message of firmness to the Middle East, in the White House Henry Kissinger was mulling over a string of arguments just adduced by his staff, which had been monitoring the company negotiations, for the U.S. government's "staying out of the dispute between the producers and the companies."

Dr. Kissinger describes these arguments as follows:

> The rise in the price of energy would affect primarily Europe and Japan and probably improve America's competitive position. To sustain a confrontation would require us to be prepared to ration oil at home to support the European economies—a difficult enterprise in a country racked by Vietnam. To improve their bargaining

position with the oil producers the oil companies might bring pressure on the government to intervene in the Arab-Israeli conflict, which was contrary to our strategy of demonstrating the limits of Soviet influence. Confrontation, it was also suggested, could weaken America's relations with the Arab world.

This was an uninspired lot of arguments to lay before one as steeped as Henry Kissinger was purported to be in the inescapable equations of power balances, power vacuums, linkages, and the like. To accept them was to accept a number of anomalies: the illusion that the contagion of higher oil prices was a one-shot affair that could be contained without being opposed and that the leader of the West could benefit by turning a profit from the economic weakening of his alliance; the disproportion inherent in the proposition that the risk of a possible political inconvenience was a respectable reason for ignoring the risk of a possible economic disaster; the upside-down inversion that the U.S. government should yield to actual, present pressure from foreign governments and so abandon the field, in order to preclude the possibility of having to resist speculative future pressure from its own companies; and the conceit that one can escape the disadvantages of confrontation by simply refusing to recognize that confrontation has already been loosed by the other side.

The insubstantiality of these arguments against government involvement, however, did not activate Kissinger's famed powers of discrimination, nor did the reminder in the papers before him that on the other side of the question lurked the danger of "a cycle of escalating producer demands shading into the political field."* If not persuaded by the arguments to "stay out," Kissinger was immobilized by them, in these circumstances the same thing. "Though I leaned to the argument for a more active government role, I concluded that I was in no position to make such a recommendation. Therefore, I took the official's time-honored way out: On January 15, 1971, I ordered a study of the national security implications of the problem."

To order a lengthy study on whether or not the government should

*There was another memorandum point that Kissinger characterizes as on the side of "deeper government involvement," but his rendering of it seems to refer to government involvement to *soften* the oil companies' stand, rather than to shore it up, so as to diminish the chance of an impasse that could cause the oil states to enforce their threat to cut off oil: "There was also the danger of supply disruptions that would seriously affect our own security as well as that of our allies."

take a hand in events a few days away from their climax, a study, in effect, into the feasibility of Undersecretary Irwin's continued support of the oil company stance on the very eve of Irwin's departure, was in fact to advocate a decision to pull back from involvement and to pull the rug out from under the Irwin mission.

On January 16, the Message to OPEC, which had emerged from the three-day oil company deliberations at McCloy's offices and the salient points of which had been endorsed by the U.S. government, was delivered to the appropriate oil state leaders throughout the OPEC world. On the same day, Undersecretary Irwin, after receiving last-minute instructions from Secretary Rogers, who had just conferred with President Nixon on the Irwin mission, took off from Washington for Teheran, the first stop on his three-nation tour. As he flew over the Atlantic, accompanied by Jim Akins, pondering the words of Rogers and the President as transmitted by Rogers and sifting through the other inputs from Department and White House, Irwin found his conception of the task before him much diminished. Only yesterday it had been the strong and simple one of voicing the U.S. government's support for the oil companies' insistence on one global simultaneous negotiation and of urging upon the oil state monarchs fairness in tactics and restraint in demands. But today his mission's purpose was (as he later described it in Senate testimony) "to seek assurances from the Gulf producers to continue to supply oil at reasonable prices to the free world."

On January 17, Irwin arrived in Teheran, where, as the guest of the American ambassador to Iran, Douglas MacArthur II, his purpose appears to have undergone a further emasculation. By now the arrival of the oil companies' joint message had created a howl of consternation at the shah's palace, and this troubled MacArthur, whose basic concern was to maintain a harmonious relationship with His Imperial Majesty. MacArthur opposed the "single negotiations" stance of the oil companies as a devious scheme to exploit the divisions among the oil states. It was in the nature of Irwin's role, however, that he ought to have anticipated and discounted this, for a President sends out representatives, instead of using his ambassador, in part to cut through this occupational confusion of identity. And certainly the instigators of Irwin's mission, the oilmen, were prompted in part by the desire to get around MacArthur's well-known partiality for the shah's side of things. McCloy regarded MacArthur as "more Persian than the Persians sometimes," and George Henry Mayer Schuler of Bunker Hunt felt that MacArthur personified the syndrome of "the representative of the United States who becomes the representative of the country to which he is accredited. His primary interests are bilateral relations between the United States and Iran and he could not care less about multinational oil negotiations."

It was, then, against a disintegrating backdrop of crumbled support for his original commission—in the White House, in the State Department, in the American Embassy—that on January 18 the newly appointed undersecretary, accompanied by MacArthur, stood before the Shah of Shahs and Light of the Aryans.

From Irwin's first utterance to the watchful shah and his retinue, he betrayed that he was there not as an advocate but as an apologist, not to insist but to importune, not even as the bearer of a fixed message but as a temporizer. He began by jettisoning altogether the original purpose behind his mission. "The United States government is not in the oil business," he said, "and does not intend to become involved in the details of the producing countries' negotiations with the oil companies." ("Neatly removing," Kissinger would later point out in good-humored, if macabre, detachment, "the one fear that might have moderated producer demands: the threat of United States governmental intervention.")

Having separated the Nixon administration from the oil company front, Irwin underlined this separation by patronizing the companies: "The United States has urged the companies to be cooperative and reasonable, and . . . the companies have already agreed . . . to negotiate the substantive demands included in the Caracas OPEC resolutions." Inexplicably Irwin chose to dwell on the great damage that would be done to Europe and Japan if oil supplies were cut off by the oil states and on the President's hope that this could be averted. ("Perhaps this is why," commented Adelman, "the Shah soon thereafter made [his] first threat of a cut-off of supply. It is hard to imagine a more effective incitement to extreme action than to hear that this will do one's opponents great damage.")

Irwin's one bow to his original mission was to explain why the U.S. government had given permission to the companies to act in unison, but even here he skewed off into a defeatism curious to indulge before one's bargaining opposite by elaborating on the worries of consuming countries over the successive rounds of price increases that would flow from separate negotiations—an argument that must have whetted the shah's famous appetite for higher and higher oil revenues.

Students and intimates of the shah have pointed out that he was of a character that alternated between moods of insecurity, self-doubt, and taking of direction from the United States when circumstances looked menacing and moods of overweening self-confidence, arrogance, and hand biting of Uncle Sam when he sensed the pendulum of advantage swinging toward him. If so, the abandonment by Irwin, one by one, of the West's trump cards in any oil impasse (the backup of the oil companies by Washington; the capacity of the oil companies and consuming nations to get along without the hosts' oil far better than the hosts could get along without dollars; the serene readiness of the United States to take the lead in mobilizing the West's alternate resources for an oil standoff) must have primed His Imperial Majesty's psyche mightily, for he commenced to lay down the law to the President's envoy in the tone of a superpower to a client instead of the other way around.

The stance taken in the so-called Message to OPEC was "a most monumental error," the shah said. The moderate members of OPEC would not be able to restrain the Libyans and the other radicals, and the price increase that would emerge would therefore be the "highest common denominator." Worse than the bad judgment of the companies' position, however, was their perfidy. "Any attempt by the companies to say that they would not sign the agreement unless other OPEC members signed it would be taken by [the shah] and OPEC as a whole as a sign of bad faith" and would lead to "serious trouble." The companies would be making a "grave mistake" if they tried to play one member of OPEC off against another, what, he said, the companies were obviously attempting to do.

Finance Minister Jamshid Amouzegar spelled the shah occasionally, at one point calling the companies' one-negotiation position "a dirty trick," aimed at throwing "the wild men" (the Libyans and Venezuelans) into the room with the moderates (such as himself) in order to stir up dissension within OPEC, at another point warning that if the companies persisted in their plot, "the entire Gulf would be shut down and no oil would flow."

How much better, the shah resumed, if the United States backed a series of separate negotiations—first the Gulf half of OPEC, then the Mediterranean members, then the non-Middle Eastern members. Why, the worries about leapfrogging were totally unfounded, the shah maintained, and he gave the solemn word of His Imperial Majesty that if the separate negotiations route were followed, the Gulf states "were prepared to sign an agreement and stick to it for the length of its term, even though the producers in other areas obtained better terms from the companies."

Irwin, as events would swiftly show, was completely won over by the shah. Akins could have told him, but apparently did not, of his own assurance to Sir David Barran, at the September 25 conference, that the Saudis would not leapfrog and demand for themselves the Libyan gains and that Barran's fear of "avalanche" was therefore baseless, but that instead the Saudis had done exactly what Barran feared. Anyone with experience in the negotiating world could have raised with Irwin the question of why, if the shah really felt that Libya's presence at the table would result in higher prices being forced upon the West, he would not be insisting on Libya's presence instead of opposing it. Irwin himself might have reflected, but obviously did not, on the situation that caused him to be in Teheran—the tearing up by Iran and the other Persian Gulf states after Qaddafi's victory of the long-term oil contracts already in effect, on so flimsy a pretext as "the doctrine of changed circumstances"—a reflection which might have led to the recognition that the only assurance of long-term fidelity to agreements rested in the power position that came from the united oil company-parent nation stance he had been recruited to speak for.

John Irwin II, however, appears to have been more a man of action than of reflection, a diplomat more given to motion than consultation. Though the two oil company negotiators were due in from London on the morrow, he did not wait for their reactions to the shah's position, and though only one of the three chief of state visits was under his belt, and there were known to be many divergences among the shah, the king, and the emir, Irwin fired off his conclusion to Washington and flew on to Riyadh just before the oilmen arrived. Irwin's cable to Secretary Rogers recommended that the United States abandon the foundation stone of the oil company strategy and "encourage the companies to negotiate with the Gulf countries separately. . . ." Ambassador MacArthur cabled his support of Irwin's recommendation, in words that could have been lifted from the shah's monologue, saying that he was "deeply suspicious and afraid that the companies intended to play OPEC members against one another."

The White House and the State Department now had before them two specimens odd enough to raise the seasoned statesman's eyebrow. An as yet untried envoy, on the basis of a first exposure, urges a 180-degree reversal of yesterday's declared policy—in the futherance of which the government has just suspended, in the larger national interest, its antitrust laws. A veteran ambassador expresses deep suspicion and even *fear* that the oil negotiators are trying to outmaneuver and divide their opposite numbers across the bargaining table—was he unaware that the oil companies' ability to "play OPEC members against one another" was and had long been the obvious key to oil company negotiating strength and that without this capacity the West had no existing oil defense?

There was a recent track record of oil relations between the shah and the U.S. government against which to measure the perhaps overwrought advice of Irwin and MacArthur.

In the three years preceding the Nixon advent there had been no fewer than five minicrises in oil diplomacy instigated by the shah, and the standard tactic in each had been pressure on the U.S. government to undermine the oil companies' position and force them to bend to the shah's demands.

As has been seen, at the time of the Mossadegh revolution, the U.S. government had played a key role in preserving the shah's rule and in persuading the U.S. majors to enter Iran and rehabilitate its defunct oil industry so as to save the country from ruin, a task the majors undertook reluctantly because they preferred to invest their money and energies in Saudi Arabian oil, which they controlled totally, and because they feared that assuming a responsibility for moving ever greater volumes of Iranian oil would conflict with and might even endanger their relations with the Saudis. But as part of the partnership with the U.S. government in oil matters, they went into Iran in condominium with the British companies, restored the industry, and promised the shah that Iran's production growth role would be kept at least equal with the average growth rate of the Middle East as a whole.

The oil companies in Iran (hereafter called the Consortium*) had kept that pledge. Between 1954 and 1966 Iranian production, after being restored, had grown by 152 percent as against the Middle Eastern average of 136 percent; however, Iran had never recovered its position as undisputed number one in Middle East oil production, which it had enjoyed before Mossadegh and which the shah, as the passing years enlarged his responsibilities and inflated his self-esteem, had come to

*Forty percent owned by American companies, 54 percent by British-based companies, and 6 percent by the French CFP.

claim as Iran's *right,* because of its large population, its comparative economic needs, and its cooperation with the West during the various Arab difficulties.

In early 1966 the shah launched a major campaign to pressure the British and American governments to force their oil companies to increase Iran's 1966 oil production growth from the projected 9 percent to 17 percent—a virtual doubling. The British Foreign Office, concerned that trouble with the shah could cost Britain the valuable balance of trade plum it enjoyed in Iran, went along with him and brought pressure on the U.S. State Department to do so, for under the voting rules of the Consortium the majority British companies could not raise output without the agreement of the minority Americans. The State Department, under Acting Secretary of State George Ball, recognizing that the allocation of production throughout the oil world by the oil companies was a key to Western control of oil prices and supply security and acknowledging its responsibility to the majors as its "instrument" in a highly complex field, backed up its companies, resisted the shah, and rebuked the British, urging them to "keep a lid on" disputes of this kind rather than exacerbate them. The American majors, in turn recognizing their responsibility to be as adaptable as they could in averting tensions, permitted Iranian output to go up about one-third of the amount the shah was demanding.

The principle being reinforced here was clear: The oil companies would try to accommodate the shah in marginal ways, but never at the risk of endangering their control system, and the U.S. government would give them full support in this, even at the cost of having to oppose its closest ally, Great Britain, and an important regional ally, Iran.

Giving up on his production demands, the shah turned in the fall of 1966 to two other money-raising schemes: giving back to Iran some of the Consortium's holdings, so that the shah could invite bids on them, and providing some oil at cost to Iran, so it could market it at its own profit. Again, the British government was enlisted to join the shah in pressuring the U.S. State Department. Again, the Johnson administration resisted this pressure, after hearing out the arguments of the U.S. majors that "relinquishment" and setting up the Iranians in oil marketing would spur like demands from all over and be a "destabilizing international precedent." To the oil companies, the State Department said, in effect, "Do your best to accommodate, but what you *can* do without destabilizing the system is a matter of company judgment." To the British government, Washington backed the arguments of the U.S.

majors that relinquishment or giving cost oil to Iran might start a "fire" in their international operations. In the end the companies worked out a compromise acceptable to them, whereby some "cost oil" was provided to Iran under the restrictions that it could be used only in barter deals with Eastern Europe, so that it could not affect the Western oil market.

In late 1967 the shah whipped up another furor by oversimplifying a complicated situation in which American oil companies operating in both Iran and Saudi Arabia could be interpreted as discriminating against Iran and in favor of Saudi Arabia in allowing overlifting—i.e., above-quota production. Here both the French and the British showed their solidarity with the shah, a situation that brought acute pressure on the U.S. government to interfere. But once more, this time guided by Undersecretary of State Eugene Rostow, it stoutly backed the U.S. majors and the world allocation system, while privately asking the majors to diffuse the issue if they reasonably could. On February 28, 1968, Washington provided its embassies in Teheran, London, and Paris with suggested counterarguments to attacks on U.S. oil company "intransigence" toward the shah. "We fear that the high visibility of Aramco may lead Iranians to concentrate their irritation on its parent companies [Exxon, Texaco, Socal, and Mobil]," said the accompanying message, "but we also fear that the French and British . . . may be encouraging this tendency by trying to shift blame to Americans when there is none to shift."

In early 1968, after the '67 Arab cutback had caused a large but, of its nature, temporary boost in Iranian production, the Consortium announced a return to the normal growth pattern. The shah, with all the fury and pathos of a betrayed ally, demanded of the Consortium that Iran maintain its oil production growth rate as its "reward" for having stood by the West, an affecting argument which His Majesty pressed upon the British and American governments. But yet again the United States, after hearing out the oil companies, backed them and, more important, their function, taking the position that "the oil companies could not set their production policies on the basis of 'reward' any more than they could allow them to be geared to the 'need' of Iran's domestic population." So Iran's rate of increase, 22 percent in the 1967 year of emergency, dropped in 1968 to a more normal 8.9 percent.

In February 1968, after two years of continuous threats, the shah dropped his insistence on a specific oil growth target and dumped upon the Consortium the general responsibility for meeting Iran's budgetary needs, however it chose to do so. The Iranian government declared that its budget would no longer be held hostage to the Consortium's production programs and that from now on the Consortium must either pump enough oil to satisfy that budget or provide the equivalent income in some other way. On February 3 the oil companies were presented with the five-year revenue requirements which they were expected to meet from 1968 through 1972 to finance the shah's Fourth Development Plan.

The American government was concerned about this, the more so because not only were the Soviets trying to penetrate the Gulf, but fears were rife that another Arab-Israeli war could erupt at any moment, bringing another Arab oil embargo, and that this time, instead of helping the oil companies break the embargo, the shah might join the Arabs, in retaliation against the American companies' refusals to meet his demands. So after a dignified wait of almost four weeks Undersecretary Rostow sat down with representatives of the American oil companies operating in Iran. Rostow explained to them the gravity of the situation from the government's vantage, and they explained to Rostow the dangers of guaranteeing long-term growth and revenue rates for Iran. If Iran were thus insulated from the realities of the world market, Saudi Arabia and the others would demand the same. Oil that could not be sold would pile up, and the international oil industry would cease to operate as a market.

Well, could they give the shah a two-year increased schedule? Rostow asked. No, the oilmen answered. Nothing could be done of that magnitude for Iran without doing the same for the other Arab countries; otherwise the resulting divisiveness would cause more harm than good to overall stability. Favoring Iran, Gulf said, would alienate Kuwait. Favoring Iran, Texaco and Socal chimed in, would endanger relations with Saudi Arabia. Rostow listened and understood. He asked the companies to come up with whatever marginal accommodating steps they could and said that in the meantime the department would be working to restrain the shah.

Three weeks later, on April 20, 1968, the representatives of the Consortium had an audience with the shah and informed him "that his five-year revenue targets were very unrealistic" and that the Consortium could not even increase the 1968 and 1969 production estimates it had given him the previous fall. However, it could expand its Abadan refinery runs for a while, and by going off the Gregorian onto the Persian calendar, it could advance payments by three months (without the Arab states' having a claim on the same) and could give the shah a small advance out of next year's payments ($75 million). These adjustments would get the shah through the first year of his plan, but that was all the Consortium could do. The shah got off some choice imperial bombast—an oil field would be seized, he would put his own people on the Consortium board, under no circumstance was the Fourth Development Plan to be curtailed, etc., etc.—but the Consortium stood its ground, with the backing of the United States. Iran's oil production growth rates for 1968, 1969, and 1970 were to stay within the pattern of the mid-1960's.

Through all five of these "crises" with the shah, the Johnson administration, like its predecessors, had been guided by its recognition that, important though rapport with the shah was, the overall equilibrium of the oil company control system was more important, that in fact, the latter was a guarantor of the former, and that so long as the United States depended on the oil company system to maintain free world oil security, it had to back that system when its vitals were threatened.

The Johnson administration was steadied, too, by its awareness that the shah, for all his agitations and fulminations, was dependent on U.S. aid in addition to oil company revenues. Iran was the recipient of five U.S. aid programs—grant military aid, substantial noncommercial credits for military purposes, P.L. 480 (food) assistance, AID development loans, and Export-Import Bank loans. The shah's imperial hopes of turning Iran into both a modern industrial state and a regional military superpower, though they drove him to squeeze as much as he could out of Oil and the U.S. Treasury, also counseled him not to cross that nebulous line of tolerable provocation.

President Nixon, too, had working experience with the financial dependency that lurked beneath the imperial façade. In March 1969, when the shah came to the United States for General Eisenhower's funeral, he made use of the occasion to offer Nixon, through Kissinger, the ten-year pact previously described for 1 million barrels of oil per day at $1 a barrel—a fire sale of Iran's resources that betokened the fundamental weakness of its bargaining position. Seven months later the shah was back in the United States with a less grandiose proposition that he broached to the President personally: He was $155 million short in his development budget and would appreciate the President's ordering increased purchases of Iranian oil to tide him over (which would mean allocating half of all our Middle East purchases to Iran). In neither case did Nixon-Kissinger pursue these propositions past the stage of bureaucratic-oil industry "shouldn't-be-dones" and "can't-be-dones," and whether they desisted out of understanding or frustration, they presumably had learned that you could say no to the shah with impunity and were likely to have frequent occasion to do so.

Such were the precedents of the immediate past that reared up to warn secretary of state and White House that what Undersecretary Irwin and Ambassador MacArthur were recommending was not only an explicit betrayal of the government's strategy commitments of the past week* but a headlong abandonment of the oil security policy of the past generation. Again, no mere haggling over commercial details was at issue here; the companies had already indicated accommodations on those. What *was* at issue—and keenly known to be so by host governments and oil companies alike—was preeminently the balance of power between the oil-consuming world and the oil-exporting world, the capacity of the oil companies to bring their united strength to bear on the demands of the precariously united host governments, and the question of whether the American oil companies had the backing of their government or whether that government was to undermine them from the rear while they faced the host governments at the front.

Yet, as soon as the Irwin recommendation to drop the unified stance and split the negotiations into separate Persian Gulf and Mediterranean sessions arrived, Secretary of State Rogers endorsed it, with no more consultation with the oilmen he had joined with three days before than Irwin had accorded the arriving cochairmen in Teheran. The collegial unity of January 15 had by January 18, without explanation, turned to peremptory pursuit of cross-purposes.

Rogers could not have made such a volte-face without the concurrence of the President, who had personally authorized the Irwin mission. On Nixon's desk on January 18 was a memorandum from Henry Kissinger which the national security adviser describes as "summing up the negotiations between OPEC and the oil companies." Kissinger explains:

*"Thus without any consultation with Piercy and Strathalmond [of BP], who arrived in Teheran the day after Irwin left, the State Department abandoned the basic strategy of seeking an overall negotiation encompassing all of the parties at the same time, a strategy which lay at the heart of the message to OPEC, the Business Review Letters and the extensive consultations with the Departments of State and Justice."—*Report of the Multinationals Subcommittee,* Senate Foreign Relations Committee, p. 131.

> If we were going to encourage the companies to stand up to the producers, I argued, six questions would have to be answered:
>
> Will the companies stick together and hold the line against the Libyan demands? . . . Will Libya stick to its extreme demands and stop the flow of Libyan oil if they are not met? . . . Will the other Arabs then stick with Libya and shut down their production as well? . . . Will the European governments panic at the potential shortages and attempt to strike their own government-to-government deals with Libya, circumventing the companies? . . . Or will the Europeans join forces to staunchly resist the Arabs? Will we then ration oil domestically to help the Europeans withstand the Arabs, and/or bring more pressure on Israel to buy off the Arabs politically?

Events do not wait upon studies, nor can policy be postponed until the outcome is known in advance. To say to Nixon, at the hour of decision and despite the commitments of the past week, that he must withdraw support from the oil companies until he had the answers to those questions, questions which could be answered only by the tortuous course of events as those events responded to the quality of U.S.-oil company leadership, was to counsel him to take the only alternate course, to "lay down" before the producers, without sketching for him the grim questions that waited at the end of *that* road. What *could be* known now was that under the well-tried policy of "standing up" in oil matters, the United States had coped successfully with such questions three years before, fourteen years before, seventeen years before.

It is unlikely that President Nixon was the hapless victim of bad advice here. Rather, it is probable that Rogers and Kissinger were telling Nixon what he wanted to hear and what they knew he wanted to hear and that the lower-level recommendations of Irwin, and before him Akins, carried the day only because they conformed to Nixon's aversion to getting into a rhubarb in the Middle East over a problem that he considered secondary and that could be bought off at a cost which could not be politically blamed on him—a position he had taken and would continue to take at every fork in the oil road. Thus, when later on in the descent Irwin and Akins submitted recommendations for tough action on energy mobilization and "leaning on" the shah to hold down oil prices, their advice got nowhere. It was only the recommendations to appease that were welcomed at the top, a circumstance which had to percolate down through the layers of an administration.

In any event, on January 18 Rogers cabled to Teheran instructions to MacArthur to work for the scrapping of the joint negotiations strategy and for putting in its place the piecemeal negotiations favored by the shah. Kissinger the historian would later write the epitaph for the policy that Kissinger the statesman had contributed to fashioning: "If confrontation was to be avoided and if our government would not involve itself in the details, the preordained outcome was that the companies must yield."

The cochairmen of the oil company joint negotiating team, George Piercy of Exxon and the younger Lord Strathalmond of British Petroleum, unaware that the plug of U.S. backing had been pulled, arrived in Teheran on the nineteenth to commence negotiation with their counterparts—Amouzegar of Iran, Yamani of Saudi Arabia, and Saadoun Hamadi of Iraq. Ambassador MacArthur had already informed the British, Dutch, and French ambassadors of the U.S. government's changed policy position, by way of lining them up behind it, and now, in the presence of all these ambassadors, he dropped the bomb on Piercy and Strathalmond. Only the Dutch ambassador supported Piercy in his shocked remonstrance that an hour before the bargaining was to begin the "essence of the combined strategy" was being gutted and that the bargaining position of all the companies, formally submitted only three days before with the approval of the home governments, was here being repudiated.

Piercy and Strathalmond, thrust from this unhorsing straight into the bargaining kickoff with the oil state negotiators, quickly decided to hold firm to the one-global-negotiations policy, the only one they were authorized by their peers to take; after all, their bargaining adversaries could not know of the U.S. split-off from the company position, and perhaps by nightfall that defection could be remedied, when Oil's spokesmen had had a chance to explain to the Nixon administration the implications of the rout it was inviting. So when Amouzegar demanded that the companies drop their stated position, that they were mere *companies,* whose positions had no official stature, whereas his side represented *governments,* Piercy refused, contending that the Message to OPEC committed the companies to a global negotiation and that *their* home governments had approved it and therefore were also committed.

Piercy had underestimated MacArthur, however. Amouzegar *did* know of the U.S. defection, had known before Piercy, and now he gently let the air out of Piercy and Strathalmond, assuring them somewhat mysteriously that they need not worry about their governments' being wedded to the "global" position. "I am quite confident," Amouzegar said, purring, "that they will agree to a regional or Gulf approach."

The attempt of the Nixon administration to mollify the shah had instead resulted—as oil company strategists had warned—in emboldening him to harden his stance against the companies. Amouzegar now served upon the crestfallen oilmen a forty-eight-hour ultimatum: Agree to the shah's "Persian Gulf only" formula by January 21 or OPEC would schedule for the twenty-fifth an extraordinary session at which it would dictate the terms of settlement, without negotiation.

Piercy could see now that there could be no appeal from Oil to the White House, no consultation whatever. The U.S. government, by moving behind the backs of its supposed oil partners, had put itself in position to say to Oil that having already stated its position to the shah's government, it, of course, could not rescind it and that further discussion on the matter would be unproductive.

In difficult circumstances the London Policy Group adopted a straddle. Strathalmond would *talk* to the Persian Gulf representatives, in that sense meeting the shah's demand, but he could not *decide* anything except in coordination with Piercy, who would conduct the negotiations with the Mediterranean section of OPEC. Amouzegar rejected this expedient as a rose by another name and promised shortly to serve on Strathalmond a letter listing Persian Gulf oil demands that must be decided separately at negotiations that must begin within a week.

At Britannic House in London, the London Policy Group, representing forces with fifty years of victories behind them, was unwilling to yield, notwithstanding the abandonment by the United States and now apparently by Great Britain. The group decided to stick to its adjusted global strategy, and it instructed Piercy and Strathalmond that Amouzegar's forthcoming letter and all other ultimatums from the producing governments were to be met by a firm restatement of the companies' refusal to negotiate a Persian Gulf settlement independently of other producing areas. What the group did not fathom, however, was that the United States not only had backed out on them but was actively conspiring against them.

In Teheran Ambassador MacArthur was demonstrating the alacrity with which—to use Irwin's phrase in vain—the U.S. government would indeed "become involved in the details of the producing countries' negotiations with the oil companies" when to do so suited its designs. Knowing that Amouzegar's "letter" was about to be served, MacArthur undertook no less than to suborn Cochairman Strathalmond into repudiating his instructions from London. MacArthur and the British ambassador, Sir Denis Wright, ganged up on Strathalmond, exhorting him to disregard his orders and to accept the letter and its procedural stipulations. Conjuring up an apocalypse, Their Excellencies worked over poor Strathalmond, assuring him that the companies back in London were only playing bargaining games and that his rejection of the shah's conditions would "finish everything here and now." Strathalmond caved in and on January 23 agreed to separate negotiations and to the shah's timetable. The "grand counterstrategy of the oil companies" collapsed with him.

Meanwhile, on January 20 in Paris, U.S. officials had sat down with representatives of the other Organization for Economic Cooperation and Development (OECD) nations to confer over the Teheran negotiations. Oil company representatives were present as "observers." Considering that the eyes of OPEC were as much on Paris as those of the OECD were on Teheran, here was an opportunity for the Western consuming nations—the mightiest aggregation of military and economic power ever seen on earth—to demonstrate that they were as prepared to resist an oil cutoff in 1971 as they had been in 1967.

But the tone of the meeting adhered to the U.S. policy of nonprovocation, and ". . . no serious attention was paid to the development of plans for oil sharing, or for increasing stockpiles, or for managing stockpiles in the event of an OPEC-wide shutdown." Even the lesser deterrence of keeping silent on the agenda and at least keeping OPEC in doubt was eschewed; a conference spokesman announced that the meeting had not discussed "contingency arrangements for coping with an oil shortage."

Bunker Hunt's George Henry Mayer Schuler, of the London Policy Group, believed that this advertised inaction in Paris had consequences: "The producing states, once they recognized that they were not going to be restrained by the joint action of the industry, and that the governments were not going to support the industry, then [realized] the only thing to hold them back was their own self-restraint."

The watchful shah, having within the week seen first the companies retreat from their strongest bargaining stance and then the industrial nations meet only to proclaim their self-paralysis, called his first press conference in twelve years to stoke the fears and feed the sympathies that were obviously prompting first the United States and now the European governments to pressure the oil companies toward surrender.

"If the companies form a powerful cartel in the belief they can put pressure on the producing countries," he said, "and if industrial nations stand behind them, we will have the most detestable expression of economic imperialism and neo-colonialism." In such a case, he would act. Unless the companies yielded soon, the shah threatened, the Gulf states would "legislate" a 60 percent tax rate, and he would personally see to it, at the next week's OPEC conference, "that the question of cutting off the flow of oil will definitely be considered."

The unfamiliar hazards of a two-and-a-half-hour press conference plus the shah's surging confidence combined to prompt him to gloat indiscreetly over the danger from which an all too fragile OPEC had just been delivered: "If the oil producing countries suffer even the slightest defeat, it would be the death-knell for OPEC, and from then on the countries would no longer have the courage to get together."

From the Oval Office things seemed to be proceeding splendidly. On January 25 Irwin returned from his triumphant Middle East tour and reported to the President ("proudly," Kissinger recalls derisively) that at each of his three stops (Iran, Saudi Arabia, and Kuwait) he had "stressed that we would follow our tradition of not becoming involved in the details of commercial negotiations." The President's representative in Teheran, MacArthur, was at that time energetically pressing the oil company negotiators to give in on price in return for pledges of assured supply—a response to the shah's hint of cutoff. The United States was pressing the oil companies, as McCloy recalled with characteristic professional euphemism, to "get across the idea that they were prepared to explore all avenues to get the proper assurances," in other words, to surrender on the solid, reliable things—money and strength in balance of power terms—in return for promises, despite Piercy's educated reminder that there was nothing in history to suggest that the promises would be kept and his warning that the Teheran negotiations were under way only because the host governments had torn up their last set of promises. Or as *Petroleum Intelligence Weekly* put it, "If such promises were worth anything, the present crisis wouldn't exist."

To veteran analysts of oil negotiations such as M. A. Adelman, the Nixon administration belief that stability could be achieved by making concessions on price was the world turned upside down. In the circumstances of the Middle East/North African areas, where prices were already fifteen times the cost of production and where, therefore, increased prices were no spur to supply, "lower prices and secure supply are the two sides of the same coin," Adelman insisted. Stability of supply was assured only for so long as the host governments were hard-up for the revenues that oil exports brought; only when they had the financial surpluses that high prices could soon pile up could they afford the kind of cutoff that the West need fear. To raise their revenues dramatically was not to buy peace but rather to put them in the driver's seat of a steamroller.

From the moment the oil companies succumbed to Nixon administration pressures to abandon the global approach to negotiations, as the Senate postmortem would conclude, "the story of the negotiations that followed is a saga of retreat by the industry negotiators before ever escalating demands by OPEC."

When the oil company spokesmen offered their specific terms on January 28, affirming a 15-cent-a-barrel increase, to rise to 22 cents by 1976, they were met with howls of derision and renewed threats to impose expropriation and an oil embargo at an extraordinary session of OPEC slated for February 3.

The U.S. diplomats in Teheran, on the whole pleased at the progress their interventions had wrought but concerned about the embargo talk being bandied about, met with the shah, in company with officials of other countries, on February 1 and February 2, reminding him of their many recent assistances and urging him to moderate his demands at the OPEC conference. Instead, on February 3 they got the back of His Imperial Majesty's hand.

The servility of MacArthur and Wright did not go unacknowledged. "The major governments happily, after my warning two weeks ago, have shown not the slightest sign of any interference or support for the companies," the shah said condescendingly. He then urged the conference to arm for war, and on the following day it voted: (1) to impose its own settlement by fiat if the companies had not offered acceptable terms by February 15 and (2) to take "appropriate measures, including a total embargo on shipments of crude oil and petroleum products" if the companies had not yielded to the imposed settlement by February 22.

Finance Minister Amouzegar summed up for the world's press the meaning of the shah's address and the OPEC resolutions in words that, foreshadowing as they did the days and years to come, might well have induced a premonitory chill in the bones of the negotiation's stage managers in Washington and the U.S. Embassy in Teheran: "There is no question of resuming negotiations. It's just the acceptance of our terms or we will go ahead with legislation."

The companies met the initial deadline by coming to agreement with OPEC's Gulf Committee on February 14. The five-year agreement called for an average increase of 51 cents per barrel during the first year, rising to 83 cents in the fifth year. The tax rate was increased from 50 to 55 percent. Host government per barrel "take," previously at $1 a barrel, was to rise by 27 percent in the first year and 54 percent by the fifth year. The increased host take in the February 14 settlement was more than three times that offered on January 28 for the first year and four times the initial offer in the fifth year. In terms of total revenue, it was projected that the agreement would add more than $10 billion to the oil income of the six Gulf members, raising their annual takes from the $4.4 billion of the prior year to $7.5 billion by 1975. The Gulf states promised five years of stability and pledged to refrain from leap-frogging in the event OPEC members elsewhere got higher prices.

The scene now shifted from Teheran to Tripoli, where the Libyans had peremptorily refused to negotiate, pending the outcome in the Gulf. Seeing their Gulf brothers catch up with them, a feat which in September the State Department had assured the oil companies need not be feared, they now charted the next great frogleap ahead. Their style of negotiation was in its rudiments the same as before, though now matured by previous success and a steadily declining fear of the United States and the oil companies. Periodic threats of nationalization and of production cutbacks were now standard fare. Joint maneuvers with the Algerians and Iraqis took on an enhanced éclat. On February 25 the three issued a joint communiqué stating that failure by the oil companies to accept the Libyans' demands would result in the curtailment of oil production by all major Mediterranean producers, a threat given further credence on March 11, when the deputy prime minister of Syria (the keeper of the pipelines) flew to Tripoli to announce Syria's solidarity with Libya.

Americans representing the oil companies in Tripoli were now being subjected to the unfamiliar usages which hostile juntas dish out to nationals of countries no longer feared or to officers of corporations whose sanctions are no longer credible. They were continually shouted at, threatened, and insulted. At times their laboriously drafted proposals, without being read, were instantly rejected, crumpled up, and thrown in their faces. Intimidations of all kinds were their hourly lot. They were ostentatiously followed everywhere by police, and their residences were surrounded by guards. They were at times denied visas to fly home, thus becoming in a sense prisoners. They would be jerked out of their beds in the middle of the night and summoned to negotiations with Major Abel Salem Jalloud, second-in-command of the RCC, who appeared for one facedown in combat fatigues and with a machine gun slung over his shoulder.

On March 21 the oil companies gave in to Qaddafi, offering him roughly twice what they had given the Gulf nations (on top of what he had already won in September)—an initial per barrel increase of 64.7 cents which was to rise to $1.14 by 1975 via the inflation adjustment: There was another 25 cents per barrel as a temporary freight premium. In another leap of the frog, it was agreed that Saudi and Iraqi oil piped to the Mediterranean for export would get the same price as Libyan oil. *In toto,* Qaddafi, who accepted the terms on April 2, was to receive $1 billion more per year than he would have received a year earlier for the same production—a jump of 130 percent since he had come upon the oil scene.

His Imperial Majesty, from the grand aloofness of the Swiss mountains, where he had taken a skiing vacation while the details were being wrapped up in Teheran, had welcomed the February 15 settlement, and he repeated his promise to honor its terms for the full five years, no matter what happened in Tripoli. Amouzegar exulted, "I was so happy I had tears in my eyes."

The Nixon administration had also hailed the Teheran settlement. The White House announced that the President was pleased. The State Department called a press conference, at which it sprinkled its hosannas throughout with the words "stability" and "durability." The U.S. government "expected the previously turbulent international oil situation to calm down following the new agreements."

But when the results of part two of the negotiations came in from Tripoli—the fruits of the separated bargaining which the administration had so resourcefully forced—the shah was "incensed," according to his spokesman Amouzegar, that the Libyans had gotten so much more than he had. Amouzegar warned that if the oil companies did not grant more benefits to Iran, it would back the "radicals" of OPEC in their next campaign. The race was already on again.

Within a very few months the claims for stability and durability, in pursuit of which the Nixon administration had virtually crippled the heretofore dominant oil defenses of the West, would be exploded, and the pledge of the oil states to stand by their new contracts would one day be characterized by Henry Kissinger the historian as "a solemn promise that must hold the world record in the scale and speed of its violation."

Chapter Ten

Options Ignored: The Irreversible Fall

The U.S. . . . had not run out of energy in the early 1970's. Rather, by a complicated series of actions and inactions, it had increasingly chosen an increasing foreign dependency.

—General George A. Lincoln, 1976
Head of Office of Emergency Preparedness
during the Nixon administration

The producing nations now have a cartel tolerated by the consuming countries and actively supported by the United States.

—M. A. Adelman, 1972, professor of economics,
Massachusetts Institute of Technology

The 1971 Agreements [at Teheran and Tripoli] were a house of cards that could come down at any time. . . . The critical issue . . . was how well the companies and the government used the two and a half years before the October War to prepare for the crisis which was clearly foreseeable at the time of the Teheran-Tripoli Agreements.

—Report of Subcommittee
on Multinational Corporations,
Senate Foreign Relations Committee, 1975

Relentlessly, if slowly, the abundance of nature and the laws of economics had for more than two decades been pressing the market price of Middle Eastern crude down, down, toward its production cost of 10 cents a barrel. By 1970, with the Persian Gulf spot price trembling at an insecure $1.25, all that stood between it and the 10 cents was the 25-cent remnant of quasi-monopoly profit left to the companies and the 90 cents in taxes levied on each barrel by the oil states, neither of which was necessarily impervious to the gravitational pull of economic reality. It had required all the diverse political interferences that have occupied the foregoing chapters to arrest that descent in 1970 and turn it around, forcing up the price against nature to about $2.50 a barrel, even higher for Libyan oil, where it now hovered precariously in the aftermath of Teheran-Tripoli at a height twenty-five times the cost of production.

The oil companies, which in the main had tried to resist the increases in income to host governments because they feared not being able to pass them on to the world's consumers and consequently getting stuck with them, now awaited with deep apprehension the reaction of their markets. It was a moment reminiscent of the return to Paris of Edouard Daladier, the French premier, following the Munich Conference of 1938. Daladier returned stricken with shame over the concessions made there to Hitler, which he had felt forced by events to acquiesce in, and when he saw a huge throng awaiting him at the Paris airport, he assumed they were there in anger over his having disgraced France, and he was fearful. But when he alighted, he found himself engulfed in tumultuous cheering from a populace exultant that peace had been bought. In such manner the chagrined oil executives were to be pleasantly surprised at the lack of complaint from consumers—the *praise,* in fact, from such as President Nixon—when they raised the price of oil to heights they regarded as unconscionable and beyond all economic reality.

This moment of recognition that oil prices could be raised outrageously, across the world, and would be paid without significant consumer resistance or government retaliation, that indeed, the oilmen would be pressed by their governments to agree to the increases, so long as the price hikes originated with the oil states and were accompanied by a hullaballoo that resembled a crisis, was a sunburst lighting up for the oilmen a new vista of changed rules—in fact, a new rulebook altogether.

The great oil companies had lost much, but they retained much. Twice in six months, in confrontations with host governments, they had been abandoned—worse, worked against—by the Nixon administration and its junior partner, the British Foreign Office. So their codominion over the world of oil had been canceled, without explanation, and they had been cut adrift, as it were, without their former responsibilities to the oil security of the West and without any reliable supporter among the governments of the world. Thereby the companies had lost the power to control production levels, transit, and prices and were about to lose their ownership of the oil they had found, developed and paid for. With these, they had lost control of the destiny of their industry, and they feared that in other hands that destiny was foreshortened and bleak. In sum, they were no longer owners and captains, merely traders and managers.

They retained, however, *operational* control of the industry. They alone possessed the technology and the capital resources; they still directed, in the day-to-day sense, the field operations, the marketing, most of the transport. If they no longer made the major control decisions, they had the means of exerting influence on them. If they were no longer exhilarated by the splendid misery of responsibility for the next century, they still could look forward to the solid pleasures of making unprecedented amounts of money in the next decade, before the roof fell in.

No longer, then, did the oil companies really resist the recurrent demands of the oil states, which resumed again in the fourth month of the five-year period of promised stability and which in the first two years of the five—before the explosions of late 1973—would cause the market price of Middle Eastern crudes to double.* Unsupported by Washington should a crunch arise, they no longer had the power to resist by insisting on observance of the Teheran-Tripoli agreements. Who would enforce their lawsuits, honor their boycotts, freeze the assets of their adversaries? No longer the architects of Oil's future, they did not have the morale to resist. And now they had made the discovery that they could make more money, at least for the short term that was their new horizon, by *not* resisting.

In 1971 the oil companies learned to swim fairly comfortably in the wake of the new monopolists, the oil states, not only smoothly passing the constantly higher prices along but tacking on increasingly higher profit margins for themselves. The prominent financial analyst Kenneth E. Hill characterized Teheran-Tripoli as "truly an unexpected boon for the worldwide industry." And when five months after the signing at Tripoli, OPEC convened at Beirut to coordinate a campaign for upward revisions, United Business Service, the American investment advisory service, opined that tax increases by OPEC were *favorable* to oil company profits.

And so they were, as 1971 turned out to be the best profit year for Oil in almost a decade. Of that doubling of oil prices between 1971 and 1973, only one-half was chargeable to increased oil state takes; the other half went into oil company profits, which leaped from 30 cents a barrel in 1971 to 90 cents in the spring of 1973! The gains oil companies could make from obliging their new masters can be seen in this: The profit split between host governments and oil companies changed from 80-20 in the hosts' favor at the time of the Teheran-Tripoli agreements to 60-40 in the hosts' favor by mid-1973, providing the hosts with a new *cause célèbre* for demanding yet higher prices.

That this conjunction of interests between the oil companies and the OPEC governments was becoming an alliance—one that supplanted the pre-Nixon alliance of oil companies and *Western* governments—was widely and publicly proclaimed. The secretary general of OPEC, Adnan Musahim Pachachi, closing the book on thousands of pages of invective, said in 1971 that he now saw no basic conflict between the companies and the producing governments. Sir Eric Drake, chairman

**More* than double when measured in dollars, almost double in other major currencies.

of British Petroleum, acknowledged that the "industry has become a tax collecting agency" for OPEC (efficiently collecting OPEC's taxes from the citizens of every consumer country in the world). Sir David Barran of Shell, apparently reconciled to the inevitable he had so prophetically warned against, spoke now of a "marriage" of companies and producer governments. Sheikh Yamani liked this analogy and embroidered on it, welcoming an "indissoluble Catholic marriage" between oil companies and oil states, united, presumably, against the consumers of the world.

Here was a symbiosis of truly frightening potential: the oil states, with the apparent clout to impose on Western governments terms they would never have accepted from mere oil companies, married to the oil companies, which wielded a magnificently sophisticated and effective machinery for transforming the commands of OPEC into accomplished facts, the proclaimed levies into actual revenues; the oil states, enjoying by default the raw power mere companies could not maintain, power to create a real monopoly if they could stick together, joined to the oil companies, able to administer so uniformly and to report so accurately and publicly as to give each oil state the assurance against the other's cheating and conspiring that till now had been the main obstacles to their unity.

The term which oil scholars adopted for the Teheran-Tripoli agreements was the "First Great Capitulation of the West." It was a phrase holding forth both despair and hope—despair in that it seemed to imply that other great capitulations were to come, hope in the implication that it was still early in the game, that time remained to redress the balance. The Nixon administration, in the aftermath of Teheran-Tripoli, was giving off signals that were grist for either interpretation. In July 1971 the State Department's Mr. Akins expressed before the House Foreign Affairs Committee the conviction that the Teheran agreement was a success and would provide stability for the full five years—an omen of an incorrigible blindness in high places that must unerringly sabotage any turnaround. Yet in the spring of 1971 the President had set all the relevant arms of his administration to the task of putting together "the first Presidential message to Congress to deal in a comprehensive way with the energy future of the country," and this offered hope of a reevaluation of the policy direction of the recent past and of a new beginning.

* * *

The putting together of an express national energy policy, to be inaugurated with a presidential message to Congress and accompanied by all the hoopla with which the President was wont to surround his "historic firsts," happily presented a Nixon White House well into its third year with the occasion, indeed the necessity, to evaluate the overall impact of a spate of Nixon foreign and domestic energy decisions. Since the President's advent in 1969 U.S. oil policy had "turned 180 degrees"—in the judgment of energy analyst Edward Mitchell, later a member of the Reagan administration. Here was a mandate to examine whether that turn had been in the right direction and should continue or should be turned around while there was still time.

What could hardly be missed was that a dozen significant energy decisions, domestic and foreign, had been made by the Nixon administration in its first two years, all with little regard to their energy security implications, each responding basically to a nonenergy stimulus, each skewed by a nonenergy goal, none coordinated with the other. All, however, were contributing to an eerily uniform result: the generating of a several-million-barrel-a-day oil shortfall that must with blindsiding effect converge upon the oil market in two to three years, unless there was a redirection of policy *now.*

U.S. offshore oil development had been essentially foreclosed by administration policy since the first weeks of the Nixon presidency, lopping off a million or more barrels a day from planned 1973 production.

The Trans-Alaska Pipeline, which under its original construction timetable would have been nearing completion now, had not even begun as a result of a series of federal interventions and negligences for which the Nixon administration bore the primary responsibility; unless the bottleneck were broken *now,* almost 2 million barrels a day counted on by the end of 1973 would not be there.

The coal-banning Clean Air Act of 1970, which according to White House energy adviser John Whitaker was signed by the President without any thought of its impact on oil consumption, was proving a major factor in the rapid shift by electric utilities and heavy industries from American coal to imported oil, a shift that, unless turned around, would add more than 1 million barrels a day to U.S. oil demand in 1973, unexpected and therefore unprepared for.

The failure to brave consumer wrath and congressional flummery by assaulting the artificially low price of natural gas long imposed by federal regulation, which for a decade had been causing a sitdown strike among gas explorers and developers that by 1969 had stopped real production growth altogether, though resources were known to be plentiful, and had begun to cause switches to oil because natural gas was unavailable—a failure not unique to President Nixon but whose consequences had for the first time reached critical proportions under him—would, unless redeemed, now lead to an underavailability of natural gas by 1973 that would increase the consumption of oil as a substitute by up to 2 million barrels a day.

A cursory look at the foreign decisions, one that left aside all the other forbidding consequences of the shift in the power balance to the host government side, could hardly avoid recognizing at least one of the costs of the botched situation in Libya, for it could be readily counted in lost barrels: A Libya in hostile hands, with lessened need of oil revenues as a result of the more than doubled prices it had won in the past year, pursuing the deliberate design of creating scarcity, had imposed an oil production cutback that would limit 1973 output to 2 million barrels a day, as compared to the 3.7 million of April 1970 and, more important, to the nearly 5 million barrels indicated for 1974 by the 1965–1970 growth trend. Here was a looming shortfall of 3 million barrels a day unforeseeable to oil planners as late as 1970.

Thus the President and his men, before they even delved beneath the surface in their labors on "America's first comprehensive energy program," were confronted with end-of-1973 oil losses, stemming largely from their own actions and conscious inactions, of 9 million barrels a day. Here was a shortfall that even the prodigious adjustment capacity of Oil could not be expected entirely to counteract, for the shortfall's political origins and uncertainties defied Oil's abilities to anticipate and react. For instance, would the Trans-Alaska Pipeline get its permit in 1971 or not?

Fortunately many of the political acts that were bringing about this scheduled shortfall were reversible; more fortunately yet, not all of them by any means *had* to be reversed to avoid a 1973 smashup, just one or two, so great and various was the abundance waiting to be drawn on. If acceptably clean air or clean water were really found incompatible with properly regulated coal burning and offshore oil drilling and were in principle deemed more important than energy adequacy (although the Nixon inner circle put little stock in the *merits* of environmentalism, in contrast with its political clout, as the memoirs of William Safire and John Ehrlichman attest), then these priorities could be honored, so long as the attendant oil loss was recognized and at least one or two of the other remedies were taken. So a vast menu, from which they could choose sparsely and conservatively enough to suit the pickiest palate, was spread before the architects of the presidential energy message.

The policy planners could resolve that the Trans-Alaska Pipeline must be finally gotten under way in *this* year of 1971 and that the procedural obstacles in which it was mired must be cleared away as a national imperative. This would assume by the end of 1973 a regular flow of 1.5 million barrels a day and an emergency capacity of 2 million —in itself a big enough add-on to make the difference.

Or if the planners found the Alaskan thicket too thorny, they could opt for cutting through the regulatory obstacles blocking offshore development—the most promising source of new oil in the lower forty-eight.

Or they could take another tack and *join forces* with the environmentalists in favor of the clean-fuel beau ideal, natural gas, and break up at last the governmental bottleneck over price regulation which had for a decade been drying up new development of what had prior to 1969 been America's fastest-growing fuel. This would save up to 2 million barrels of substitute oil a day by crisis time. But to win on this one, Nixon would have to be willing to face the protests of gas users and to use the presidential pulpit to explain to the nation that a price rise of pennies now could save a rise of dollars later—for all fuels—and at the same time keep the nation energy-secure. The President was known to believe in gas deregulation philosophically, but he was also quick to caution his speech writers, "You can't explain economics to the American people."

Or if stopping the government-forced decline of natural gas were too prickly, they could stop the decline of *coal's* share in the national energy mix by stretching out the pollution control timetable, if Congress would permit it, or by using the administration's existing power to refuse to waive further the import restrictions on heavy fuel oil.

Or they could gingerly dip their big toe into the waters of conservation, as White House aide S. David Freeman was urging, for instance, by setting energy efficiency standards for new automobiles and future buildings. If the automobiles on the drawing board for the mid-1970's could be held to the miles-per-gallon performances of the mid-1960's —by no means a stringent vise—the daily savings in oil would be almost three-quarters of a million barrels, and even greater savings could be had by establishing energy-saving standards in building construction, an industry long conditioned to codes.

Or if mild conservation be found too bold, they could follow the well-trod path of tax incentives to stimulate domestic energy production. There were many proposals: reversing the focus of oil production incentives, which continued to reward Middle East production better than U.S. production, or incentives to attract the capital needed by the nuclear power industry that since 1969 had been falling behind its timetable for assuming in the 1970's a large percentage of the nation's electricity generation.

Or if the planners decided to let "nuclear" stew in its own troubled juices, they could turn to the defensive stratagems like oil stockpiling which had been worked out by the Office of Emergency Preparedness and the Cabinet Task Force on Oil Import Control in 1969. (Even the shah of Iran had proposed one to Nixon, assuming from his remote vantage that it was such a plausible solution to what must be so grinding a concern to Americans that it was an ideal vehicle for selling cut-rate oil to us.) A year's worth of imports from "insecure" areas could be purchased on a spot market that was still in glut and, stored in salt domes or steel tanks, held in readiness against embargoes or extortion attempts. And there were alternate types of "strategic petroleum reserves"—hiring oil companies to find oil on federal lands and keep it in readiness for pumping on short notice or paying oil companies to develop spare capacity of their own which would be kept available.

Or they could opt in favor of just cutting consumption under a standby rationing system to be imposed in time of trouble. According to a government contingency plan already suggested, a 10 percent reduction of oil use—more than enough to free the United States altogether from the need for Eastern Hemisphere oil—could be borne indefinitely by the oil-wasting American economy without major dislocation, a critical factor not only in enabling vigorous American leadership in time of oil crisis elsewhere but also in halting the inroads of American imports on Middle Eastern supplies.

If reading this abbreviated listing of alternatives causes the eyes to glaze over, it only reinforces the point: As of the spring of 1971, two and a half years before the denouement, though much had been thrown away, the means of retrieval were still immense. No one could be sure, of course, that Congress would follow where the President led, but the prerequisite to all was that he now lead.

That on June 4, 1971, he did not seize the moment is a fact of history. "The message was intentionally silent about the short term . . ." said a study by the Brookings Institution. "Congress ignored the energy message and within the Administration, too, it was quietly laid aside."

Staff proposals for the message that really came to grips with the sources of the shortfall were intercepted and culled out on reaching the inner circle. Nothing that could cause a stir or provoke a battle survived in the final draft. The Alaska pipeline was not mentioned. The offshore oil blockage was not seriously confronted. The gauntlet was not thrown down on the deregulation of natural gas prices. The hemorrhaging shift from coal to oil was left undisturbed. The rise of oil imports was not challenged but accommodated. Even tax incentives, that old standby of would-be visionaries without visions, were ignored. Protective devices such as stockpiling, sounding as they did an alarmist note, were not invoked.

What remained of the presidential message—an emphasis on nuclear energy without a program to effect it; an aspiration toward more offshore leasing but no measure that could in fact realize it; modest research aid for remote demonstration projects in the fields of coal gasification, the nuclear breeder reactor, shale oil production, and thermal energy (rather than the crash programs suggested by the Cabinet task force); the relaxation of import quotas against foreign oil—was delivered to Congress in an envelope, not in person, perhaps in recognition that here was not the stuff of a televised clarion call to the nation.

Henry Kissinger indicates history's judgment of this episode by making it clear that he had nothing to do with the preparation of the message. "There was no sense of urgency, much less of international crisis," he concedes. "Nixon's program addressed a problem of the more or less distant future."

The Church inquiry of the Senate Foreign Relations Committee was less tactful: "The critical issue . . . was how well the companies and the Government used the two and a half years before the October War to prepare for the crisis which was clearly foreseeable at the time of the Teheran-Tripoli Agreements. The time was not used very well . . . the U.S. Government had no energy policy."

The key to the problem of administration nonperformance could be found inevitably at the top, in the President's reluctance to recognize the onrushing danger no matter how many aspects of it were thrust upon him for decision and in his related refusal to expend political capital or take political risks for oil security or even to make ordinary exertions in its behalf if it meant tangling with some determined special interest group of left, right, or center, however small. (I am indebted to John Whitaker, a White House energy counselor under Nixon and later his undersecretary of interior, for some illustrations of this.)

The U.S. government owned a vast area of Alaska known as Naval Petroleum District Number Four which had been carefully insulated from the entanglements attending Alaskan statehood. Pet Four, as it was called, was known to have as much oil under it as Prudhoe Bay. But development of this great resource had never gotten under way because the admirals feared that if Pet Four oil were developed, it would be *used,* whereas they wanted to keep it untouched and in reserve for World War III, which would have to last a number of years before this oil could materialize from scratch and come to the rescue of the fleets. Before the area could be opened up, the admirals and the Navy's congressional sidekicks would have to be persuaded to loosen their death grip, a task not beyond the resources of an involved Commander in Chief. But President Nixon would not make the effort.

By mid-1971, given the progressive collapse of the Alaskan situation, the virginal Pet Four had become a bit long-range for the near-term emergency. But the Navy was also sitting on another huge reservoir—the Elk Hills Naval Reserve in California—and this one was already developed and capable of soon pumping several hundred thousand barrels a day. Beginning in 1969, the then Interior Secretary Walter Hickel had pressed the White House to start production at Elk Hills as an antidote to the Nixon-Hickel shutoff of offshore oil development. After all, in the modern world the industrial superpower role of the U.S. Navy was to keep the sea lanes open and assure our access to fuels and other resources from around the globe, not to use up what domestic resources we had on *itself.* But the admirals were as adamant about Elk Hills as they were about Pet Four, and Nixon was loath to brook them; not now, not even when the disaster actively hit in late 1973. (It remained for President Ford to start production at Elk Hills, but the year was 1975, years after the point of no return.)

Broad electoral politics, like its companion special interest politics, dominated the Nixon oil security decisions.

Some of the President's economic advisers were urging that if nothing formal was to be done to increase domestic oil output or to stop the shift from coal and natural gas to oil, oil production could be boosted informally but effectively by allowing the price of domestic oil to rise moderately. They were particularly concerned that with domestic oil prices standing still against a background of rising inflation in general, the economy was naturally turning more and more to oil and overusing it because of its relative cheapness.

This had contributed to an alarming development. Prior to the Nixon years, there had been a slow long-term decline in the use of energy per unit of the gross national product, a beneficent trend that portended rising prosperity—we were getting more product for less energy input. But in the Nixon years this trend had turned around (it would resume again after the Nixon presidency). Now the use of energy was rising faster than the gross national product, a fact that portended both a decline in real prosperity and explosive pressure on oil prices caused by overuse of oil.

The scientific solution was to cool down the overuse, the waste of oil —imported oil now because with domestic production falling, all increased use of oil came out of imports—by restoring the price of oil to its former relationship with that of other products. A small increase now would restore balance and help to stave off out-of-control prices later.

This would happen in the natural order of things; with domestic production inching downward as the demand for oil rose, its price would rise also, causing lessened consumption and more production, as hitherto unprofitable-to-produce oil came on the market—a double-barreled counterattack on the widening gap between U.S. production and consumption. That is, unless the administration intervened to thwart the natural process.

But the political assessment was something quite different. Oil prices were the brightest spot in the politically sensitive inflation picture and were keeping the consumer price index from rising as fast as it otherwise would.

There were three political ways readily at hand for the President to hold domestic oil prices in place until election day, while his foreign policy was pushing them up abroad: to jawbone and intimidate; to impose formal price controls; and to adopt a policy of unlimited imports of foreign oil, which, cheaper than American oil for yet a while, would suppress its price.

The President had already moved on the first of these fronts. When Gulf had raised its prices by about 8 percent in November, 1970, Nixon, who had just concluded pressuring the oil companies in Libya to surrender to the Qaddafi price increases of October, mounted a public campaign against Gulf and like-minded domestic companies, which included pulling oil produced offshore (and therefore under federal jurisdiction) out of the prorationing compact by which the state commissions had long set domestic prices.

The other two steps were of graver moment. To freeze the price of oil by government fiat would put an end to any hope of restoring the growth of domestic oil production and would instead speed its decline. Oilmen, looking at a world scene characterized by rising prices, simply will not develop oil here at a frozen price. They will leave it in the ground and wait for a better day.

And letting in all the foreign oil that Americans wanted, which would have to come largely from the Middle East, would mean totally scrapping the only oil security program the United States had—the oil import quota system. Only sixteen months before, Nixon's Cabinet Task Force on Oil Import Control had reported to him that to allow Middle Eastern oil imports to rise above 5 percent of domestic consumption before the late 1970's, and above 10 percent *ever,* would endanger oil security, but if he now opened the floodgates, that first level would be breached in a matter of months, the second in a year. Not to let the oil in, however, would result in shortages, would cool down the economy that ideally should be heating up over the next year and a half toward an election day peak.

In August 1971 the President froze oil prices as part of a general price control program that would last until early 1973—that is, until after the reelection campaign had been safely concluded in November 1972 with a resounding victory. Even after the other controls had been lifted, however, controls on oil were left in place. The anticipated impact on domestic production would be borne out by a steadily declining curve which would not be interrupted until oil price controls were lifted several years later.

The President also did away with oil import restrictions, although here he acted on a piecemeal basis for almost two years, simply raising the quota each time imports reached its limit, until in April 1973 the dodge had become such an embarrassment that he abolished the quota system altogether.

Starting in 1970, when Nixon began his waiver policy in earnest, the upward creep, and then the leap, of oil imports furnish a graphic representation of a policy that has yielded to every politically potent interest that has clashed along the way with the constituentless interest of energy security. By 1972 all of the President's previous accommodations had fused to encourage an unprecedented surge in the use of oil —which had to be imported oil. Pump priming of the economy aimed at producing an election year boom was causing a feverish hike in the demand for energy. That energy couldn't be coal, in decline because of recent restrictions on its use. It couldn't be natural gas because nothing had been done to break the long standoff between government regulators and gas producers. It couldn't be nuclear power because since 1969 its schedule for assuming an ever larger share of power generation had been allowed to founder. It couldn't be new domestic oil—that had been choked off by tax disincentives, environmental disincentives, and price disincentives. It wouldn't for much longer be Canadian oil, our largest foreign source; the Canadians were estranged by Nixon's refusal to include Canada in the Alaska pipeline routing and in a genuine overall energy partnership with the United States, so Canada would before long be announcing the end of its oil exports south of the border.

It could only be imported oil, pouring in without restriction, without thought for any system of national security, its volume rising dizzily, out of control. On President Nixon's inauguration day, oil imports stood at 19 percent of domestic oil consumption, a figure that had remained fairly stable for some time, that was more than offset by American spare capacity, and that had stood the test of safety in the 1967 embargo confrontation. The figure for 1971 would be 25 percent; for '72, 28 percent; for September 1973, 38 percent—6.2 million barrels a day, as opposed to 2.5 million in 1968. And there was no longer any American spare. As for imports from the Eastern Hemisphere, the United States was still within that 5 percent safety margin in 1971. But in 1972 "insecure" imports would rise above 7 percent; in September 1973, above 14 percent! So far as oil was concerned—the basic ingredient of industrial strength—we were now drifting at the mercy of chance events.

Chapter Eleven

Unconditional Surrender

It is now obvious that this decision [the OPEC price rise of December 1973] was one of the pivotal events in the history of this century.

—Henry A. Kissinger

The decisions against any immediate national effort either to boost domestic energy production or to restrain or put safeguards around leaping oil consumption, decisions taken piecemeal in the early Nixon years and then reaffirmed across the board in mid-1971—by the stands not taken in the presidential energy message and the stands taken in the "New Economic Policy"—left the United States nowhere to turn but abroad for one-third of its oil needs in the early seventies and an estimated *half or more* in the late seventies. This prospect of onrushing insecurity was heightened by the fact that some regions were far more dependent than the nation as a whole; the East Coast, already importing more than 95 percent of its utility and industrial oil needs, was as vulnerable as Japan or even Hong Kong.

Coping with this dependence would be the more difficult because the foreign policy decisions of the same years had crippled the system of Seven Sister-parent government control that until 1971 had been able to assure the free world a plentiful, uninterrupted supply at predictable prices. But coped with it must be. If the American-led West could no longer control the decisions about oil production and price on which its prosperity and cohesion depended, it could at least influence those decisions, put limits on them. The Atlantic Community's military power and diplomatic capacities, and the linkages it was in position to exact from oil states dependent on it for protection, modernization, and mediation, left aside, its most obvious and pacific path was in the direction of at least a loose oil affairs unity among the consuming industrial nations, an arrangement to share the oil and coordinate the bidding in time of supply interruption, so that they would not out of isolated insecurity work against one another and so that their oil companies would in time of emergency be under Western control and coordination, rather than pursuing helter-skelter advantage and survival in an oil game that now had new masters and new rules.

In short, the question before Richard Nixon in mid-1971 was: Having definitively cast aside the national policies of energy independence at home and international oil dominance abroad that were still in place on his inauguration day, could he now in part compensate by leading the world's strongest, most politically sophisticated and historically cohesive nations into a minimal consumers' unity capable of matching the producers' unity being fearfully and stumblingly pieced together by the world's weakest, most politically backward and historically divided nations?

Though the West had abandoned the mechanisms and strategies that had assured oil security, it still maintained a psychological ascendancy too great to be wholly dissipated by a year or two of folly, an ascendancy impossible to quantify but surely significant. It stemmed from the long-ingrained perception by the third world oil states of their weakness and incompetence compared to the strength and know-how of the West and from the fogged-in horizons that, because of their unfamiliarity with the things that had always been managed by oil, hovered low over them, shrouding the limits of what was possible. How high could the price of oil really go? Five dollars? Six dollars? Could they really master the intricacies of selling it on the world market if they had to? Were the oil companies as indispensable as Sheikh Yamani kept saying they were, and had the all too familiar strength of the West in defending its interests really been fatally diminished or was it just taking a catnap, soon to rise up reinvigorated?

This was the insecurity and defeatism that Nasser had so long despaired over ever conquering, and this must have been troubling the shah when, on the very eve of victory at Teheran, he speculated about how easy it would be for inconstant OPEC to fall and break into a dozen pieces and never rise again. To preserve such helpful forebodings, it was incumbent on President Nixon not to stumble into collateral conflicts with the oil states unless he intended to wage them vigorously, lest they become accustomed to victory, and not to make them presents of glimpses at the shining, undefended heights that they had not imagined were there and that were just waiting to be occupied.

Immediately one of these admonitions was to be ignored; soon all would be. In August 1971 the President unwittingly initiated a no-win one-sided conflict with OPEC when he ordered the devaluation of the dollar without giving any thought to its impact on the five-year oil agreements of Teheran and Tripoli, in which so much had been given up for the promise of "stability." OPEC promptly demanded an upward revision of the Teheran payment schedules. Its case was far from ironclad since it was debatable whether it would be really disadvantaged* and since during the Teheran negotiations the companies had offered periodically to readjust payments to account for inflation, and the oil states had refused, taking instead a fixed amount increase of 2.5 percent *plus other benefits.* With the other benefits safely tucked away, it was contestable whether they should now be allowed the flexible inflation adjustment they had rejected. But it was not contested by the newly docile oil companies, which agreed to an 8.5 percent hike and to quarterly adjustments henceforth and which, instead of protesting this increase as a violation of the Teheran-Tripoli accords, chose to call it a "supplement." The first hole had been punctured in the five-year agreement before it was a year old; before it was two and a half years old the host government take per barrel would have been doubled instead of stabilized and the oil states' historic self-doubt would have been assuaged by a quarterly diet of small, uncontested, almost automatic victories over the oil companies and the oil-consuming nations.

Among those victories of 1972 and 1973 was that of participation—nationalization of the companies in all important aspects but the nitty-gritty of operating them—by which the companies agreed to give up to the host governments, in stages, their ownership and control of their oil holdings, yet stay on to run them. Under these participation agreements, the oil companies turned over to the hosts gradually increasing amounts of crude oil, which they wanted so that they could learn the ins and outs of doing their own marketing, refining, transporting, and thereby find out how far they could ultimately go with their new-won dominance. The more oil they "moved" by themselves, the more confi-

*Adelman argues: "Devaluation of the dollar . . . was [only] an incident in the worldwide price inflation to which the Teheran and Tripoli agreements had adjusted by providing for periodic escalation. Moreover, Persian Gulf revenues were mostly *not payable in dollars.* The new element in the situation was not the increased dollar cost of imports to the producing countries, but the fact that prices in dollars increased, especially in Germany and Japan [italics added]."

dent they became of their ability to take over the whole show. But until it suited them to do so, they had the oil companies to translate their wishes into accomplished facts.

In early 1972 there was a stirring of alarm at some of the lower levels of the Nixon White House and the State Department over the visibly mounting dangers to oil security and the realization that the presidential energy message of June 1971 had been a nonresponse. A confidential State Department report urged a "crash program" to develop energy sources at home, along with a "major international effort" to put together a unified response among the oil-consuming nations in such areas as oil sharing in emergencies, stockpiling, and the avoidance of panic bidding. And in the summer of 1972 Undersecretary of State John Irwin began a round of visits to every Cabinet officer and agency head with an energy responsibility, seeking to alert the highest level of the American government to the danger that increasing dependence on foreign oil would "inevitably result in losing our freedom of action in foreign policy."

General George Lincoln, head of the Office of Emergency Preparedness in the White House and a member of the National Security Council, later made the astonishing observation that Irwin's effort was "the first time that energy received top institutional attention from U.S. foreign policy machinery." It comes as a bit of an anticlimactic letdown to read Lincoln's conclusion that "nothing spectacular came of his efforts."

At a number of gatherings of the Organization for Economic Cooperation and Development (OECD) during 1972 and early 1973, representatives of the consuming nations discussed the sharing of available world oil supplies if an Arab shutoff occurred, so that through a fair distribution of the two-thirds of free world oil production that was non-Arab and the bulk of Arab oil that would likely not be involved in a cutback, all might function at reasonable levels for many months and none could be picked out—like Occidental—for strangulation. Under such safeguards, the Western nations could deal with OPEC coolly and unitedly, as they had in 1967, when the oil majors performed this function for them; the alternative course would be for each industrial state to try to wheedle separate oil deals at panic prices, with every sign of desperation heightening the adversary's confidence and raising his estimate of what it was possible to extort.

But no agreement emerged. If the Europeans were poor followers, the United States showed itself a poor leader, in fact, a nonleader.

Partly this was traceable to a loss of American credibility on the subject among the Europeans. The United States was more and more looked on as a seeker of competitive advantage and as the *cause* of problems rather than as a self-abnegating coalition leader bearing aloft the solution. Nixon's dollar devaluation of 1971 (and a second in February 1973) was seen as a sabotage of European exports in competition with U.S. exports, making the former costlier and the latter cheaper. The Nixon administration's support of higher oil prices for Qaddafi in 1970 and at Teheran-Tripoli in 1971 were now seen not as efforts to keep Europe's supplies flowing but as suspect maneuvers aimed at bringing European oil prices up to those paid by American manufacturers for domestic oil. The sudden surge in American imports of Libyan and Middle Eastern oil—which before Nixon had been left largely to the European and Japanese markets—had created a new long-haul tanker shortage that was driving up tanker rates to new record highs (from 74 cents per barrel in April 1972 to $4.09 in August 1973), making the Europeans and Japanese feel that they were paying an extravagant price for American indiscipline. And so far as many European representatives were concerned, the danger of an Arab cutoff of oil directed at the West arose generally from American overidentification with Israel, specifically from the present administration's refusal to pressure Israel into some kind of phased withdrawal from its 1967 conquests, a refusal they attributed to petty considerations of domestic electioneering.

The suspicions that wreathed the heads of the U.S. representatives at OECD oil-sharing discussions were reinforced when they insisted on an allocation formula that gave advantage to the United States—oil-rich though it was—as against oilless Europe and Japan. The United States insisted that only "waterborne imports" be shared in the oil-sharing formula, which involved only 14 percent of the American supply, ignoring the unignorable fact that America still produced 70 percent of its oil and virtually all of its coal and gas. The French, strong in some energy respects, advocated a sharing formula based on each country's "vital needs"; the Japanese, weak in all respects, wanted a scheme based on "total energy requirements." So each session ended in deadlock.

It might be argued that the Americans were no more guilty of a self-seeking divisiveness than the others, but to argue so would be to miss the point. The Western nations could unite against the oil threat only if the United States rose above the pack, as it had been doing rather consistently since 1940. Only the United States bore the responsibility for leading the Western alliance and had the incomparable resources that enabled it to "give." By coming forth with a formula that in time of crisis would reduce its own energy supplies by only 3 percent, while Japan's, for example, would fall by five times as much, the Nixon administration abdicated leadership and assured the failure of oil sharing, setting up an "every man for himself" scramble toward disaster.

If concert among the nations was out, cooperation in crisis prevention from American oil companies, which still had much to gain and lose in Washington, ought to have been more easily achievable, but it was not sought.

It is notable how many remnant pockets of old-system resistance to the new oil order were left in the 1971–73 period and that had their appeals to Washington and other Western capitals for reinforcements been met, they might not have been mopped up so easily after all and the end might have yet turned out differently.

Libya, though it had permanently cut production to about half the April 1970 level, was unable to move even this reduced amount in 1972 because it was overpriced. Iraq was also having serious trouble selling its overpriced crude (i.e., $2.50 a barrel), so much so that to head off bankruptcy, it was offering to sell at "reduced and competitive" prices —an offer that threatened to undermine the newly bulldozed OPEC price hikes. And despite calculated long-term production cutbacks by such major producers as Libya, Kuwait, and Venezuela, aimed at creating an exploitable oil shortage, the indomitable glut periodically welled up during 1971 and 1972. But by and large the oil companies were not taking advantage of these opportunities to resist and destabilize their new masters, and whenever one of them did resist, it soon became an object lesson when Washington failed to back it up.

On December 7, 1971, Qaddafi nationalized British Petroleum's operations in Libya. BP fought back, urging its sister oil companies to blackball expropriated BP oil that Libya was trying to market and filing suits of enjoinment, as in the good old days. Unable to sell hot BP oil except in barter deals with the Communist bloc that, contrasted with Western gold, grew unappetizing, Libya ordered the American company Bunker Hunt, BP's concession partner in Libya, to take over BP's former production share and market it, threatening nationalization, cutbacks, and the like, if Bunker Hunt refused. But Bunker Hunt did refuse and was consequently nationalized.

The sanctions were working, and the more assertive of the oil states faced a test. Both BP and Bunker Hunt requested the U.S. government to play its traditional role by ordering American oil companies and oil purchasers not to buy Qaddafi's hot oil. But in both instances the United States refused to involve itself beyond lip-service requests that were ignored by oil companies which no longer would honor mere requests and no longer felt tied to one another by a common bond or strategy.

At an earlier stage in Libya, when Qaddafi's acts against the oil companies had fallen just short of nationalization, the Nixon State Department had justified its nonresponse by arguing that only outright nationalization of an American firm would justify U.S. diplomatic and economic intervention. That stand was now revealed as hollow; it had become obvious that in the eyes of the Nixon administration *nothing* justified an active defense of the oil security structure.

Iraq, with much of its oil not moving in 1972 because of overpricing, a drop in demand, and a reappearance of glut, responded first by demanding in King Canute fashion that its oil consortium (IPC) move it nonetheless—the one demand it could not oblige—secondly by nationalizing the IPC for its disobedience; and thirdly by offering to cut prices on this seized oil below those of competing oil states. Iraq, with lessened leverage on the oil companies since cheaper oil was plentiful and without the financial reserves that had fortified Qaddafi for his 1970 challenge, tottered on the brink of a financial collapse that could bring to pass a new turn in oil affairs.

But the desperation of radical Iraq that threatened to let loose floods of price-breaking oil was cured by the disunity of the awkward West and the unity of the sophisticated Arabs. France broke ranks with the Western nations and oil companies and in the nick of time rushed in to form an oil partnership with Iraq, under which it was to buy the expropriated oil not at cut rates but at Iraq's premium price. And Libya and Kuwait came forward to lend Iraq enough money to obviate its need to turn to cut-rate oil sales. "Behind the Arab nations' action," reported the trade press, "lies an offer by Iraq to sell its newly nationalized oil at a cut-rate which would have driven down the revenues received by the other countries for their oil."

The oil majors were rapidly adjusting to their new role as dependencies of the OPEC governments. In 1972 the companies began cutting back on their sales of crude to independent refiners, utilities, and other "third party" users, so that this oil would be available to the host governments as participation oil if they demanded it. The former buyers of this crude were thus suddenly cut off from the supply sources they had long depended on and were no longer able to obtain the long-term supply contracts that are the guarantors of stability. Their only recourses were to enter the spot market and to seek contracts with the host governments for the participation oil the hosts were trying, experimentally, to market. In both instances these independent purchasers were bidding against each other for oil that often meant their survival.

Here was a situation tailor-made to destroy what stability remained, but here also was a situation amenable to U.S. intervention. The obvious remedy was for the United States to regroup the oil companies, most of them American, and to order the majors to continue their sales to the independent users, as the only antidote to panic buying that would blow the market sky-high. But as Washington would not intervene to defend the companies against the host governments, neither would it intervene to defend the small companies against the big ones. Nothing was done, with the inevitable result that the spot market began a steady rise and the oil states found eager buyers for their participation oil, dramatically widening their horizons in regard to both the vulnerability of the consumers and the exploitability of their new power.

By May 1972, as Dr. Weisberg points out, "Saudi Arabia had sold all of its participation crude available for 1973 and part of that to become available in 1974 and 1975." By August 1973 spot market prices for crude, driven up by the scramble of insecure purchasers, had risen above the posted prices in the oft-revised Teheran-Tripoli pacts to $4.90 a barrel, so Algeria, no longer tolerating even the charade of negotiations, decreed a price of $5 a barrel.

There had as yet, in the summer of 1973, been nothing that could be termed a crisis—no Arab-Israeli war, no anti-Western jihad, no OPEC-wide cutoff or embargo. Yet since 1969 the market price of oil had more than quadrupled. What would happen if there were a crisis?

Five-dollar-a-barrel oil in mid-1973 was not the $1.20 oil of 1969. A fourfold rise in the cost of the energy that went into turning the world's wheels and making its goods constituted a hurtful brake on the growth of prosperity in the developed world and on the inching up out of poverty in the underdeveloped world. Yet, in retrospect, perhaps $5 oil would have been a tolerable place to stop the escalator. A price fifty times the cost of production might well have been high enough to remedy waste and revive competition, to penetrate the guilt-indifference complex that had in the recent past caused such reluctance to fight for cheap oil. Now that the United States had gotten itself into the fix where oil was no longer a consumer price index boon for politicians, where imported oil cost more than domestic oil, and where Americans were suddenly dependent on it for one-third of their supply, soon to be one-half, maybe it would again become fashionable, if there were time enough, for statesmen to take an interest in holding the oil price line and keeping open the oil supply line.

And since early 1971 a momentous, unlooked-for, unrecognized demarche toward the Western side had been developing in Cairo and Riyadh, which had the potential both of stabilizing the oil price and supply situation and of removing the long-imminent threat of another Arab-Israeli war that could blow it to bits.

Egypt was in the throes of attempting to turn its posture inside out —from Moscow-armed confrontationist state committed to an eventual war of reconquest against Israel to an American-subsidized conservative-leaning partner of the oil monarchies, committed to peace with Israel and to an American-led diplomatic solution of their disputes. Saudi Arabia was moving toward a proposal of a permanent commercial partnership with the United States, rooted in oil, under which the Saudis would triple oil production, meet all U.S. foreign oil needs in the future, at stable prices, and invest its surplus revenues in the United States—the tie that would bind—in return for a privileged commercial relationship regarding such things as import duties and investment access, especially to oil refining and distribution. Moreover, Cairo and Riyadh, after a decade of hostility, were working in tandem at Egypt's reorientation and were moving toward a formal rapprochement, an entente if you will, that for Arab affairs represented a far more significant swing of the pendulum toward the West than the Qaddafi revolution had meant in the other direction.

Anwar el-Sadat came to be a candidate for greatness along a highly circuitous route. Terrorist and assassin in his earlier years, he slowly climbed the rungs as a champion of violence, a specialist at anti-Western propaganda, Cairo's expert behind the Yemen War debacle. For a time he was known as Nasser's yes-man and was sometimes referred to by Nasser as "my black donkey," yet ten months before he died, Nasser named him to the place of succession. Considered by most a transition caretaker when he assumed the presidency in September 1970, Sadat quickly displayed a mastery of intrigue and survival that would have made *El Rais* proud.

It did not take President Sadat long in office to conclude that neither he nor Egypt could last much longer unless the shame and dislocation of Israeli occupation of the Sinai were ended, peace restored in place of half war, and the economy revitalized; otherwise dissension would before long tear the regime and the country asunder, the already dangerous dependence on Soviet economic aid and military presence would deepen, and the future of Egypt must become a race between dissolution and Communist takeover. The best hope for achieving cooperation from Israel, economic uplift, and a weakening of the Soviet grip lay in the West, in particular with the United States. Sadat well remembered Eisenhower's role fourteen years earlier in clamping severe pressure on Israel to force its withdrawal from Suez and the Sinai. Only the United States could influence Israel now to surrender the conquered lands peacefully, provide the economic investment and know-how, and replace the aid now coming from Moscow. But Sadat recognized that to obtain U.S. involvement, Egypt must drastically change its national tune.

And so it did. Beginning in early 1971, Sadat made a series of dovish overtures toward a settlement with Israel—his own peace proposals, his acceptance of the UN's Jarring proposal that involved the signing of a peace treaty, his acceptance of the U.S. initiative of Secretary of State Rogers. (The Israelis, not so anxious to liquidate a great victory as the Arabs were to undo a great defeat and justly determined to exact strong security guarantees as the price of partial withdrawal, took a dim view of the efforts of so many parties to give back *their* conquests.) In the spring of 1971 Sadat purged the pro-Moscow clique in Egyptian politics by way of an early-morning police roundup and entered into a formal entente with King Faisal, who with Sadat's permission was keeping the White House informed that Sadat was systematically moving toward the West and was even considering ousting the Soviet military forces from Egypt. (Kissinger had suggested publicly on June 26, 1970, that if Egypt wanted the United States to pull its coals out of the fire, why not "expel" the Soviet military presence, to create the proper atmosphere?) As the year progressed, Sadat deemphasized Egypt's twenty-year policy of Pan-Arabism and took various steps to make the socialist economy of Egypt more attractive to Western investment and more open to domestic free enterprise.

But a year passed without any response from President Nixon. The "Rogers initiative" for partial withdrawal of Israeli occupying forces foundered amid obvious signs of lack of White House support for it. Sadat's April 1972 attempt to open up another "track" by establishing secret contact with Henry Kissinger was rebuffed when Kissinger kept postponing the initial meeting with Sadat's representative, Hafiz Ismail. Angered, mystified, yet unwilling to abandon the hope of cooperation with the Nixon administration, Sadat resolved on a bold gamble calculated to remove all doubts in Washington as to his bona fides.

On July 13, 1972, he sent a preliminary message to Dr. Kissinger through King Faisal's "secret channel," Kamal Adham, advising that he would send a representative to Washington if the United States had something to propose. Then, on July 18, Sadat ordered the 15,000 Soviet military "advisers" out of Egypt—giving them one week to vacate—a stunning defeat for the Soviets that rocked the chancelleries of the world.

While Sadat was awaiting the response of Nixon and Kissinger, his Saudi partners were readying their proposal for an oil partnership with the United States that would, as its by-product, constitute what Sheikh Yamani called a "foundation for real cooperation" between the United States and the newly realigned Arab world. On September 30 Yamani made a public proposal in Washington, reinforced by a personal letter from King Faisal to President Nixon endorsing that proposal, that called for the following: The Saudis were prepared to raise their oil production to 20 million barrels a day by 1980, enough to meet the most extravagant projections of Western demand, and offered to meet all the import needs of the United States, at stable prices, even if they reached the 8 million barrels a day of some estimates. In return, the Saudis would expect a privileged commercial relationship with the United States that would assure a stable market for Saudi oil, a hospitable environment for the investment of mounting Saudi treasury surpluses, and access for the Saudis to all "downstream" activities of the international oil business—refining, shipping, marketing—that would enhance Saudi return on its oil resource.

Thus in the summer and fall of 1972 a gratuitous array of unsought opportunities offered Nixon and Kissinger the chance to avert the consequences of almost four years of reverses in oil affairs. Without the encouragement or even the interest of the U.S. government, Faisal and Sadat had created a duumvirate in Arab affairs that wished to form a multilayered partnership with the United States. Sadat had defied the Soviets, turned to the West, and was prepared to banish the specter of an Arab-Israeli war if the United States would commit its great leverage to bringing about what was just and in the not so long run inevitable —the Israeli withdrawal from Egyptian land—and, as if to make the cup run over, Faisal was offering a *modus vivendi* in oil that, by involving the Saudis in an economic partnership profitable to them, provided its own self-enforcing cement.

The United States did not, of course, have to accept whole hog the propositions of Sadat and Faisal to profit by them. It was necessary merely to respond to them interestedly, to engage seriously in the process they opened up—accepting this, reserving judgment on that, postponing the other, buying time, building a rapport that served the interests of all.

But neither the gigantic gestures of Sadat nor the visions of safe oil proffered by Faisal were sufficient to elicit a timely or interested response from the White House. The top-level meeting Sadat wanted between his man, Hafiz Ismail, and Nixon's man, Kissinger, was put off by Washington until October, then put off again and not held until February 1973—seven months after the expulsion of Soviet troops and almost two years after Sadat, through Faisal, had begun to seek a new relationship with President Nixon. By the time Ismail was permitted to visit Washington, Sadat had long before decided that U.S. involvement was unattainable and that war was his only option. The letter from King Faisal to President Nixon about the oil partnership was not to be answered for three months, and the answer displayed such a lack of interest as to be a terminal rebuff.

For the explanation of the first of these oddities I am indebted to Dr. Kissinger's memoirs: ". . . Nixon thought Middle East diplomacy was a loser from the domestic point of view and sought to deflect its risks from himself."

What Nixon meant by this, Kissinger explains, was that progress toward a peace settlement would depend on the United States pressuring Israel to yield up much of its 1967 conquests and that if he applied such pressure, it would stir up reelection problems for him with American Jews.

Richard Nixon's initial strategy for postponing this political inconvenience until his distant reelection was past was to leave the Arab-Israeli dispute to the State Department, on the assumption that nothing much would happen there, with his personal stimulation absent. The plan went awry in the third year, when Secretary Rogers, pursuing a plan for Israel's eventual withdrawal to its pre-1967 borders, began to provoke a backlash from Tel Aviv that was ricocheting off the American Jewish community. The President did not want personally to call off Rogers, for he did not want his long friendship with the secretary to be strained any further than it had already been by Kissinger's eclipse of him, so he began to seal off Rogers's effort surreptitiously, as earlier he had isolated Walter Hickel. ". . . the White House acted as if the State Department were a foreign sovereign power," Kissinger writes.

But Sadat's insistent efforts to involve the United States in a leadership role were pumping life back into Rogers's scheme and causing matters to get out of hand, so Kissinger was called in to "drag things out through the Presidential election," as he puts it. "Late in 1971, Nixon began shifting responsibility to me. . . . My principal assignment was to make sure that no explosion occurred to complicate the 1972 election—which meant in effect that I was to stall."

Stall he did. By letting Sadat know that he was the man who must be met with before there could be any definitive action from Nixon, but by not making himself available for such a meeting until the second term, Kissinger touted Sadat away from Rogers yet prevented anything from transpiring on his own track, or "back channel" as he called it. As Nixon and Kissinger saw it, 1974 would be a good time to get cracking on the Middle East. For after Nixon's reelection campaign ended, in November 1972, the Israelis would be having their own election campaign coming up in the fall of 1973, which had to be waited out. It was a highly effective system for delay, but history does not always stifle its explosions until it is politically convenient for campaigners to deal with them, and from the middle of 1972 on a disillusioned Sadat was committed to war and actively preparing for it.

Kissinger half defends Nixon's policy of delay and snub on the ground that Sadat was sending confusing signals, one day booting the Russians out, the next day signing an arms and friendship pact with them, one day proposing peace with Israel, the next preparing for war; besides, delay and distancing fitted in with Kissinger's notion that a few more years of Arab frustration over not getting anywhere via Soviet help were necessary before conditions would be auspicious for U.S. intervention. Given the election campaign timetable that is the clear reality here, Kissinger's other justifications have the scent of sophistry. How could confusion over Sadat's true aims be a valid justification for avoiding the chance he was so patiently and spectacularly seeking to establish the contact that could enable him to make his aims unmistakably clear? And was not the obvious issue for Sadat, as 1971 passed into 1972 and 1972 into 1973, the one of: Could he get anywhere via *American* help? After all, he was at least getting the arms from the Russians which made possible his war option if the peace option failed. From the Americans he could not get even an appointment.

Kissinger's first meeting with Sadat's representative would have thus had to be a stunning success to cause the recall of war plans already unfolding. But alas, Ismail's trip to Washington was to confirm Sadat's discarding of the American option, not to arrest it. Ismail also had appointments with President Nixon and Secretary Rogers, and he gave the President a stark message from Sadat: "The situation in our region has deteriorated almost to the point of explosion." Kissinger recalls: "Within the space of forty-eight hours he would encounter the American President uttering generalities, the State Department pushing an interim agreement without White House backing, and Nixon's national security advisor discussing principles of an overall settlement at a secret meeting without State Department participation." It is a measure of Sadat's despair over this apparently endless runaround that even war with Israel seemed to offer a better chance of success.

King Faisal went along with Sadat's decision for an autumn strike at Israel, for though he judged it would turn out badly, he was satisfied that—given the deteriorating internal situation in Egypt, where mass discontent was joining army discontent over the apparent passivity toward the Israeli occupation, and the complete failure of both Sadat and Faisal to activate Washington—Sadat had no other choice. Faisal agreed to play two important roles: to be the partial bankroller of the Egyptian war effort and to mount an ambitious effort to pressure the United States against coming to the military assistance of Israel.

By early 1973 Faisal, stung by his inability to lure the Nixon administration into a pro-Arab tilt by his "positive use of oil," abandoned his long opposition to the "negative" use of oil as a weapon of foreign policy. In what was for him an unprecedented series of personal exhortations—letters to Nixon, dispatchings of Yamani and other personal message bearers to Washington, audiences with oil company executives in Riyadh and Geneva, even a televised appearance on NBC and an interview with *Newsweek*—Faisal pushed the line that Saudi Arabia's willingness to continue to meet the rising needs of the West for oil would depend on a U.S. contribution toward a Middle East settlement and, above all, on the United States' ceasing its one-sided support of Israel.

It was not necessary that the United States get results in order to ward off the split between itself and the conservative oil states that Faisal said was imminent. "Even a simple disavowal of Israeli policies would do something to change the anti-American sentiment in the region," the king told Aramco manager Frank Jungers.

The funny part of it was that all the while Nixon believed he *should have been* moving toward Sadat and Faisal. In October 1972, when Defense Secretary Melvin Laird in a memo after Sadat's expulsion of the Russians recommended that the President should open up secret contacts with Sadat and move closer to the Arab position, Nixon penned in the margin, "K— I lean to Laird's view." Again, in early 1973, the President said to Kissinger, "I have delayed through two elections and this year I am determined to move off dead center." But he did not.

By the spring of 1973, as the outbreak of war approached, Faisal was leaning hard on his new servitors, the American oil companies that made up Aramco. In May he told local Aramco officials that its concession was in danger unless the United States changed its Middle East policy. In Geneva on May 23, after a meeting with Sadat, Faisal called in the Aramco directors and ordered them to inform the U.S. government that unless such a change were made, American business interests in Saudi Arabia "will not be preserved. . . . Action must be taken urgently. You can lose everything."

The oil companies did their best to get across the king's message. They put ads in the papers and sent letters to their stockholders urging "evenhandedness," and delegations of oil chiefs called on high officials at State, Defense, and the White House to present Faisal's warnings. The messages were not taken seriously. According to their later testimony, a typical response was: "Faisal is calling wolf when no wolf exists except in his imagination." Another: "There is little or nothing the U.S. government can do or will do on an urgent basis to affect the Arab/Israeli issue."

On October 6, 1973, the day of Yom Kippur, Egypt and Syria launched their surprise invasion into the occupied territories, scoring initial successes and inflicting heavy losses of men and matériel. Almost immediately Israel called on the United States for a massive airlift of supplies, especially tanks and planes, which it had to have if it was to counterattack. On October 8 OPEC met in Vienna for a previously scheduled appointment to dismantle what was left of the Teheran agreements; by the ninth the focus had shifted to an emergency meeting of the Arab oil ministers to consider oil action against the West if it intervened in Israel's behalf.

Faisal held out against precipitate use of the oil embargo, wanting to avoid action against the United States if Nixon left him any choice. His esteem for the United States as anti-Communist protector and economic partner was second only to his need to maintain solidarity with his Arab brothers. There is evidence that Faisal was resigned to some American aid to Israel. Much depended on the strings attached to it and on Nixon's avoidance of public ostentation in its provision (Faisal remembered how a needless, far-in-advance Nixon announcement of Phantom jets to Israel—aimed at the Jewish vote with such crassness as to embarrass Kissinger—had helped convince Sadat that the American option was hopeless). Again, Faisal relied on his Aramco messengers.

On October 12 John J. McCloy hand-delivered to Alexander Haig in the White House a joint memorandum for President Nixon signed by the chairmen of the four American partners in Aramco. The memo warned that further U.S. action in Israel's behalf would provoke a major cutback of Arab oil:

> We have been told that the Saudis will impose some cutback in crude oil production as a result of the United States action taken thus far [support of Israel in the United Nations]. A further and much more substantial move will be taken by Saudi Arabia and Kuwait in the event of further evidence of increased U.S. support of the Israeli position. . . . We are convinced of the seriousness of the intentions of the Saudis and Kuwaitis [who would between them account for more than half of the oil cutoff] and that any actions by the U.S. government at this time in terms of increased aid to Israel will have a critical and adverse effect on our relations with the moderate Arab producing countries.

The oil companies, which once influenced host governments in behalf of U.S. views, were now trying to influence Washington in behalf of the oil states. But no matter. If Nixon read the oilmen's memo, he did not act on it. Not only did he order an air resupply of Israel, he did so in the most public way possible, sending a formal message to Congress requesting $2.2 billion in aid for the embattled Israelis, an act probably intended to bolster his popularity at a desperate time in the Watergate affair and one which fell on Arab capitals like a declaration of war.

King Faisal responded immediately by throwing his enormous influence behind an oil embargo-and-cutback that was accompanied by a dictated price increase of 70 percent (93.8 percent in Libya). The cutback was administered with high efficiency by the oil companies, which notified Washington of their intentions to abide by Arab orders and encountered no objection from a U.S. government that had no plan of its own. This time the United States and its allies had no defense, having so thoroughly dismantled the former one and having so nonchalantly refused to put anything in its place, and though the amount of oil involved was relatively small, panic broke out across the oil-consuming world.

At its peak the amount of oil cutback was a little over 4 million barrels a day, counting offset production increases in Canada, Iran, and Indonesia (but not the United States, now without that all-important spare). The shortfall in available oil, at its lowest point, was only 7 percent below the September preembargo level. Had there been in place a system for sharing oil and for controlling the bidding practices of companies purchasing oil, most of them American—in other words, a system to prevent panic buying—all could have proceeded smoothly.

But as it was, uncontrolled bidding became panic bidding. In early November there were 200 bidders for a single lot of Nigerian oil selling at $16 a barrel—more than three times the preembargo price. The shah tested the market in December with an oil auction that brought a staggering revelation—a price of $17.40. The vulnerability of a completely disorganized and demoralized West was illuminated by daily demonstrations of this sort, and led by the shah, OPEC set its sights on permanent prices that before it would never have dreamed of.

Obviously the market would bear a great deal more than the 70 percent hike of October, so on December 28 OPEC met in Teheran to raise the price (the oil companies were no longer present at these price-fixing sessions, even as straight men). The shah, again playing an active personal role, proposed a price of $14 (some accounts say $20). The Saudis, with a 100-year supply of oil to get rid of and a fear that too high prices could destroy oil as a product before they had gotten rid of one-tenth of it, suggested a $7.50 price. In the maneuvering before the Teheran meeting, the Saudis and the new American ambassador to Riyadh, Jim Akins, tried to get Kissinger to put pressure on the shah, using U.S. leverage as Iran's arms supplier; but nothing came of it beyond a routine letter from Kissinger, which the shah ignored, and the Nixon administration kept unsullied its record of never having used muscle to keep down foreign oil prices. A "compromise" figure of $11.65 was agreed to at Teheran.

Fittingly the shah made the announcement of the group's decision.

There would be a 130 percent increase, on top of the 70 percent increase of two months before, and smaller hikes in the months before that, which added up to a cumulative increase of 438 percent in one year. The annual bill of the oil-consuming nations would be $95 billion higher in 1974 than in 1973, a first installment on the greatest transfer of wealth in human history. Of course, this didn't count what the oil companies would add to their margins.

The increase was a "minimum," the shah said. More hikes would follow. "The industrial world will have to realize that the era of terrific progress and even more terrific income and wealth based on cheap oil is finished."

When Richard Nixon resigned his office in August 1974, the United States, which had dominated the oil world when he became President, was reduced to the position of not only accepting whatever OPEC decreed but also working out the credit arrangements through which the impoverished majority of the free world could pay the bill. The price, which had been dipping toward a $1 a barrel on Nixon's arrival, was $12 a barrel on his leaving. That was 120 times the cost of production—and going up.

Henry Kissinger continued as secretary of state under the new President, Gerald Ford, until Ford left office in January 1977, having been defeated by Jimmy Carter—primarily because the chronic buffetings of inflation and recession stemming from the oil price explosion, and its proliferating reminders of America's lost control of its destiny, prevented the well-liked Ford from ever establishing whatever leadership credentials he may have possessed.

During these years Secretary Kissinger was occupied largely with trying to piece together a sustainable order out of the debris left by Sadat's war and Faisal's oil action. He and Ford would leave office with a few organizations set up but no significant defense in place, at home or abroad, against the next offensive by OPEC. Meanwhile, the OPEC states—somewhat jittery over having plunged the Western world, their customers, into the worst recession of the postwar era and over having, by their price shock, stopped in its tracks the previously steady rise in world demand for oil—held in abeyance their next move pending the West's recovery, lest the goose be killed with many a golden egg yet to lay.

Chapter Twelve

The Crippling Legacy

The consensus was that the public acknowledged my intelligence and integrity, my ability to articulate problems and to devise good solutions to them, but doubted my capacity to follow through with a strong enough thrust to succeed. Most of this doubt about me had risen from the struggle over energy, with my repeated exhortations and lack of final action by Congress. It was not pleasant for me to hear this, but I felt their analysis was sound.

—Jimmy Carter, 1982
Keeping Faith: Memoirs of a President

Sometimes it seems as though our political leaders have no consistency, no coherence, no staying power, no theory that puts it all together—only ambition and anxiety.

—Meg Greenfield, 1980
Newsweek columnist

A world-upending revolution, wholly triumphant because wholly unopposed and wholly acquiesced in, had placed in the hands of surprised OPEC regimes the power to reduce oil output below normal demand, which quickly became the power to raise prices to whatever level the panicked bidding of a leaderless, oil-dependent West would bear. And once jacked up, the price that had been gained by contrived shortage could be substantially maintained, even at 120 times the cost of production, by the restriction of supply to the level necessary to support that price, though this forced the closing of factories and shops all over the consuming world and the progressive stifling of economic growth which had been geared to energy costs one-eighth of OPEC's 1974 price.

In the beginning unreflective wits had quipped, "We have merely swapped monopoly by oil companies for monopoly by oil states." But in fact, OPEC rule was a complete turnaround from the comparatively benign dominion of the oil companies. Long mellowed by circumstance, the oil companies had rested content with trying to avoid surplus, with shaping supply to a constantly rising demand and looking to *rising volume* for increased profits. Thus they had accommodated unfailingly, and at *declining* prices, a world thirst for oil that rose annually at 7.5 percent a year (11 percent for Middle Eastern oil). The new rulers of the oil world had begun by cutting supply below demand to drive up prices and, having quadrupled the price in one year, were determined henceforth to choke off supply to a level low enough to maintain the new price and pave the way for yet higher prices. Under the oil companies, steadily rising oil supply and gradually falling price had been the spurs to economic growth for industrial and third world society alike; under OPEC, static oil supply and rising price would be the brakes on growth.

Purists contended that the system of connivance among the OPEC states was too loose to satisfy their definition of a monopoly operation. Of course, there *were* differences in the character of this new collaboration among sovereign, disparate, politically rivalrous regimes and that of textbook cartels of mere corporations united in a single, limited purpose. But the key factor—purposefully restricted production that forced up the price and kept it up—was indisputably there. Whatever its deficiencies, the cohesion among the OPEC states to limit oil output first to a no-growth level and then to a shrinking level, so as to support astronomical prices, was effective enough to work. Americans grown accustomed in 1982–83 to an OPEC production level hovering for many months around 15 million barrels a day were in position to appreciate the extent of its working during OPEC's first decade.

In September 1973, the last month of the old order, the oil companies operating in the OPEC states were producing 33 million barrels a day. The productive capacity at OPEC's command in 1974 (which included reserves hooked up to the production and transportation apparatus and ready to pump on short notice) was 38 million barrels a day,* a capacity scheduled by the oil companies to rise to well over 50 million barrels a day by 1980, in order to keep pace with the normal consumption growth pattern of past decades.

But once the OPEC oil states assumed full authority, they never permitted their post-1973 production even to return to its September 1973 level, let alone rise incrementally to the levels planned for the future. OPEC production was choked off at 31 million barrels a day. "Supply is no longer determined predominantly by economic considerations," observed oil analyst Walter J. Levy, "but rather by the production policies of OPEC members."

A Central Intelligence Agency analysis noted that the OPEC countries with the largest reserves had "taken steps that have lowered current oil production and limited investment in the expansion of productive capacity. Saudi Arabia, Kuwait and Abu Dhabi have imposed restrictions that have kept output below capacity. Iraq is moving in the

*According to the January 2, 1975, report of the Senate Multinationals Subcommittee.

same direction. . . . Keeping production below existing capacity . . . also serves as a disincentive to invest not only in new capacity but even in maintenance of existing capacity. Kuwait, for example, which had substantial excess capacity for years, has let capacity erode."

Each successive oil supply forecast by the CIA reflected the effects of what private analyst John H. Lichtblau called "the imposition of production ceilings by several OPEC members below their existing technical potential." In 1977 the CIA estimated that enough momentum remained from the oil company expansion programs of the pre-1974 era to assure an OPEC productive capacity for the early 1980's of between 43 and 47 million barrels a day, but in 1979 the CIA was to lower that figure to 35 million, while a consensus of oil companies placed it in the high thirties.

OPEC's actual production in the 1974–78 period was stunted at a ceiling of 31 million barrels a day because that was the maximum the consuming world could afford to buy at $12 a barrel, a price OPEC was determined to exact. To produce more would result in a surplus unsellable at $12, a surplus that would soon force the price down. The expansion blueprints and investment plans OPEC inherited from the oil companies (Saudi Arabia alone was slated for a 10-million-barrel-a-day jump in the latter half of the 1970's) were thus not only unnecessary but inimical to the new scheme of maintaining double-digit prices by holding down production.

After the second great price explosion, of 1979–80, when the $12 price tripled to the $34–$41 range, the amount of OPEC oil the consuming world could afford was to drop in half, to 17 million barrels a day, to 14 million, and OPEC was to drop its production to those levels —at least 20 million barrels below its deliberately shrunken production capacity—in a desperate effort to maintain the new price.

When one considers how excessive the profit percentage of Middle Eastern oil would still have been at even a quarter of this price (for instance, a $9 selling price for a barrel of oil the production cost of which had risen to perhaps 25 cents after years of regional inflation and falling efficiency), it is clear that had any degree of free market competition been able to penetrate the production restrictions of the OPEC states, it would have called forth many millions of barrels a day in increased output from that locked-up capacity. That this did not happen shows the effectiveness of the OPEC monopoly, not because the cartel was strong—far from it—but because it was never challenged by the consuming West during its first decade. In those millions of daily barrels, which at a cheaper price could have been purchased to turn profitably the idled wheels of production but were not, would lie a basic cause of unemployment for millions in the industrial West and deepening destitution for hundreds of millions in the third world.

Although by 1974 the United States had abandoned its role as guardian of Western oil security, the sheer size of its resources and appetites made it the key, for good or ill, to the regaining by the industrial nations of some sort of handle on oil stability before the circumstances that could set off the next price explosion again meshed. In the Nixon era the leap in U.S. oil imports from 2.5 to 6 million barrels a day and the diplomatic encouragement given by Washington to the OPEC price hawks had been the basic destabilizing factors. For the remainder of the 1970's the question of oil stability turned on whether the United States could cut back its imports of OPEC oil or whether those imports would worsen from year to year, canceling out the conservation successes of the other industrial democracies and thwarting all efforts to reestablish an equilibrium between the power of consumers and exporters.

On the hopeful side were America's underexploited resources in oil, natural gas, and coal, its vast technological capacity that, if mobilized, could either restore its nuclear energy timetable* or transform its limitless supplies of shale rock or coal into millions of barrels a day of oil equivalent. Moreover, America's spectacular wastage of oil left a huge margin for conservation gains in a nation that burned up one out of every nine barrels of the world's oil as gasoline on its highways and where 6 percent of the world's population consumed more than 30 percent of the world's energy. On the downside was the reluctance thus far of American politicians to do anything decisive or unpleasant, an unwillingness either to set priorities between energy development and other national goals or to impose minor consumption disciplines for the years ahead.

It could be said with reasonable confidence that by the advent of the twenty-first century technological breakthroughs in any of a number of fields—nuclear fusion, the breeder reactor, solar energy, geothermal energy—would put humankind in comfortable sight of oil's obsolescence as a necessity and render the question of how many hundred

*The nuclear power program of France, for example, was being pushed so effectively in the 1970's that by 1982, 39 percent of France's electricity was nuclear-powered, and plans to raise that to 70 percent by 1990 were on schedule.

billions of barrels were left in the earth's crust and what would happen to them, in Professor Adelman's phrase, "not only irrelevant but uninteresting" (except to their owners). But whether or not the period between 1974 and then was to be a lost era of disintegration depended largely on America's political capacity to solve what was at bottom a problem of self-discipline.

In political affairs, catastrophe is the great teacher. The basics of what needed to be done were obvious enough and were recognized by the immediate successors to Richard Nixon—Gerald Ford and Jimmy Carter. To enact these lessons into law, however, two great obstacles would have to be overcome: the Nixon energy legacy and the stalwart defense of that legacy by the congressional Democrats.

The achievable goal, given the posture of nonconfrontation with OPEC in any form which the consuming West had adopted in the Nixon era, was not a dramatic rolling back of 1974 OPEC prices but a slow whittling away at them—through the gradual effects of reduced oil purchases that could keep the OPEC market loose enough so that oil prices henceforth would rise more slowly than creeping inflation. Meanwhile, routine commonsense security steps had to be taken to contain the effects of another supply interruption that otherwise would surely blow the price through the roof: the creation of a six-month stockpile, a standby rationing plan to stretch that stockpile, an improved oil-sharing scheme, and controls on panic buying of oil.

Slowly to cut back dependence on OPEC and chip away at its price through consumer restraint had the virtue of being doable within the passive mode the United States had imposed upon itself—doable, that is, if American politicians were up to imposing austerities openly now, which could be blamed on *them,* in order to forestall calamities later, which could be blamed on accidents or foreigners or the oil companies.

Here the Nixon inheritance cast its encompassing shadow. The turnaway from new domestic energy production toward foreign imports in the Nixon term saddled President Ford, and President Carter after him, with a trajectory that called for a 50 percent jump in oil imports in the immediate years after 1974. In the first three post-Nixon years, 1975 to 1977, oil imports would rise from 6 million to almost 9 million barrels a day. On Nixon's exit one-third of U.S. imported oil was Eastern Hemisphere oil; by Carter's first year its share would be two-thirds.

The nonconfrontationist way for the United States to avert this stupendous increase in jeopardy was to cease hiding from its consumers the extent of the defeat by OPEC and to cease shielding them from the full impact of OPEC's quadrupling of prices in 1973–74—by letting the price of oil in America rise to the world level. A sudden, staggering increase in the price of oil products would force on most consumers the necessity to cut down on oil use, through lowered thermostats, smaller cars, insulation, and switching to cheaper fuels, and would make it irresistible for industries—big energy users—to pursue to the hilt energy conservation and the technology of energy efficiency. And it would do more than anything else to stimulate increased exploration and production of domestic oil, by providing both the profit incentive and the investment capital.

Raising the price of domestic oil from $5 to $12 a barrel was not in any way an agreeable course. Radically higher oil prices would stoke inflation and end in reduced economic activity, which spelled hardship for millions. In a sense, it was ratifying OPEC's dirty work and making it permanent. But for a nation that rejected rationing, that pooh-poohed conservation, that would not resolve its standoff between environmental safeguards and the need for new energy, that distrusted the oil companies but had evolved no alternative to the profit incentive for finding oil, $12 oil *now* was the only defense left against $40 oil a few years hence, when continually deepening U.S. dependence on OPEC had to reduce Western bargaining power to zero. Cutting imports from OPEC by using less and producing more was bound to take a long time to restore the loose "buyers' market" that would force OPEC to cut its prices. Being a monopoly of sovereign states, OPEC could cut production as demand fell, thus preserving scarcity and high prices and leaving the consumers playing what for years was likely to look like a mug's game. But sooner or later, no matter how great their cash surpluses were in the mid-1970's, or how few their people to spend them on, OPEC governments were bound to go the way of *all* governments and begin spending more than they had, thus restoring their dependence on the Western consumer and reestablishing the basic workings of economic law.

President Ford, and President Carter after him, acknowledged this reality and seemed willing to risk the immediate political penalties that went with it. But here again the Nixon legacy, now guarded by a Democratic Congress, blocked the way. The reason that domestic oil prices had not automatically risen to the world level set by OPEC was that President Nixon had imposed price controls on oil in 1971 as part of a sweeping preelection price freeze and had kept them on oil after they had been generally lifted following the election, breeding in the consuming public the notion that it was entitled to protection from the world oil price. At the time of Nixon's exit two-thirds of the oil consumed in the United States was produced at home, and the price on most of it was frozen at less than half the world price. The impact of the OPEC price explosion thus reached the American consumer in greatly diluted form, the more so in the case of gasoline since a large element of its price at the pump was comprised of state and federal taxes and refining and distribution costs, which remained relatively stable. As a result, the OPEC price rise was a lesser impediment to oil consumption in the United States than it was in most industrial nations.

Through most of his first year in office Gerald Ford tried to get Congress to decontrol domestic oil prices, but he was rebuffed. Jimmy Carter, in his turn (and by 1977 oil imports had risen to almost half of consumption), tried to boost U.S. oil prices to the world level by means of a tax (so that the Treasury, not the oil companies, would reap the windfall), but once more the Nixon price control system prevailed. Carter's plan was rejected, and U.S. oil imports continued to rise in 1977 by a rate of 1 million barrels a day, negating the reductions painfully achieved in the rest of the industrial world.

Neither Ford nor Carter demonstrated the leadership skills and the single-mindedness necessary to prevail over the congressional traffickers in consumer politics, who could gain applause today for appearing to hold the price line and disappear into the woodwork tomorrow when the politics of palliatives and postponement blew the price of oil sky-high.

A more determined Ford could have won out since the oil price controls were due to expire in 1975, and as a last resort he could have let them expire and used his veto to prevent their reenactment. But to do so might have cost him the passage of a package of less controversial energy legislation which included long-range gasoline efficiency standards for autos and would have caused him to stand alone before the electorate as the culprit in the oil price rise; in the end Ford signed a bill extending oil price controls until 1981—and the imports from OPEC continued to surge.

Carter's 1977 effort to raise domestic oil prices by taxation instead of decontrol incited the opposition of Senator Russell Long and his oil state delegations, as well as of many proconsumer tribunes, and failed. "When you add up all the people against part of it," explained Representative Abner Mikva (D-Ill.), "you end up with a majority."

Though Carter accepted intellectually the necessity for allowing consumer oil prices to rise to the world level and espoused it in Washington speeches and at international conferences, his support of that goal was diverted and diluted by conflicting aims that assured a confused struggle ending in defeat. He wanted to make the price transition relatively painless by stretching it out over a period of years, instead of inflicting the sharp jolt that was required to turn around consumer habits. And he wanted to raise the price of the oilmen's oil, without raising their income from it, by the tortuous, messy process of taxation instead of the simple, direct one of decontrol. Timid gradualism cost him much of the intellectual force of his argument and all of its urgency and mocked his slogan "The moral equivalent of war." The taxation approach queered the "production incentive" half of the argument for world-level prices and made bitter-end foes out of the natural supporters of a price rise, and it consigned the energy program to that exitless labyrinth wherein oil state strength was strongest—Russell Long's Senate Finance Committee.

Even after the defeat of the tax approach, with all its waste of irretrievable time, President Carter continued to spurn the only other route, decontrol, though the reverberations of the second great price explosion were already audible. At the end of his second year in office, in late December 1978, a presidential energy spokesman announced, "We're not yet far enough along in the decision process to completely rule out decontrol, but it is not an attractive option because of the transfer of dollars from consumers to producers."

Carter did make an effort, in 1978, to decontrol prices gradually in the long-stultified natural gas industry, prices frozen at a small fraction of the cost of equivalent oil, and the effort ended up rather like Ford's attempt to decontrol oil prices—a snail's pace easing of controls that would not fully expire until distant 1985.

In their largely futile grapplings with the energy problem Ford and Carter, it must be said, were much the victims of their inheritance and of congressional paralysis. The governmental process of the United States had become mired in a rut of small-mindedness and chronic obstructionism in which it could no longer act on grave, urgent matters by enacting new corrective laws but only by providing that old culpable laws could finally expire in the next decade.

The bottom line was that as of early 1979—the sixth year since the economic Pearl Harbor of 1973—the United States was still unmobilized, still had taken no timely basic step to reverse the trend of annually deepening dependence on OPEC oil, and was therefore less able than ever to resist another price explosion that was always a clear and present danger, deferred only by the caution of an OPEC unsure of its horizons.

Since the 438 percent rise in oil prices in 1973–74, OPEC had experienced an understandable difficulty in gauging the true extent of its further opportunity: With oil prices already at the unreal height of 120 times the cost of production there was no system of market logic or economic law in command of the process. At this uncharted altitude there was no guide to how much more overpricing the consuming world would tolerate, or could absorb, except the guide furnished by its own responses to events—unplanned events. The OPEC of the 1970's was finding its way up the price ladder by Western reactions to accidents.

For example, not long before the Yom Kippur War, the shah of Iran was given the results of an exhaustive study by his oil technicians on the highest possible oil price that could be achieved and maintained. One of the architects of that study told me that the figure agreed on and reported to the shah as the optimum price was $7 a barrel, which they estimated to be the cost of the substitutes for Persian Gulf oil that could be developed by the West in sufficient quantity. But within a year the shah had learned from unexpected events—such as the $17 bid received at an Iranian oil auction—that the West would pay more than twice $7 if it were sufficiently panicked. Thereafter the shah's minimum price target was in the teens. Had the Western governments made precautions before 1973 for calmly riding out the sort of temporary, partial shortage a nervous OPEC had always restrained itself to, it is unlikely that oil prices would have risen above single digits, despite all the West's other mistakes. Could not this lesson be applied to the post-1973 scene? Since nothing significant was being done to reverse the long-term trend, was it not all the more urgent to take precautions against accidents or emergencies that could explode at any time?

Recent history was littered with warnings of the near certainty of another circumstantial oil interruption before the end of the 1970's—a 1950's-style Iranian shutdown, a 1956 Suez confrontation, a 1967 embargo, a 1970 Libyan-Syrian squeeze, a 1973 war eruption. From Western reaction to these periodic episodes, the oil regimes had historically adapted to what was revealed as possible: sometimes bowing to Western strength, sometimes capitalizing on Western disarray, sometimes fleecing Western panic. If during the next emergency a U.S.-led

West could respond calmly and unitedly and keep under firm control its penchant for self-revelatory fear, there was no reason why the $12 to $13 price of the middle 1970's should not last indefinitely, its real cost slowly eroded by the inflation it helped spawn, the more so since $12 oil had long ago stopped the growth of world demand for OPEC oil.

But if the United States and its allies were again caught unprepared and divided and permitted themselves a repeat of the 1973 oil-at-any price performance, nothing could keep the lid from being blown up onto a new plateau of extortion that would henceforth become the new base to defend, a plateau at a height that would otherwise never have been reached or dreamed of.

The lessons of 1973–74 directed U.S. policymakers to four obvious preparations: (1) oil stockpiles capacious enough for several months' substitution of imports; (2) a ready-to-go rationing program, able to reduce nonessential use and stave off hoarding and the insecurity that breeds it; (3) a tightened-up oil-sharing program among the industrial democracies so that none would be on its own; (4) a control system on price bidding for oil, to snuff out the panic buying that had such calamitous results in 1973.*

But when the inevitable interruption began in late 1978—in the form of an oil strike in Iran's Khuzistan Province, which heralded the fall of the shah and caused a brief, limited oil shortage in early 1979, not a single one of these elementary precautions was in place.

A stockpiling program, first urged upon Nixon in 1969, had won congressional authorization under Ford in 1975; but in the subsequent four years neither Ford's nor Carter's administration had produced the stockpile, and it would be 1983 before a reserve of middling consequence had been amassed.† There was no rationing program standing by when the Iranian stoppage struck the United States in early 1979, and one year later, on January 28, 1980, Hazel Rollins, administrator of the Energy Regulatory Administration, admitted, "It would take from seven to twenty-one months to get a stand-by gasoline rationing

*A number of bidding-control plans, the basic principle of which was set forth by the famed international oil consultant Walter Levy, were advanced during the 1970's: At home, importers of oil who exceeded the quotas set for emergencies by paying excessive prices would be required to resell their overage *at a loss.* Abroad, within the framework of the oil-sharing agreement, any member nation that exceeded its predetermined fair share of imports could have to sell that excess at a loss.

†Three hundred million barrels, about sixty days' worth of imports at early 1983 levels.

plan into operation if an emergency arose now." The holes in the foreign oil-sharing agreement of Kissinger's day were still gaping wide; consequently, insecurity inflamed the oil scene, and the stampede toward panic bidding that had climaxed the disaster of 1973 recurred in early 1979, for no controls had been established at home or abroad to prevent fear from creating its own reality. The Iranian shortfall was by no means unmanageable, the less so since more than half the loss was made up by increased pumping by other OPEC states. During its severe stage, the first quarter of 1979, total OPEC production was down only 2.2 million barrels a day from the fourth quarter of 1978 and was actually higher than for the corresponding first quarter of 1978. In the United States, where March, April, and May were the months of critical impact because of the several-week length of tanker voyages from the Persian Gulf, Americans consumed 17 million barrels a day, compared to 18.1 million during the same months of the previous year. Total U.S. oil imports for all 1979 were 8.4 million barrels a day, as against 8.2 million in 1978.

Yet the United States and its allies turned a mild shortfall into a crisis out of fear, as if they had lost the capacity to learn from past debacles.

Professor Dankwart Rustow has described how, with no element of any policy of damage control in place, the interruption of early 1979 was allowed to snowball into an epic disaster: "Anxious buyers and hopeful speculators, expecting that the shortage would send prices soaring, bought whatever oil they could get their hands on and paid whatever price was asked. The effect was that of a self-fulfilling prophecy. The frantic buying aggravated the shortage, speculation drove up prices and attracted additional speculators, and rising prices caused yet more panic. As the exiled shah made his way from Egypt to Morocco and Mexico, spot prices doubled and tripled."

OPEC's "marker crude"—Saudi light—$12.70 a barrel in late 1978, rose on the spot market to $16 in January, to $22.50 in February, to $35 in June, on its way to $41 in November—a rise of more than 300 percent.

In the real world of oil production things were back to normal by July, but in the world of fear—where President Carter and Energy Secretary James Schlesinger kept warning of growing shortages—"panic and hoarding each time quickly re-opened the gap that higher production was beginning to close." By the middle of 1980 so much excess oil had been bought and hoarded—much of it stored in idle tankers built for a world the oil trade of which was in the process of diminishing by half—that the excess inventory held twice as much oil as had been lost by the Iranian shutdown the year before.

Had that oil been in public inventory in 1979 as a result of government precaution, instead of in private hoards in 1980 as a result of fear, the Iranian "pause" would have caused scarcely a ripple in the world market. This would be demonstrated in the fall of 1980, when the outbreak of war between Iraq and Iran caused a 3.5-million-barrel-a-day loss of production with no serious effect on the market.

Finding out, again by accident, that the United States was still defenseless and that the West would tolerate $40 oil—would beg for it, in fact—OPEC members in short order hoisted their contract prices onto the mid-thirties to low-forties plateau, which now became the new norm, the floor to be defended. Some terrible enervation of common sense and ordinary resolve in Washington had permitted a containable fluctuation to become transmogrified into an incalculable disaster. Congressman Richard Bolling, retiring out of frustration from his powerful post as chairman of the House Rules Committee, told reporters, "Nothing has worked. Our policies are blowing up. We're factionalizing. We're unable to address ordinary dilemmas. I think it is absolutely predictable that something dreadful is going to happen unless we change."

Once the second oil crisis took off, only the boldest and luckiest leadership could have atoned for the squandering of the years between 1973 and 1979; having defaulted in the arena of strategic preparedness, only leaders using the most brilliant tactics could have avoided the consequences. But there was to be no leadership, and such tactics as there were only compounded the disaster.

An emergency Western "summit," an accord among the key consuming nations that forbade scrambling for oil, that fairly shared in the Iranian cutback, that put a unified lid on price offers, could have done the trick, given the providential smallness of the overall shortage. But there was no grand initiative, no following, no unity. In May 1979 oil authority Walter Levy described what in fact occurred:

> . . . We have been stymied by the fear that OPEC, or some of its members, will not welcome a joint policy of importing countries

> and will characterize such an approach as confrontation. But OPEC has never questioned its own right to act as a group on oil prices. . . . We have until now been afraid to cope effectively as a group with oil and related problems. Instead, importing countries have pursued their selfish interests to save their own skins by attempting to make special deals at the expense, if necessary, of all other consumers—even though such an approach will surely prove futile in the end.

The Carter administration proved no more masterful at the tactics of crisis management than it had at the strategy of crisis prevention. In the early weeks of the Iranian shutdown the White House seemed to see not disaster unfolding but a needed corrective of complacent public attitudes. Americans must "use less oil and pay more for it," Carter said primly, preaching yesterday's sermon when the need was for urgent action to head off a truly calamitous price increase. Pipe-smoking Energy Secretary James Schlesinger had for some time been bemoaning the lack of a crisis or "pinch" that would wake up the country and speed the adoption of his energy policies; as his deputy, John O'Leary, put it: "The pain to the consumer is not sufficiently severe." And so, as the pain increased, the White House seemed for a time to regard it as an ally that would vindicate its policies rather than as a potential bull on the rampage that would chase it to cover.

By March 1979, with spot market prices up from December's $13 to $23 a barrel, an awakened administration tried to turn back the tide by persuading the larger U.S. oil companies not to buy oil on the spot market and not to pay extra price "premiums" anywhere. Had this been enforceable on *all* U.S. purchasers, and synchronized with other Western governments, it would have been a salutary move. Had the administration proved staunch enough to persevere in it, even alone, when its effects reached the local gas station, it might still have been salutary, such was the weight of the American market in world scales. As it was, however—partial, uncoordinated, temporary—it was a debacle. The big American oil companies bowed to the Carter request, perhaps cynically, and dropped out of the spot and premium price markets, to which more and more OPEC oil was being diverted. Citing their patriotic compliance, they reduced supplies to the United States and cut them off altogether to many of their smaller customers, leaving them without any source of oil; predictably, as the alternative to going out of business, the stranded firms surged desperately, alongside some Europeans and Japanese, into the spot market, where even small purchases can be decisive in setting the pace.

The results of this maneuver were analyzed later in an internal White House memorandum, dated November 7, 1979, prepared for the President's "chief inflation fighter," Alfred E. Kahn, by his energy analyst, Terence O'Rourke:

> The bulk of foreign oil traded in international markets and imported into the United States is controlled by a handful of major international companies. Other companies buy all or most of their foreign oil from them. In recent months, as the world oil supplies became tight, these major companies reduced their third-party sales to other companies in order to meet their own needs and/or divert supplies to take advantage of high spot market prices. At the same time, they greatly expanded their mark-ups on remaining third-party sales. Their customers who were cut back were driven into very thin spot markets for oil where they bid up prices to extraordinary levels. They imported this oil at vastly inflated prices into the United States where it has had the double impact of driving up prices for both domestic crude oil and refined products.

By early May the spot price had risen to $25 a barrel, but the Europeans and Japanese had reined in their bidders (France, for example, put ceilings on import prices by controlling the prices of products made from them, while the United States, which controlled domestic prices, allowed the prices of imported oil to be passed right through); oil production had returned to precrisis levels; the spot price had stopped spurting; and market equilibrium seemed to have been restored. But by May Americans were feeling the loss of the purchases its great oil companies had forgone at the administration's request.

I was at the time enjoying access to the "for your eyes only" transcripts of meetings of the Carter Cabinet. At the April 23 meeting Secretary Schlesinger had assured the President that their decision to reduce purchases was having a beneficial effect, as was their decision to have refiners cut back on this summer's gasoline production in favor of heating oil production for next year's winter. "The Energy Department," assured Schlesinger triumphantly, perhaps with a thought to Carter's interest in the New Hampshire presidential primary, "will have enough oil available to meet New England's heating needs this coming winter."

But by the May 7 Cabinet session these decisions looked less brilliant. "The President opened the meeting by noting that his trip over the weekend to Iowa and California was useful and successful." Yet all was not quite roses. "The major issue on each stop was energy. . . . Iowans are very concerned about adequate fuel supplies for their agricultural economy. In California there was substantial anxiety by motorists over limited gasoline supplies."

Coincidentally or not, on or about May 7 the administration reversed its policy on the spot market and urged the big American oil companies to plunge back in, providing refiners a special $5 a barrel incentive to do so. The result was soon recorded in the journals devoted to oil affairs. On May 14 the *Petroleum Intelligence Weekly (PIW)* reported: "The . . . spot crude oil market is spinning totally out of control with prices now rapidly escalating toward the $30 a barrel mark—at premiums of $8 to $10 above official OPEC levels. This latest round of sweeping increases has been generated by the large scale entry of important U.S. refiners into the market as buyers." *Platt's Oilgram* on May 22 declared: "Prices of spot sales are now about $30 to $35 a barrel, almost double the price of term sales." On May 28 *PIW* found that: "The return of U.S. buyers to world markets in recent weeks sparked a dramatic rise in Europe's markets which in turn is spilling over to east of Suez markets." And on June 4, *PIW* reported on the diplomatic ramifications:

> Hopes that consumer countries will show solidarity in the face of oil supply shortages and rising prices now look illusory. The U.S. announcement of a $5 per barrel incentive to U.S. refiners to import distillates brought a swift and incensed reaction from European governments last week. They saw the move as a callous attempt to hog limited supplies. European Community Energy Commissioner Guido Brunner called in the U.S. Ambassador to the EEC to protest that the U.S. was "exporting its problems" only days after a ministerial meeting of the 20-nation International Energy Agency stressed the need for international cooperation.

During June thoroughly panicked American refiners sent OPEC a signal it could scarcely believe, but one clear enough to reveal the oil price horizon of the future. "The United States set a dismal new spot price record for the world's oil consuming nations during June," reported *PIW* on July 2. "A short-lived stampede on the U.S. Gulf Coast spot market drove refined product prices high enough to 'support' an OPEC market price of almost $41 a barrel for Arabian light crude." In the past OPEC price hawks had watched for these peak prices, "voluntarily" offered *in extremis,* as guides to the ultimate price levels they could seek to make permanent, settling for a price somewhat lower than that which had been offered in the depths of panic. By December 1979, contracts notwithstanding, the price of 20 percent of OPEC oil had risen to $40 a barrel. When "normalization" returned in 1980, the new plateau would be at an average of mid-thirties for Persian Gulf oil and low forties for African oil—300 percent higher than its level before the "Iranian pause."

Although the American public could not follow the intricate details, by late June 1979 it had grasped the results. President Carter was in Tokyo on June 28 attending an economic summit, when he received an alarum from his principal aide for domestic affairs, Stuart Eizenstat, that back home the sky was falling. Citing public wrath over gas lines and oil prices, Eizenstat warned, "Nothing which has occurred in the Administration to date . . . has so frustrated, confused, angered the American people—or so targeted their distress at you personally."

Prominent among the victims of the ensuing Cabinet shake-up was Energy Secretary James Schlesinger. In his farewell address on August 16, perhaps reflecting awareness that for all his strenuous efforts over two and a half years he had been the mouse in the OPEC cat-and-mouse game, Schlesinger made an ever-so-fleeting, half-disguised reference to the salient factor behind all of the West's oil distress: "The United States, after all, is currently producing as much oil as Saudi Arabia. But we are exploiting our own proven reserves about six times as rapidly."

This was a nonconfrontational way of saying that Saudi Arabia was limiting its oil production to one-sixth of what was physically feasible. At one-fourth of what was feasible, the price of oil would have been back in the single digits, and the economic progress of the West could have been rolling forward.

In 1981 OPEC let its oil production fall from the 31-million-barrel-a-day level of the 1970's to 24 million, to support the new price plateau. When that proved insufficient, OPEC lowered its output to below 17 million barrels in 1982 and to 14 million in early 1983—20 million barrels a day under its productive capacity, which in turn had been purposely stunted by another 20 million barrels a day below the planned output scheduled by the oil companies in the early 1970's. By early 1983 the Saudis were at one-twelfth the production rate of the United States relative to reserves, the oil price had dropped several dollars but was still holding near $30 a barrel, the 1984 unemployment projection for the industrial democracies was 35 million, and the question was: How much more of this could either side stand for the sake of $30 oil?

The failure of government either to avert the two oil price plagues through preventive precautions or to ameliorate their onslaught by apt diagnosis and timely treatment left the Western world somewhat in the predicament of a stricken person dwelling in the prescientific age: The victim must simply suffer the full ravages of the disease, his recovery depending on how much punishment his system could endure, how vigorously its unaided antibodies could rise to the occasion, and how strong and durable was the disease itself.

As the West grew ever weaker and more idle, it was able to consume less and less OPEC oil, and its defensive mechanisms learned to stretch oil further; as the high-price virus depressed the appetite, it activated the reactive search for substitutes, especially for non-OPEC oil, that could be relied on, slowly reversing the balance between OPEC and non-OPEC energy production. In combination the two processes had so reduced the consumption level of cartel oil by the early 1980's that even at a profit of 3,000 percent, OPEC income was getting too low to support its regimes in the manner to which they had so soon grown accustomed. How would the cartel respond to the first great challenge to its self-discipline, a challenge occasioned not by Western policy but by its own profligacy and overreaching?

The second OPEC price explosion accelerated the trends set in motion by the first, the more so because in the United States the political placebos which had previously masked the impact had lost their capacity to defer reality. It was not just that consumers were forced to retrench. Improved energy efficiency in everything from cars and planes to factories and light bulbs, caused basically by high fuel prices and the technology they inspired but with help from legislation mandating improved auto mileage, were by 1982 causing energy use to grow only half as much as economic output, a dramatic turnaround. In 1980 the United States consumed 8 percent less oil than in 1979, and the decline was just beginning. Between March 1979 and April 1981 U.S. gasoline consumption dropped 16.5 percent. With the United States no longer bucking the trend, the free world recorded a 14 percent drop in oil use in 1980–81. As the recession deepened into the worst numbers in several decades, the decline in oil use deepened apace. In 1982 oil use in the United States dropped by 11 percent and the consumption of residual oil by American industry dropped 41 percent! Overall, oil demand in the non-Communist world shrank from 52.4 million barrels a day in 1979 to 45.5 million in 1982.

Meanwhile, enforced higher prices were stimulating the oil production growth in the West that government policy had not accomplished. Non-OPEC output rose from 15 million barrels a day in 1977 to 22 million in 1982.*

How long would OPEC persist, as it watched its customers sinking deeper and deeper into economic sickness and resistance to its oil, in enforcing an oil price the world could not afford? How long *could* OPEC persist?

By 1983 it had persevered, after its fashion, for three years in supporting its mid-thirties price while its sales were dropping from 31 million barrels a day in 1979 to about 14 million during early 1983. This was a cut much steeper even than the industrial world's cut in oil consumption, for OPEC oil was not only being done without but being replaced. In 1979 OPEC had supplied the West with two out of every three barrels it used; by early 1983 the figure was less than one out of two, and falling. Now OPEC's resolve began to crack here and there as cheaters shaded prices a few dollars in order to raise exports by a total of about 2 million barrels. Was it possible that the breaking point of OPEC unity had been staggered into, possible that OPEC, which in 1970 had had an oil income of $7 billion, and in 1974 of $72 billion, and in 1980 of close to $300 billion, which in 1980 had enjoyed a leftover trade surplus with the West of $109 billion after buying all its hearts desired, was spending its income so wildly, had become so addicted, so rapidly, to that level of spending that it was on the verge of restoring its old dependence on Western gold?

There were signs that this was so. OPEC's $109 billion trade surplus of 1980 plummeted to an $18 billion *deficit* for 1982. Iran, Libya,

*In the "contiguous forty-eight" American states (a formulation designed to exclude Alaskan oil) higher prices and the final petering out of government price controls ended a decade of 4 percent a year production decline—a yearly slippage of between 300,000 and 400,000 barrels a day—and sparked slight increases over the previous year in 1980 and 1981.

Nigeria, Indonesia, and Venezuela, anxious for more revenues, all cut their prices from $2 to $6 a barrel below the OPEC floor so as to raise their exports above the OPEC ceiling, cheating at both ends in their anxiety to raise revenues. Historically this is the point where monopolies start to fall apart.

Up until early 1983 the Saudis and their neighboring satellites—Kuwait and the United Arab Emirates—kept cutting their own production (rather than cutting prices to be competitive) in order to offset both the cheating on output and the ongoing decline in Western purchases, until after eighteen months these cutbacks had reached a level of 10 million barrels a day. Thus the Saudi bloc was holding the price line at about $30 a barrel, but could it go on indefinitely losing revenues of this proportion?

If any of the OPEC states could long stand the gaff, it would be Saudi Arabia, Kuwait, and the UAE, with their bloated treasuries and investment reserves and their tiny populations to care for. But in late January 1983 the United Arab Emirates oil minister Mani Said al-Otaiba announced a "decision to boost output by 45 percent to 1.6 million barrels a day," explaining that the UAE had a $1.2 billion deficit in its current budget and that the next year's shortfall would double if the current daily production rate of 1.1 million barrels were maintained. Kuwait's budget was also running in the red, Otaiba revealed. The Saudis, who by 1981 had managed to raise their expenditures to $96 billion a year while enjoying oil revenues of $120 billion, faced the prospect of making do with oil revenues one-half their annual budget; that meant dipping into investment reserves and making politically difficult cuts—or cutting oil prices sharply to restore lost exports and revenues.

Even the richest and most sparsely populated among OPEC members, then, despite revenues that had multiplied 100 times in ten years, had joined the goodly company of deficit spenders, raising the hope that the industrial democracies might regain some measure of control not through their own performance but through the incredible carelessness and profligacy of their exploiters.

How apt the Reagan administration would prove at rising to the opportunities remained clouded in doubt. By its third year it was certainly not pursuing a policy aimed at actively exploiting the chance for a steep decline in oil prices. Pressure groups that had a vested interest in the high-price status quo—oil-producing countries, major oil companies, the international banks, and the multinational corporations, all afraid of the impact on themselves of either falling oil prices or falling credit-worthiness among oil countries—mobilized to keep oil prices from falling below the $29 benchmark price OPEC leaders were seeking to establish as their final fallback line, and they appeared to have immobilized the administration. "U.S. Seems to Straddle Issue of World Oil Price," capsuled *The New York Times,* on January 15, 1983, adding, "The United States Government is expressing increasing anxiety as energy conservation and recession bring a continuing decline in oil prices and talk of a precipitous drop this spring." Its degree of involvement was measurable by the fact that the Department of Energy post of assistant secretary for international affairs remained vacant for several months. When in March the OPEC governments pulled themselves together, at least temporarily, in defense of the $29-$30 price floor, and solemnly pledged to keep down oil output to the level necessary to sustain that floor, the reaction of Washington and London was not to attempt to undermine the blatantly monopolistic pact but rather to support it by declining to rock the boat.

President Reagan had begun with some forceful steps in the energy field. One of his first acts as President was to decontrol domestic oil prices in January 1981, eight months earlier than they would have finally expired under the Carter implementation of the Ford compromise legislation of 1975. He moved aggressively into oil development offshore and on federal lands. And his new team took up the old task of filling the Strategic Petroleum Reserve with a sense of drive (and at a cost about thirty-six times higher than that proposed to Mr. Nixon by the shah of Iran in 1969).

But in early 1983 his budget called for cutting back the filling rate of the Strategic Petroleum Reserve by more than 50 percent, postponing eventual completion until the 1990's. Oil companies and industries were cutting *their* private stockpiles way down, a reprise of the error made just before the 1979 price explosion. Reagan's Interior Secretary James Watt had repeated the mistake of the early Hickel by causing such a furor among environmentalists over his oil exploration plans as to provoke massive opposition that after two years was beating him back on project after project. Reagan had waited until his third year to propose amelioration of the natural gas regulatory muddle that by then had only two more years to run anyway. The Alaskan natural gas pipeline project was foundering as badly as had its famous predecessor for so many years. Nothing of consequence had been accomplished to rescue coal or nuclear power from their doldrums or to subsidize conservation programs, such as home insulation, against the day when falling oil prices and economic recovery might spark the next surge in oil consumption. And Reagan had slowed down and miniaturized the Carter program for the subsidized development of synthetic fuels, for which the windfall profit tax was passed in 1980. Moreover, the Reagan administration, because of its antigovernment regulation ideology, opposed most of the steps needed for damage control during unforeseen emergencies—rationing authority, allocation planning, tight oil-sharing accords abroad, controls on panic bidding. Thus under Reagan the United States was in many ways as unprepared to cope with a sudden interruption in oil supplies as it had been under Nixon, Ford, and Carter.

As the oil cartel suffered through the winter of discontent that comes at some point to all monopolies, as the Saudi bloc deliberated how much longer it could carry the burden of supporting the oil price by suppressing its own production to a small fraction of its capacity, as the other members vacillated between the joys and dangers of cheating on price floors and output ceilings, the OPEC governments were looking for their deliverance to another panicked response, to another "accident," another round of appeasement, another bailout by the West that might hold the line and set the stage for the next plateau.

Notes

INTRODUCTION

1. p. ix "... thirtyfold leap in oil prices...": If "thirtyfold" surprises the reader, it is probably because most journalists and commentators have typically used less alarming comparisons—"tenfold" or "twelve times" or "fourteen times"—to describe the leap in the world oil price—i.e., the FOB price charged at Persian Gulf ports. For example, Alan Greenspan, in an excellent piece in the *Wall Street Journal* (March 11, 1983), speaks of the "twelvefold increase in prices between late '73 and mid 1981." This is due to their use, for their "floor" figure, of the price as it stood in 1973, or in 1972 or 1971, after the first doubling or tripling had already occurred, or because they used the "posted price" rather than the much lower actual price at which oil was being traded, whereas *our* floor price is the arm's length price oil was actually sold at in the first Nixon year, before Nixon administration interventions in oil pricing began to push it upward.

M. A. Adelman, in his monumental work *The World Petroleum Market* (Johns Hopkins University Press, 1972), establishes on pp. 183–91 the 1969 price of Persian Gulf crude at no more than $1.20 per barrel, and as low as $1 during some periods of that year. Robert Stobaugh and Daniel Yergin, in *Energy Future* (Random House, 1979), confirm Adelman's assessment (see Stobaugh and Yergin, "References," p. 276): "The price of oil in the Arab/Persian Gulf at the end of the 1960's was $1.00 to $1.20 per barrel." And that was a price still in the midst of a long decline, a decline all the more impressive when contrasted with the general price level. From 1960 to 1969 the world price of oil dropped by one-fifth, the fact that the general price level rose by one-third notwithstanding.

At the upper end, the "spot market" price for the benchmark oil, Saudi Arabian light, stood at just under $40 in late 1979 and again in late 1980 and early 1981. (See *Time* magazine's chart of figures taken from *Petroleum Intelligence Weekly,* March 7, 1983, p. 62.) The rest of OPEC oil was generally more expensive as the various Mediterranean and African crudes reached levels several dollars higher. The "official" or contract price for Saudi light followed the spot price upward (it was raised no less than nine times in the 1979–1981 period) until it stabilized at $34 in late 1981, while other OPEC crudes leveled off at $2 to $7 higher.

2. p. ix "... the industrial democracies ... ": Present and projected unemployment statistics came essentially from an analysis published by the Organization for Economic Development and Cooperation (OECD) as reported by the Washington *Post* (Hobart Rowen, "Worst Slump in 50 Years Stifles Global Economy," January 9, 1983). The estimates worsened slightly as 1983 progressed.

3. p. ix " . . . steel mills were operating at 31 percent of capacity . . . ": *The New York Times,* November 11, 1982. See also Joseph Kraft's syndicated column, January 11, 1983. Capacity utilization can vary considerably from month to month. The American Iron and Steel Institute reports that *for the entire year of 1982,* steelmakers used an average 48.4 percent of capacity, the lowest level in nearly forty-four years. Reported by Warren Brown, Washington *Post,* April 24, 1983.

4. p. ix "Automobile sales . . . lowest in twenty years": Warren Brown, Washington *Post,* January 6, 1983. Also, Motor Vehicle Manufacturers Association, reported by Washington *Post,* March 27, 1983. "Business profits had fallen. . . . Bankruptcies were at the highest level . . . growth rate . . . had fallen below zero." U.S. Department of Commerce figures, reported by Caroline Atkinson, Washington *Post,* November 21, 1982.

5. p. x The inflation rate, growth rate, and unemployment figures for Western Europe: From analysis of OECD data reported by William Drozdiak, Washington *Post,* January 1, 1983.

6. p. x " . . . crude oil rose from $1.20 a barrel to $41 . . .": See Introduction, Note 1.

7. p. xi " . . . hundreds of billions of dollars . . . drained out of the economy by the 'OPEC tax' ": "The cumulative transfer of income to the OPEC countries made possible by the price increases that began in 1973 exceeds $1 trillion." Daniel Yergin and Martin Hillenbrand, eds., *Global Insecurity* (Houghton Mifflin, 1982), p. 20.

8. p. xi " . . . 1973 marked a statistical watershed": A comparison of inflation, unemployment, and economic growth in the years before and after 1973 is given, up to 1981, in Yergin and Hillenbrand, *op. cit.,* p. 6, showing a post-1973 deterioration that in most respects has intensified since 1981.

9. p. xi "In 1982 . . . the volume of world trade actually contracted": Analysis of IMF and OECD reports by Washington *Post* economics writer Hobart Rowen, January 9, 1983.

10. p. xi Robert Strauss quotation: From an article by Strauss in the *Wall Street Journal,* January 24, 1982.

11. p. xi "Peter Drucker reminds us": From an op-ed piece by Drucker in the *Wall Street Journal,* January 26, 1982.

12. p. xii "Developing nations . . . are getting the lowest prices in thirty years . . .": From statement by the World Bank, reported by Judith Miller, *The New York Times,* January 16, 1982.

13. p. xii ". . . lowest growth rates in *several decades.* In Latin America . . ."

(inflation and unemployment rates): Report of United Nations Economic Commission for Latin America, reported by Jackson Diehl, Washington *Post,* January 9, 1983.

14. p. xii *Time* magazine quote on the role of oil prices in international debt: From the issue of January 10, 1983, p. 45.

15. pp. xii–xiii "As Henry Kissinger told us . . .": The two Kissinger quotes are from his memoirs, *Years of Upheaval* (Little, Brown, 1982), pp. 858, 859.

16. p. xiii ". . . a minimum 20 million barrels a day in unused production capacity . . .": The sustainable production *capacity* of the OPEC states in the early 1980's was 35.3 million barrels a day (*International Energy Statistical Review,* June 24, 1980, p. 3), opposed to actual production of only 22 million barrels a day in 1981, 18 million in 1982, and less than 14 million in early 1983, as production was successively slashed to prop up the price (*Time* magazine chart based on OECD figures, February 7, 1983, p. 43).

This 21-million-barrels-a-day spread between output and existing capacity in early 1983 is only half the story of withheld supply; the 35.3-million-barrels-a-day capacity figure is an estimate of what the OPEC governments *have permitted* to exist "rather than of what may be physically possible." Craufurd D. Goodwin, ed., *Energy Policy in Perspective* (Brookings Institution, 1981), p. 644.

In the first half of the 1970's, when OPEC production capacity reached 38 million barrels a day and actual output reached almost 33 million (U.S. Senate Report of Hearings Before the Subcommittee on Multinational Corporations, January 2, 1975; *Fiasco,* Chapter Twelve, Note 4), the routine development plans and schedules under which the international oil companies were proceeding called for an early 1980's OPEC production capacity of well over 50 million barrels a day, a figure which already-found Middle Eastern reserves could sustain by themselves for decades. Saudi Arabia alone was scheduled for a 12-million-barrels-a-day increase, from 8 to 20 million, by 1983 (J. B. Kelly, *Arabia, the Gulf, and the West* [Basic Books, 1980], p. 372). According to a 1974 analysis by Professor Richard B. Mancke, "Recent experience and engineering data both indicate that output from *any* of the large Persian Gulf sources could easily be doubled or even tripled within just a few years and with no appreciable rise in unit production costs" (Richard B. Mancke, *The Failure of U.S. Energy Policy* [Columbia University Press, 1974], p. 18).

What has happened in the past decade, however, has been the progressive scuttling of exploration and development in an attempt to maintain high oil prices based on an artificial scarcity and a steady retreat from the original output goals for the 1980's. By 1977 the Central Intelligence Agency estimate for mid-1980's production had narrowed to a level between 43 and 47 million barrels a day; by 1979 the CIA projection had fallen to 35 million. (The CIA's 1979 analysis "The World Oil Market in the Years Ahead" elaborates on these figures and furnishes country-by-country information on how oil development had been suppressed by official policy.)

Thus, when one encounters commentators who still describe the present oil situation in terms of the "forces of supply and demand," it is pertinent to recall that supply is being suppressed to less than half of capacity and that capacity itself has been suppressed to about three-fifths of what was planned.

17. p. xiii ". . . so as to maintain the price at 120 times the cost of production": Under the dominion of the oil companies, in the late 1960's and early 1970's the price of crude oil in the Middle East/North Africa region—$1 to $1.20 per barrel—was 10 to 12 times the average cost of production, which was established by Dr. M. A.

Adelman at about 10 cents a barrel (Adelman, *op. cit.,* p. 76, Table II-8). Adelman's work, Dr. R. C. Weisberg assures us, is "widely accepted" and "independently substantiated" (Richard C. Weisberg, *The Politics of Crude Oil Pricing in the Middle East 1970–1975* [Institute of International Studies, University of California-Berkeley, 1977], p. 16). The 10-cent production cost figure was still valid in 1974, according to Dr. Mancke (*op. cit.,* p. 18). Thus, when the cost of crude reached about $12 in 1974, it stood at 120 times the cost of production.

By 1980, when the spot price for Arabian light had reached $40, and the price of Mediterranean crudes a much higher level, the average production cost of Middle Eastern oil was 25 cents a barrel, according to the British historian J. B. Kelly (*op. cit.,* p. 427), which meant that Middle Eastern oil was selling for at least 160 times the cost of production. By 1982, when the crude price had settled into the mid-thirties range, the ratio was 140; the March 1983 average OPEC price of about $30 is 120 times the cost of production.

It is possible that since 1980 the production cost has risen above 25 cents a barrel (though ABC News, for one, has continued to be satisfied with that figure into 1983), in which case the 120 ratio would be a bit high. However, given the factors that (1) OPEC has cut its production in half since 1979 and is thus able to center on its cheapest-to-operate wells and that (2) OPEC members have starkly deemphasized development activities, which have traditionally constituted *half* of production costs, it is unlikely that there has been any valid increase in production costs of late.

18. p. xiii "... between 1950 and 1973 the world's oil reserves expanded eightfold while demand was expanding only fivefold": Yergin and Hillenbrand, *op. cit.,* p. 2.

19. p. xiii "As an American diplomat told *The New York Times* ...": Thomas L. Friedman, "After Lebanon: The Arab World in Crisis," *The New York Times,* January 22, 1982.

20. p. xiii The quotation from Eqbal Ahmad: From Ahmad's article in *The New York Times,* "On Arab Bankruptcy," August 10, 1982.

CHAPTER ONE

1. p. 5 "Too much in too many hands ...": At the end of 1967 the proved reserves of the free world totaled 371.7 billion barrels, as opposed to consumption in 1967 of 10.6 billion barrels (*Basic Petroleum Data Book* [*BPDB*], American Petroleum Institute, Vol. II, No. 3, September 1982, Section II, Table 1 [for January, 1968] and Section IV, Table 1).

This imbalance was the more striking in that "proved reserves" represent only a fraction of the oil that has been found by exploration—oil in place. "Proved reserves are that small part of oil in place which has been developed for production by the drilling and connecting of wells and associated facilities. It is the total of planned production from facilities already installed and paid for" (Adelman, *op. cit.,* pp. 26, 27). Moreover, proved reserves were growing each year at a rapid pace. Thus, five years hence, at the end of 1972, free world proved reserves stood at 566 billion, as against 1972 consumption of 15.3 billion barrels (*BPDB,* Section II, Table 1 and Section IV, Table 1).

The end of 1967 free world figures represented an "inventory" of thirty-five years' supply, about three times as large as the ten- to fifteen-year reserve the industry considers optimal (because the industry wants reserves large enough for efficient extraction and security of supply but not so large as to incur excessive maintenance costs for oil not to be sold for decades [Adelman, *op. cit.,* pp. 71, 72]. But the excess of reserves versus consumption so far as Middle Eastern/African oil was concerned was even more striking. There proved reserves were 291.5 billion barrels, against output of only 4.8 billion, a sixty-one-year inventory (*BPDB,* Section II, Table 1 and Section IV, Table 1).

Inordinate proved reserves caused the aspiring monopolists among the sheikhs and colonels of Islam to be vulnerable in the more immediate arena of production capacity: the sustainable output an oil company speedup could achieve in a short time. In 1967 the oil companies maintained a ready-to-pump "shut-in" capacity of 7 million barrels a day, 4 million of it in the United States (Weisberg, *op. cit.,* p. 72, figure 3). This "spare" equaled more than half the actual production of all the Middle Eastern and African producers put together (*BPDB,* Section IV, Table 1), enough to nullify even a durable supply shutoff by several oil states acting in concert—a feat the oil states had never been able to persevere in.

And the larger the time frame considered, the stronger the grip of the oil status quo. Out of the abundance of proved reserves, the oil companies could double and triple the output of any of their larger Islamic oil provinces within one to three years (Mancke, *op. cit.,* p. 18). Thus the dream of a collective uprising against the oil companies was futile so long as even a small minority of Middle Eastern/African concessions remained in the Western orbit.

2. p. 5 " . . . 13 million barrels of oil a day": *BPDB,* Section IV, Table 2, revised combined production figures for Middle East and Africa, 1967.

3. p. 6 "Did not Syria force a shutdown . . . while proclaiming a war against cartels . . . only to eat its holy vows": A thorough account of this episode is furnished by Christopher T. Rand, *Making Democracy Safe for Oil* (Little, Brown, 1975), pp. 88–93.

4. p. 6 ". . . in 1967 the obstreperous faction was driven from control, the terms of the company were accepted . . .": *Ibid.,* pp. 178–80, and Adelman, pp. 209–10.

5. p. 6 ". . . Nasserite[s] . . . halt oil production . . . by autumn the agitators were in prison and oil production was headed for another record year": See Joe Stork, *Middle East Oil and the Energy Crisis* (Monthly Review Press, 1975), pp. 114, 154, and Rand, *op. cit.,* Appendix 1, which show that in 1967 Libyan oil output rose 15.7 percent above 1966 and reached a new high of 1.7 million barrels a day.

6. p. 6 The rise and fall of the 1967 oil cutback and embargo are described in detail in Chapter Six of *Fiasco.*

7. p. 7 ". . . host government revenues had doubled . . .": *More than doubled,* in fact. Middle East production rose from 1.6 million barrels a day in 1958 to 3.6 in 1967 (*BPDB,* Section IV, Table 1) while host government "take" per barrel was rising from 69 cents to 88 cents (Weisberg, *op. cit.,* p. 20, Figure 2).

8. p. 7 "They accepted . . . the rationale that Too Much Oil meant that the price per barrel could not rise and that revenue gains must come mainly from increased production as demand rose . . .": "By 1967 OPEC had given up the attempt" to pursue revenue gains through forcing up the price per barrel by limiting production (Anthony Sampson, *The Seven Sisters* [Viking, 1975], p. 165). "Most countries saw higher revenue as a function of higher production" (Weisberg, *op. cit.,* p. 25). "For the conservative oil producing regimes, the goal was the establishment of stronger political and eco-

nomic links with the companies, along with a 'more equal' role in the partnership" (Stork, *op. cit.,* p. 101).

9. p. 7 "The radicals . . . blamed conservatives . . . for betraying all their dreams": For example, Syria refused to attend the Arab summit at Khartoum on that ground. "Syria publicly attacked the sheikhdoms, calling the oil there the 'rightful property of the Arab people as a whole' (Stork, *op. cit.,* p. 113). "For their part, the radicals . . . were crestfallen at the outcome of Khartoum. They complained that . . . the Khartoum signatories had weakened before the imperialist oil monopolies. . . . The conservatives, they maintained, were capitalists" (Rand, *op. cit.,* pp. 99–100).

10. p. 8 ". . . the radicals were everywhere in defeat by the end of 1967 and the conservatives in clear ascendancy": ". . . a political shift advantageous to pro-Western interests in the Arab world"; ". . . the Arab nationalist wave seemed . . . to be receding" (Rand, *op. cit.,* pp. 98, 153). The 1967 alignment represented a "shift in political power in the Arab world"; "the defeat of Egypt and Syria left Saudi Arabia in a dominant political position among the Arab states . . ." (Stork, *op. cit.,* pp. 116–17). Chapter Six of *Fiasco* enlarges on this theme.

11. p. 9 ". . . in violation of the commonly accepted norms . . .": "The nationalized Iranian oil industry lacked the legal right to sell abroad the oil produced from freely and willingly negotiated concessions made previously between government and company and recognized by customary international law" (Peter R. Odell, *Oil and World Power: A Geographical Interpretation* [Penguin Books, 1972], p. 73).

12. p. 9 "That defense was pacific but effective . . .": "The Anglo-Iranian Oil Company's threat to take legal action against any entity 'buying' and using its oil was sufficient in the circumstances to insure that over this period no Iranian oil moved on to world markets" (*Ibid.,* p. 73).

"During this period, Iran exported no oil at all. British Petroleum made effective use of its weapon, which was to intimidate foreign buyers, threatening them with law suits all over the Western World should they endeavor to lift any. And American firms, among most others, refused to lift Iranian oil" (Rand, *op. cit.,* p. 137).

13. p. 10 ". . . in 1938 Mexico nationalized the oil companies. . . . Before long Mexico . . . had dropped off the charts . . .": Odell, *op. cit.,* pp. 78–79.

14. p. 10 ". . . Mossadegh became dependent on the Tudeh Communists": ". . . he came to depend more and more on Tudeh support in the streets of Tehran" (Rand, *op. cit.,* p. 138); "The Communists clearly had become the best-organized and most-disciplined force in the country. They had infiltrated many government departments. . . . The American Embassy estimated their strength in Tehran at 8,000 to 10,000 activists, with an equal number in other cities and a great many nonmember supporters. The Shah was the only other political force with a strong political base in the country" (Barry Rubin, *Paved with Good Intentions: The American Experience in Iran* [Oxford University Press, 1980], p. 80); "Iran [is likely to] pass under a dictatorship which the Communists support as a precursor to their own tyranny" (Editorial, *The New York Times,* August 4, 1953).

15. p. 10 "British Petroleum . . . made up for most of the lost Iranian oil through increased liftings from its wells in Kuwait and Iraq": Rand, *op. cit.,* p. 141.

16. p. 10 "As for the American companies . . . they would just as soon have seen Iran's oil permanently gone from the market. . . . When asked by their government . . .

they at first resisted . . .": See Report of the Subcommittee on Multinational Corporations of the Committee on Foreign Relations, United States Senate, January 2, 1975 (hereinafter called Multinationals Subcommittee Report), pp. 102, 173, Note 2.

17. pp. 10–11 "Iran . . . had by 1953 sunk without leaving a ripple on the oil waters. . . . its disappearance was considered a general convenience": *Ibid.,* p. 2.

18. p. 10 " . . . not-so-covert aid of CIA agents . . . ": For a detailed account of the fall of Mossadegh and the restoration of the shah, see Rubin, *op. cit.,* pp. 79–90.

19. p. 11 "In terms of payoff . . .": For volumes, *BPDB,* September 1982, Section IV, Table 1. For prices and percentages, Rand, *op. cit.,* pp. 81–82; Weisberg, *op. cit.,* p. 20, Figure 2; and John Blair, *The Control of Oil* (Pantheon, 1976), p. 78.

20. pp. 12–13 "The shah created a propitious atmosphere . . . the new consortium explicitly and firmly retained those operating controls . . . deemed essential": See Multinationals Subcommittee Report, pp. 102, 173; Rand, *op. cit.,* p. 145, observes: "Iran's ability to influence the Consortium was never strong."

21. p. 14 ". . . from 1954 to 1967 Iran's growth rate exceeded Saudi Arabia's ten times . . .": Rand, *op. cit.,* Appendix I, a table compiled from Annual Reports, *Platt's Oilgram,* and *Petroleum Intelligence Weekly.*

22. p. 14 "The shah's annual oil revenues grew . . .": Rubin, *op. cit.,* p. 96; and Stork, *op. cit.,* p. 108.

23. p. 14 " . . . the [Kuwaiti] revenue spigot was promptly turned up": For 1965–1967, the years of the tiff, Kuwait's oil production growth rates were 2.2, 4.8, and 0.1 percent; after good relations had been restored in 1967, the growth rates for 1968–1971 were 5.6, 6.4, 6.2, and 7.0 percent (Rand, *op. cit.,* Appendix I).

24. p. 15 ". . . was changed to a pro-oil-company policy which frustrated collective action and soon diminished OPEC . . .": Stork, *op. cit.,* Chapter 5, establishes this in detail, concluding: "Saudi oil policy steadfastly supported the position of the international oil companies, while occasionally increasing the price of that support by way of financial demands on Aramco." (p. 111); "Saudi Arabia sabotaged all efforts to establish a joint Arab oil policy" (p. 112); and "OPEC's political credibility stood at near zero among the masses of the oil producing countries and the companies as well" (p. 98). Anthony Sampson, *op. cit.,* p. 176, says, "As for OPEC, the whole basis of its unanimity now seemed to be in ruins."

25. p. 15 ". . . jumped from 5 percent to 18 percent . . .": Rand, *op. cit.,* Appendix I, years 1964 and 1965.

26. p. 16 ". . . ceasing further development and minimizing production gains. The oil growth of Iraq was henceforth to be stunted . . .": The Multinationals Subcommittee Report, pp. 100–01, quotes State Department oil spokesman James Akins on the 1960's period: "The companies are not investing in Iraq now and have not for years. In fact, they are disinvesting. . . . There is no question of growth or new facilities."

27. p. 16 "Every significant scheme of the Iraqi government to lure in foreign developers was broken up [by U.S. State Department and the American oil companies]": See Multinationals Subcommittee Report, pp. 101–02, which describes how the State Department, at the behest of oil majors, "warned off" American independents and intervened to dissuade foreign companies as well.

28. p. 16 The production comparisons among Saudi Arabia, Iran, Kuwait, and Iraq in 1960 and 1967: From Rand, *op. cit.,* Appendix I.

29. p. 17 ". . . the convenient submersion of Iraq . . .": In an interview with Anthony Sampson, Exxon's Howard Page said, ". . . sometimes they make it easy to cut down by breaking an agreement, as in Iraq; then we could tell 'em to go to Hell" (Sampson, *op. cit.,* p. 168).

30. p. 17 The figures for Middle East oil reserves and production and years of supply: Taken from, or calculated from, *BPDB* September 1982, Section II, Table 1 and Section IV, Table 2.

31. p. 17 "Only 10 cents a barrel . . .": See Introduction, Note 17.

32. p. 18 ". . . promptly activatable production capacity far greater than current demand . . .": See this chapter, Note 1.

33. p. 18 The per barrel prices, profit shares, and tax takes given at the top of page 18: From Adelman, *op. cit.,* pp. 183–91, and Weisberg, *op. cit.,* p. 20, Figure 2.

34. p. 19 ". . . the unfolding plans of the great oil companies to reduce overconcentration. . .": See Odell, *op. cit.,* pp. 105–06; and Sampson, *op. cit.,* pp. 177–80, who describes, among other planned diversifications away from OPEC, the plan of Exxon alone for "a massive programme for exploration. It was to cost $700 million in the three years from 1964 and it concentrated on territories outside of OPEC."

35. p. 19 ". . . routine development of fields already operating . . . would enhance present proved reserves by about 60 percent over fifteen years . . .": See M. A. Adelman, "Is the Oil Shortage Real?" *Foreign Policy* (Winter, 1972–73), p. 74, Table 1.

36. p. 20 The Anthony Sampson quote: From Sampson, *op. cit.,* pp. 185–86.

37. p. 21 ". . . occasionally toppling or preserving a government . . . withdrawing income from one and adding to another . . .": "Any country that made too many difficulties for the oil companies provided a useful excuse to cut back. . . . Iraq was held out as a new warning to the others, like Iran under Mossadegh, to 'cooperate or else' " (*Ibid.,* p. 168).

38. p. 21 ". . . the oil Sisters had larger revenues than . . . concessionary states, more tonnage . . . than a superpower's navy . . .": *Ibid.,* p. 6.

39. p. 22 "To provide *too little* oil . . . would provoke political interference . . .": "The international oil companies, acting individually, were not in a position [to stagnate production and hike prices] because of their need to remain competitive with each other. Had they tried to do so collectively, or even given the appearance of such action, they would have run into major political and legal trouble in their home and other consuming countries" (John H. Lichtblau, *World Oil—How We Got Here—And Where We Are Going* [Petroleum Industry Research Foundation, Inc., 1979]).

40. p. 23 ". . . their profits would remain steady in the aggregate, even as they declined per barrel": Joe Stork, an unsympathetic chronicler of the oil industry, records (*op. cit.,* p. 120) that total profits for oil companies operating in the Middle East "remained just about the same [between 1963 and 1969] while output more than doubled from 2.6 billion barrels in 1963 to 5.5 billion barrels in 1969."

41. p. 23 ". . . the oil-endowed states . . . all demanded higher and higher oil

production quotas . . .": Howard Page, elder statesman of Exxon, testified: "The Iranian government was as every other government . . . always trying to get more . . . than their agreement called for" (Multinationals Subcommittee Report, p. 105).

42. p. 23 "The Seven Sisters had conspired in every way this side of prison gates to keep out the independents . . .": "My job was to keep 'em out of jail" (John J. McCloy, lawyer to the oil majors, quoted in Sampson, *op. cit.,* p. 166). The case of the oil majors against the independents is elaborated on in Chapter Eight of *Fiasco.*

43. p. 23 ". . . owned exclusive oil rights to *all* the acreage in an entire country . . .": "Aramco has an area [in Saudi Arabia] the size of Texas, Oklahoma, Arkansas and Louisiana thrown together" (Richard Funkhouser, Multinationals Subcommittee Hearings, Part 7, p. 169).

44. p. 25 ". . . every postwar President had made difficult interventions at critical junctures": Truman's role in quashing the antitrust prosecution, on national security grounds, was testified to by the Justice Department official in charge of the prosecution, Leonard J. Emmerglick (Multinationals Subcommittee Hearings, Part 7, pp. 102, 103). The history of the Truman administration's foreign tax credit is recounted by John Blair, *op. cit.,* pp. 195–203. For Eisenhower's refusal to bail out Mossadegh unless Iran resolved its conflict with British Petroleum, see Rubin, *op. cit.,* pp. 81, 87. The Oil aspects of Eisenhower's intervention in Lebanon are discussed by Stork, *op. cit.,* pp. 81–82. Kennedy's efforts to thwart the penetration by Soviet oil into the Western sphere are recounted in the preface to Charles H. Breecher, *Oil—the Big Ripoff* (Carlton Press, 1975). For Johnson and Peru, see Sampson, *op. cit.,* pp. 191–92. The Johnson administration's repeated backstoppings of the international oil companies regarding the shah are detailed in Chapter Nine of *Fiasco.*

CHAPTER TWO

1. p. 29 ". . . spare capacity of almost 4 million barrels a day . . .": Weisberg, *op. cit.,* p. 72, Figure 3.

2. p. 30 ". . . which kept domestic oil prices 40 percent higher than the world market . . .": Douglas R. Bohi and Milton Russell, *Limiting Oil Imports* (Johns Hopkins University Press, 1978), pp. 284–85.

3. p. 30 ". . . tax write-offs . . . to one-sixth the stated rate . . .": Blair, *op. cit.,* p. 187, says that "in the face of a corporate tax rate of 48 percent, the federal income taxes paid in 1974 by the nineteen largest oil companies amounted to only 7.6 percent of their income before taxes." For the *very* largest oil companies, the effective U.S. tax rate was even lower. Multinational Subcommittee Hearings, Part 4, p. 104, reveals the following average tax rates from 1962 to 1971: Exxon, 7.3 percent; Texaco, 2.6 percent; Mobil, 6.1 percent; Gulf, 4.7 percent; and Socal, 2.7 percent.

4. p. 31 ". . . about $1.5 billion each year . . .": Because of the differing combinations of tax write-offs that could be used, estimates vary on this point. An excellent article by Murray Seeger in the Los Angeles *Times* on December 29, 1968, points out that U.S. Treasury studies put the 1968 depletion allowance subsidy at $2.5 billion. The Joint Congressional Committee on Internal Revenue set it at $1.5 billion. The CONSAD research group arrived at $1.2 billion per year for *the mid-1960's* (*The Economic Factors Affecting the Level of Domestic Petroleum Reserves* [CONSAD Research

Group, 1969], p. 83). Expert testimony at the Multinationals Subcommittee Hearings (Part 4, p. 19), placed the combined *1972* tax loss from depletion plus expensing of cost allowances at $2.35 billion (after the 1969 reduction in the depletion allowance).

5. p. 31 "... Nixon ... promising oil millionaire John Shaheen": This promise was duly reported in "The Washington Merry-Go-Round" on May 7, 1969.

6. pp. 31–32 "... as Nixon analyzed it, *he* ran out of television money and Kennedy didn't": William Safire, *Before the Fall: An Inside View of the Pre-Watergate White House* (Doubleday, 1975), pp. 57, 80.

7. p. 32 "... the Chicago miasma of division and ineffectuality ...": Hubert Humphrey to Theodore White: "I felt when we left the convention we were in an impossible situation. . . . Chicago was a catastrophe. My wife and I went home heartbroken, battered and beaten. I told her I felt just like we had been in a shipwreck" (Theodore H. White, *The Making of the President 1968* [Atheneum, 1969], p. 303).

8. p. 32 "There was no up-front Texas money, usually a factor in Democratic campaigns ...": "There simply was no money ... there was no Texas money, as there usually is in a Democratic campaign; there was no "smart" money from the great corporations and operators who usually bet on both sides, if both sides have a chance" (*Ibid.*, pp. 339–40).

9. p. 33 "... Eugene McCarthy ... had picked up $40,000 in donations": David H. Davis, *Energy Politics* (St. Martin's Press, 1974), p. 65.

10. p. 33 "... Humphrey's fund raisers ... flew to Houston ... to beard the oil moneymen for a quick loan of $1 million": This incident is recounted by Jeno Paulucci, the head of the Humphrey mission, in an interview with Murray Seeger of the Los Angeles *Times,* February 16, 1969.

11. p. 33 "... the Houston oilmen ... [asked] one blunt question: What was Humphrey prepared to do about the oil depletion allowance?": *Ibid.*

12. p. 35 Humphrey's gradually diminishing role in the oil depletion struggle was keenly apparent to this reporter, who missed the cut and thrust of Humphrey's earlier *tours de force.* It is alluded to in the autobiography of Senator Paul Douglas, *In the Fullness of Time* (Harcourt Brace Jovanovich, 1972), p. 429.

13. p. 35 "... Nixon would state the rule ...": Safire, *op. cit.,* p. 509.

14. p. 36 "Some of Humphrey's seminars were held in secret ...": Douglas, *op. cit.,* pp. 423–35.

15. p. 37 "The percentage depletion allowance was Oil's creative leap ...": Excellent descriptions of the oil depletion allowance, how it originated and how it worked, are given by Mancke, *op. cit.,* pp. 77–87; and Blair, *op. cit.,* pp. 189–92.

16. p. 37 "Oil production ... was being held down to one-half and sometimes one-third ... by state-federal prorationing regulation ...": Bohi and Russell, *op. cit.,* p. 31, Table 2–4.

17. p. 38 "... *no* congressman ... could hope to gain appointment ...": The exclusion of antidepletion congressmen from the House Ways and Means Committee is discussed in a "Special Report," "The Year Oil Gets Its Lumps," *Business Week,* May 17, 1969.

18. p. 39 "... for years thereafter Douglas was an isolated pariah . . .": Douglas, *op. cit.,* pp. 204–05. Another antidepletion senator to get on the Finance Committee, William Proxmire (D-Wis.) encountered the same obstacle. " 'Just let a Senator try to get on the Finance Committee if he is a dedicated and outspoken critic of the oil industry,' Proxmire said after he was repeatedly passed over for assignment to the committee. 'I was told confidentially by Senators from oil states that my oil depletion views was the reason. They said this was a fact of life in the Senate. This is why the Senate has not acted on the oil depletion question' " (Los Angeles *Times*, December 29, 1968).

19. p. 40 "... 27½ percent depletion . . . had ceased to have any significant effect on domestic oil exploration": See *Fiasco,* p. 88.

20. p. 42 ". . . he had begun to feel guilty about exposing his comrades to the retributions of the oil lobby . . .": Douglas's change of tactics to spare his Senate allies was explained to the author by a key Douglas aide of those years, Kenneth Gray.

21. pp. 42–43 The Frank Moss and Eugene McCarthy incidents are from Douglas, *op. cit.,* pp. 211, 433.

22. p. 43 The Evans and Novak quote: From Rowland D. Evans and Robert D. Novak, *Nixon in the White House: The Frustrations of Power* (Vintage Books, 1972), p. 214.

23. p. 44 Paulucci's conversations with the oilmen and with Humphrey are taken from his account, given to Murray Seeger, Los Angeles *Times,* February 16, 1969.

24. pp. 45–46 The several ways in which the Humphrey campaign was crippled for lack of financing are detailed by White, *op. cit.,* pp. 340–41, 354, 356.

25. pp. 45–46 The comparisons between Nixon and Humphrey campaign outlays come from *ibid.,* pp. 329, 357, Note 2.

26. p. 46 The comparisons between Oil's financial contributions to Nixon and Humphrey are taken from Professor Bruce Ian Oppenheimer, *Oil and the Congressional Process* (Lexington Books, 1973), p. 40.

CHAPTER THREE

1. p. 49 "... the combined recoverable energy resources of the United States were greater . . . than any conceivable demand . . .": This claim assumes that our vast surplus of coal will be suitably exploited, that the oil shale resource will be developed if needed, and that the great reserves of natural gas will be pursued with vigor; but it does not rely on strenuous development of nuclear, solar, or other resources. As for the conditional sufficiency of the individual fuels, see the conclusion of the landmark 1972 report of the National Petroleum Council: "No major source of U.S. fuel supplies is limited by the availability of resources to sustain higher production."

2. p. 50 ". . . the world's greatest single producer of crude oil . . . capable of producing just about as much as it used": This U.S. place in oil production is shown in *Energy Perspectives 2* (U.S. Department of the Interior, June 1976), p. 33. U.S. 1968 capacity to produce as much oil as it consumed is demonstrated by total import figures

26. p. 56 ". . . gasoline consumption . . . was lately accelerating at twice that level . . .": Darmstadter and Landsberg, *op. cit.,* p. 28.

27. p. 56 ". . . less than half the oil that was pumpable . . . was being pumped": Bohi and Russell, *op. cit.,* p. 31, Table 2-4.

28. p. 56 "This had the side effect of retarding oil exploration . . .": "In 1963, for example, the large, efficient, low-cost fields in Texas were cut back to under 30 percent. . . . The result was a dampening of profit incentives for further exploration and development" (James McKie, "The United States," in Vernon, *op. cit.,* p. 74).

29. p. 56 ". . . to deduct a dollar from their U.S. taxes . . .": Blair, *op. cit.,* pp. 193–94.

30. p. 59 " 'the highest return to the government . . .' ": John C. Whitaker, *Striking a Balance* (American Enterprise Institute-Hoover Institution Policy Studies, 1976), p. 266.

31. p. 60 ". . . as OPEC held most of its price gains by halving its oil production against a background of . . . unemployed . . .": "There is still an oil gap; only now it is expressed in the number of unemployed" (Ulf Lantzke, executive director of the International Energy Agency, "Energy Vulnerability and the Industrial World," paper presented at twentieth anniversary meeting of the Atlantic Institute, October 1981, p. 9).

32. p. 64 "My partner, Drew Pearson, now led the attack on Hickel . . .": Hickel later wrote in his memoirs: "The Pearson column, initially at least, did more than anything else to generate opposition to me. Each time Pearson (or Anderson) launched an anti-Hickel balloon, its language would be picked up and repeated by other journalists and by members of Congress . . ." (Walter J. Hickel, *Who Owns America* [Prentice-Hall, 1971], p. 14).

33. p. 64 "An unprecedented volume of wires and letters . . . poured in . . .": *Ibid.,* p. 14. Also, Chairman Henry M. Jackson, Committee on Interior and Insular Affairs, on the Confirmation of Governor Walter Hickel to be Secretary of the Interior, January 15, 1969, p. 3.

34. pp. 64–65 The stories and speculations on how Nixon came to appoint Hickel are based on the accounts in (1) Safire's memoirs of his years in the Nixon White House, *Before the Fall, loc. cit.,* p. 192; (2) Hickel's autobiographical recollections of his early meetings with Nixon, *Who Owns America, loc. cit.,* pp. 7–8; and (3) John Ehrlichman's assessment, *Witness to Power: The Nixon Years* (Simon & Schuster, 1982), pp. 97–101.

35. pp. 65–70 The episodes and excerpts from Hickel's confirmation struggle are taken from Hearings on the Nomination of Governor Walter J. Hickel of Alaska to Be Secretary of the Interior, Committee on Interior and Insular Affairs, U.S. Senate, Ninety-first Congress, First Session, January 15–18, 20, 1969.

36. pp. 69–70 "Thoroughly whipped by now, Hickel . . . abjectly submitted . . . the Nixon administration had lost the authority to act decisively . . .": This judgment of the authors seems inescapable and is supported by an aide who assisted Hickel through his ordeal—James Watt, later himself to be secretary of the interior: "I'll get personal on this. I was an aide to Governor Wally Hickel of Alaska in 1968. Nixon selected the governor to be his secretary of interior, and I brought him through the hearings. He was subjected to such a *brutal,* harsh, ugly attack that it changed his

whole personality and style of behavior. . . . He raced to the left to prove to the environmentalists that he was one of them. He ran so far to the left that Nixon had to fire him. . . .

"Here we use the phrase, 'Don't Hickel-ize yourself.' We came in determined not to let committees of Congress take away our Constitutional duties . . ." (Interview given by Secretary Watt to James Conaway, the Washington *Post,* April 27, 1983).

37. p. 71 The Whitaker quote: From Whitaker, *op. cit.,* p. 269.

38. p. 71 ". . . mid-transformation to a new identity . . .": See this chapter, Note 36.

39. p. 71 "The claims of vast and permanent damage did not square with the accumulating evidence": Whitaker, *op. cit.,* pp. 268–69.

40. p. 72 The Whitaker quote: *Ibid.,* p. 272.

41. p. 72 ". . . the administration rejected pleas [for] environmental assessment of the frontier areas of the Atlantic . . . erased the offshore oil option . . . for the decisive period": *Ibid.,* p. 334.

42. p. 73 The account of the May 1981 lease auction, the initial oil strikes thereafter, and the reactions of Beinicke, Zierold, and Schuyler are from *Time* magazine, November 29, 1982; *The New York Times,* August 10 and November 21, 1982; and the *Wall Street Journal,* June 25, 1982.

CHAPTER FOUR

1. pp. 77–78 The quotes of oil executives are taken from a *Wall Street Journal* roundup of oil industry opinion, James Tanner and Norman Pearlstine, "Oilmen Expect Better Treatment by Nixon Than They Got from His Predecessors," November 20, 1968, p. 36.

2. p. 78 ". . . Johnson Treasury Department . . . had offered as ransom . . .": Oppenheimer, *op. cit.,* p. 106, quotes a former Treasury official as saying, "The real pressure [for depletion repeal] came in 1967 when Treasury had to commit itself to further tax reform in 1969 in order to get Congress to go along with the surtax."

3. p. 79 ". . . politically explosive application . . . build a refinery . . . at Machiasport . . . 10 percent below the going price . . .": "The Year Oil Gets Its Lumps," *loc. cit.,* p. 99.

4. p. 79 ". . . Mills came away from the Pierre satisfied . . . no Nixon interest whatever in tax reform . . . 'Pretty close to dead last' ": Evans and Novak, *op. cit.,* pp. 215, 212.

5. p. 79 ". . . snatched back from the Interior Department all powers of decision over oil imports . . . creating a Cabinet Task Force to undertake a study . . .": "On February 20, 1969, Nixon announced that he was assuming direct responsibility for oil import policy and that the Mandatory Oil Import Program was to be reviewed by a cabinet task force" (Goodwin, *op. cit.,* p. 400).

6. p. 80 ". . . one group he didn't owe a thing to . . . was the Jewish voters": Sampson, *op. cit.,* p. 206.

7. p. 80 "Over at Treasury, more mail . . . than in all the previous year . . .": Evans and Novak, *op. cit.,* p. 215.

8. p. 81 ". . . Walker urged the White House to get aboard . . . Nixon agreed . . . now favored tax reform . . .": *Ibid.,* p. 216.

9. p. 81 "He simply declared oil depletion not to be a loophole . . . other things to reform": Drew Pearson and Jack Anderson, "The Washington Merry-Go-Round," May 7, 1969. Also, Nixon's secretary of the treasury, John B. Connally, told the Senate Finance Committee, "I don't consider the depletion allowance to be a loophole" (Washington *Post,* February 29, 1972, p. 23).

10. p. 81 "Some of their colleagues . . . had been assured by Nixon himself . . .": "The Washington Merry-Go-Round," May 7, 1969.

11. pp. 81–82 The *Business Week* quote and the examples of lopsided congressional support: From the issue of May 17, 1969, p. 99. Also, Oppenheimer, *op. cit.,* p. 104.

12. p. 82 " 'There was Johnny Byrnes acting just like Bill Proxmire' " and the quotes from Mills and Ikard: *Business Week,* May 17, 1969, p. 69.

13. p. 83 " 'stick out like a sore thumb' ": Oppenheimer, *op. cit.,* p. 123.

14. p. 83 " 'It's a new group . . . couldn't care less . . .' ": *Business Week,* May 17, 1969, p. 108.

15. p. 85 The speech by Senator Stevens: *Congressional Record—Senate,* December 1, 1969, pp. 36218–20.

16. p. 86 The testimony of Browne and Wright and the conclusion of Dr. Blair: From Blair, *op. cit.,* pp. 4–10.

17. p. 87 ". . . rising from 2.2 million barrels a day in 1968 to 4 million in 1973": "U.S. Net Trade in Energy" tables, *Energy Perspectives 2, loc. cit.,* pp. 190–205, converted from yearly to daily barrels by authors.

18. p. 88 "In 1980 . . . U.S. reserves ceased their decade-long decline and began to hold even . . .": *BPDB,* September 1982, Section II, Table 2. (Moreover, in 1981 oil production in the contiguous forty-eight states, which since 1970 had been falling annually at an average rate of 250,000 barrels per day turned around and rose by 3,000 barrels per day over 1980 [UPI dispatch in *The New York Times,* December 8, 1981]).

19. p. 88 " '. . . oil companies do not reinvest the depletion incentive . . .' ": Oppenheimer, *op. cit.,* p. 166, Note 2.

20. p. 90 ". . . it cost American consumers an extra $5 billion a year . . .": Blair, *op. cit.,* p. 181, figured at the 1968 consumption level.

21. p. 91 ". . . White House advisers, led by John Ehrlichman . . . pulled some of its sharpest teeth . . .": Evans and Novak, *op. cit.,* p. 218.

22. p. 91 " '. . . this doesn't change depletion, does it? . . . I'm committed on depletion' ": *Ibid.,* p. 219.

23. pp. 91–92 ". . . Nixon's fundamental posture was hopelessly unacceptable . . . he couldn't get one without the other": "There was unanimous agreement among

those [congressmen] we interviewed that a tax reform bill containing no change in depletion would not be accepted as a reform bill" (Oppenheimer, *op. cit.,* p. 123).

24. p. 92 ". . . it was counting on to be the premier Nixon legislative achievement . . .": Evans and Novak, *op. cit.,* p. 212.

25. p. 92 The proposals by Boggs, Byrnes, Proxmire, Long, and others would be offered during the House and Senate committee deliberations: Oppenheimer, *op. cit.,* pp. 123–26.

26. p. 92 ". . . independents . . . discovered 80 percent of the oil and natural gas . . .": See Ruth Sheldon Knowles, *America's Oil Famine* (Coward, McCann & Geoghegan, 1975), p. 45: ". . . the most basic oil industry group is the 10,000 independent producers, who traditionally have found, and continue to find, three-fourths of our new oil and gas." And p. 46: "All the decrease in domestic petroleum exploration and development since 1956 is due to the decrease in independent oilmen's activities. In 1956, independents spent $2.5 billion for domestic exploration and development. By 1971 they spent less than $1.2 billion." (The big companies, which often financed the independents, had shifted their financing to offshore and overseas projects.)

27. p. 93 "Cohen and his Treasury aides . . . came up with a tough proposal": ". . . it would keep the depletion allowance at 27.5 percent but require that every last cent of tax benefits from the allowance had to be plowed back into petroleum exploration" (Evans and Novak, *op. cit.,* p. 219). Also, "Plowbacks would allow depletion to the extent that the taxpayer 'expended an equivalent amount for exploration or development' of a natural resource" (Oppenheimer, *op. cit.,* p. 177, Note 66).

28. p. 93 ". . . Treasury raised the matter again on July 17 . . . elaborate memorandum": Oppenheimer, *op. cit.,* pp. 102–03, 177, Note 66.

29. p. 93 ". . . George Bush . . . 'the most articulate spokesman for the industry position' . . . a cut to 23 percent": *Ibid.,* p. 124.

30. p. 94 The mixed fate of the Boggs amendment: *Ibid.,* pp. 1, 126.

31. pp. 94–95 ". . . Secretary Kennedy flew to San Clemente . . . 'I made a promise, and I intend to keep it' ": *Ibid.,* p. 126.

32. p. 95 " 'Let's just accept it and go on from there' ": Evans and Novak, *op. cit.,* p. 221.

33. p. 96 "John Ehrlichman: . . . Kennedy confused . . . Dick Nixon would *never* back a cut . . .": *Ibid.*

34. p. 98 "Thirteen years of unbroken decline in domestic drilling . . . slide in home investment . . .": "From an all-time peak of 16,173 wildcats drilled in 1956, the number dropped to 7,693 in 1970, with a corresponding decline in oil reserves" (Knowles, op. cit., p. 44). For the "slide in home investment," see Chapter Three, Note 23, of *Fiasco.*

35. p. 98 ". . . has since been cited by journalist and academic alike . . .": For example: Evans and Novak, *op. cit.,* p. 216: ". . . the President was never the master of the situation. Not since the Landrum-Griffin Labor Reform Act in 1959 had major legislation been drafted in such great part by Congress without domination by the administration." And Professor Oppenheimer's entire study of the tax reform bill of 1969 is a study of the absence or irrelevance of administration input.

36. pp. 100–101 Long's efforts to sway the Senate Finance Committee: See Oppenheimer, *op. cit.,* pp. 67, 126; also *Congressional Quarterly Almanac,* 1969, p. 620.

37. p. 102 "More natural gas had been used up than found . . . second year running": *BPDB,* September 1982, Section XIII, Table 2.

38. p. 102 ". . . construction timetable for nuclear power plants . . . slipped a year . . .": Knowles, *op. cit.,* p. 27.

39. p. 102 "Demand for electric power . . . 8.1 percent instead of 6.9 . . .": *Ibid.*

40. p. 102 ". . . capital exploration investments . . . declined in absolute terms": *BPDB,* September 1982, Section V, Table 8.

41. p. 102 ". . . down to half the $2.5 billion . . . U.S. oil reserves were 1 billion barrels below . . . January": Knowles, *op. cit.,* p. 47, and *BPDB,* September 1982, Section II, Table 1.

42. p. 103 ". . . domestic . . . investment in oil . . . would fall by 12 percent . . .": *BPDB,* September 1982, Section V, Table 6. The University of Pittsburgh anecdote is from the *Wall Street Journal,* April 16, 1981, p. 29.

43. p. 103 "exploration wells . . . would continue to decline . . .": *BPDB,* September 1982, Section III, Table 1A.

44. p. 104 "The [stockpiling] program was opposed by the oil lobby. . . . Nixon went along with the oil merchants . . .": "A storage program was considered in 1969. . . . The industry, however, preferred to stick to the existing quota system. . . ." (McKie, *op. cit.,* p. 79). (McKie was chief economist of the President's Cabinet Task Force on Oil Import Control.)

45. p. 105 "The task force . . . established the danger point above which Middle Eastern imports should not be allowed to rise . . .": General George Lincoln, Chapter 2, in Joseph Szyliowicz and Bard O'Neill, eds., *The Energy Crisis and U.S. Foreign Policy,* p. 38.

46. p. 106 "The oil lobby was dismayed by the report . . .": "Coming Crisis in Oil: Battle over Bigger Imports," *U.S. News & World Report,* January 12, 1970.

47. p. 106 " 'Don't put the President in a box on this' ": Evans and Novak, *op. cit.*

48. p. 108 "If a Democrat were in power . . .": Oppenheimer, *op. cit.,* p. 41.

CHAPTER FIVE

1. pp. 117–118 The misadventures attending the construction of the first haul road and the quotation "a proof . . . last remaining wilderness": From Mary Clay Berry, *The Alaska Pipeline: The Politics of Oil and Native Land Claims* (Indiana University Press, 1975), pp. 94–97.

2. p. 120 ". . . not ready with any proposal until June, 1969 . . .": *Ibid.,* p. 65.

3. p. 121 The Goldberg-AFN-Alaskan bar contretemps was reported in "Justice Goldberg Quits AFN," the *Tundra Times,* May 16, 1969, p. 1; also, it is covered by Berry, *op. cit.,* pp. 62–64, 72.

4. p. 122 ". . . white Alaska . . . was insisting on taking for itself . . . let it come out of the feds' 10 percent . . .": An excellent summary of white Alaskan sentiment appears in Berry, *op. cit.,* pp. 69–75.

5. p. 123 The governor's position is summarized from his letter of November 18, 1969, to Senator Henry M. Jackson (files of Senate Committee on Interior and Insular Affairs).

6. pp. 124–125 The dispute between Senators Stevens and Gravel, its resolution, and the consensus reached by the two with Jackson and Allott: Berry, *op. cit.,* pp. 77–80.

7. p. 124 ". . . Gravel . . . spoke to the press . . . brought only acrimony . . . all previous understandings were off . . .": *Ibid.,* pp. 80–81.

8. p. 126 " 'You've just ruined yourselves . . .' ": *Ibid.,* p. 105.

9. p. 126 ". . . arrive . . . unheralded and to depart unnoticed . . . a swath of destruction . . .": *Ibid.,* pp. 83–84.

10. p. 127 "A gleeful Saylor . . . Train was sent away rebuffed . . .": *Ibid.,* p. 113.

11. p. 128 ". . . Interior . . . had gotten away to a promising start . . . had not lagged behind TAPS but kept ahead": From Anderson staff interviews with Hickel and Train.

12. pp. 128–129 Assessment of Hickel's working philosophy: *Ibid.*

13. p. 129 "I have always had a healthy respect for America's oilmen . . .": Hickel, *op. cit.,* p. 87.

14. p. 129 " '. . . they had to have the permit by the first of July or . . . disaster' ": Anderson staff interview.

15. p. 130 " 'They approached the whole thing as . . . routine . . .' ": Anderson staff interview with Train.

16. p. 130 " 'We'll elevate the pipeline where we have to' ": *Ibid.*

17. p. 130 ". . . TAPS insisted it could bury the pipe even there . . .": From Train's testimony before Senate Interior Committee, cited by Berry, *op. cit.,* pp. 112–13.

18. p. 130 " '. . . We know how to do it; now leave us alone' ": Anderson staff interview with Train.

19. p. 132 " '. . . They had no engineering plan . . . but they wanted a permit . . . like they were going to build a pipeline from Lubbock, Texas, to the Gulf . . . it seemed like they did not have an interest . . .' ": Anderson staff interview with Hickel.

20. p. 132 ". . . occasional reports surfaced of foot-dragging accusations . . . a lawsuit allegedly threatened . . .": For example, William Moore, "Two Oil Companies Opposed Pipeline," San Francisco *Chronicle,* April 30, 1974; Sampson, *op. cit.,* pp. 178–79; and Rand, *op. cit.,* pp. 242–45.

21. p. 133 " 'We were faced with "decision by committee" . . .' ": Hickel, *op. cit.,* p. 125.

22. p. 133 The Baker quote is from *The New York Times Magazine,* August 9, 1981, p. 14; the Ehrlichman quote is from his book *Witness to Power, loc. cit.,* p. 208.

23. pp. 134–135 The report of Nixon's shifting attitude toward Hickel is based on Hickel, *op. cit.,* pp. 3–8, 202–15; Ehrlichman, *op. cit.,* pp. 97–99; and Safire, *op. cit.,* pp. 192–93.

24. pp. 135–137 The description of the National Environmental Protection Act—its content, the people who shaped it, its uncontested passage into law—is based in part on the authors' interviews with Mitchell Melich and Russell Train, in part on the study by Richard A. Liroff, *A National Policy for the Environment: NEPA and Its Aftermath* (Indiana University Press, 1976), and in part on Whitaker, *op. cit.,* pp. 48–50.

25. p. 139 ". . . a directive from President Nixon . . .": Whitaker, *op. cit.* pp. 6–7.

26. pp. 140–143 ". . . a turnaround in the performance of the pipeline consortium buoyed hopes . . .": This account is based on the on-the-scene observations of a member of the Anderson staff and on contemporary reporting by a number of publications, including, Ruth Edmondson, "North Slope Problems and Progress—A Status Report," *World Petroleum,* January 1970; Sterling G. Slappey, "America Must Have Alaskan Oil," *Nation's Business,* September 1971; and Patricia Delaney, "Alaska's Frustrating Freeze on Oil," *Time,* June 26, 1971. The quote by Dr. Victor Fischer is also from *Time,* June 26, 1971.

27. pp. 143–145 The second abortive attempt to construct a haul road draws on the reporting of James E. Bylin, "Road Dispute Bottles up Alaska's Oil," *Wall Street Journal,* July 3, 1970, op-ed page, and on a June 18, 1970, *Wall Street Journal* article, "Boom to Gloom: Oil Field Development Is Stymied in Alaska," p. 1.

28. pp. 146–147 ". . . Haley . . . suffering a slow burn . . . 'The lobbyists will write this bill over my dead body' ": Berry, *op. cit.,* p. 84.

29. p. 148 " 'The only thing we haven't had is the funeral' ": *Ibid.,* p. 134.

30. p. 149 ". . . suspicion had fallen on Pollock . . .": *Ibid.*

31. pp. 149–150 The account of the House Interior Committee's wind-up session, including the quotes from Edmondson, Aspinall, Saylor, and Pollock, is drawn from *ibid.,* pp. 134–37.

32. p. 154 The exchange between Whitaker and Ryan: From Hickel, *op. cit.,* p. 250.

33. p. 155 "Mitchell claimed to know nothing": *Ibid.,* pp. 276–78.

34. p. 156 " 'Hickel is through' press leaks started up again": As in the Washington *Evening Star,* November 20, 1970: ". . . it is no coincidence that the name of Walter J. Hickel heads just about everybody's list of presumed expendables. This ubiquitous highly reliable anonymous official has been at it again, whispering sweet somethings in the ears of selected pundits."

35. pp. 155–156 The account of Hickel's last days as secretary is based on Hickel's memoirs and on Ehrlichman, *op. cit.,* pp. 98–101.

36. p. 156 "In March he began to explore a radically different pipeline route . . .": "Oil Companies Should Review Entire Alaskan Project, Says Morton," Anchorage *Daily Times,* March 3, 1971, p. 1.

37. p. 157 "By early June . . . 'I don't favor the damn thing much at all' . . .": Anchorage *Daily Times,* June 11, 1971, p. 1.

CHAPTER SIX

1. p. 163 ". . . oil imports, especially from the Middle East, which between 1970 and 1973 were to rise by 500 percent": Imports from the Middle East proper rose by almost five times, from 61 million barrels in 1970 to 292 million in 1973; imports from Africa rose by more than six times, from 44 million barrels to 285 million (*BPDB*, September 1982, Section IX, Table 3).

2. pp. 163–164 The Henry Kissinger quotes: From his memoirs, *Years of Upheaval, loc. cit.*, p. 855.

3. p. 165 ". . . brought home to the local despots at Nasser's elbow how thin was their writ": For an illustration, see David Holden and Richard Johns, *The House of Saud* (Holt Rinehart & Winston, 1981), pp. 188, 189: "King Saud, at his side, was utterly overshadowed. For that day, at least, Nasser was King. . . . Until Nasser's visit to the Kingdom the Saudis had assumed that money would buy him much as it bought tribal loyalty. But this reception made it obvious that he had established an independent source of power through his own actions. . . ."

4. p. 167 "Nasser had proved a durable rallying point for key groups in most Arab states": Miles Copeland, *The Game of Nations: The Amorality of Power Politics* (Weidenfeld and Nicolson, 1969), pp. 216–218; and Holden and Johns, *op. cit.*, p. 187.

5. pp. 167–168 "Though upright enough in many respects, he was amoral when it came to the use of subversion and even terrorism . . .": For an assessment of Nasser's character, see Robert Stephens, *Nasser: a Political Biography* (Simon & Schuster, 1971), pp. 561–62; and Copeland, *op. cit.*, p. 239. For his open defense of terrorism and subversion, see Copeland, *op. cit.*, pp. 172–76.

6. p. 168 "The basis of Nasser's public policy, Florentine though it was in execution, was simple in theory . . .": See Stephens, *op. cit.*, pp. 562–66, and Copeland, *op. cit.*, p. 185. (Copeland's book is devoted in large measure to the complicated execution of Nasser's foreign policy.)

7. p. 169 ". . . periods of breathtaking expectation . . . the aspect of a rising crescendo . . .": "Expectations . . . were higher than ever in the aftermath of Suez . . . the advantage now lay almost entirely with Nasser. Riots and demonstrations in his name became regular happenings from Lebanon to the Gulf" (Holden and Johns, *op. cit.*, p. 190).

8. p. 180 ". . . Nasser's designs to redistribute oil wealth Arab-wide and to drive out the Western oil companies": Stork, *op. cit.*, p. 92, cites the elements of the radicals' oil policy as voiced in the resolutions of the Nasser-dominated second congress of the Federation of Arab Labor Unions held in Cairo in late 1960: "The resolutions of the congress included demands that (1) Arab oil wealth be considered the property of the whole Arab nation; (2) Saudi Arabian oil revenues be allocated to meet the needs of all the Arab people; and (3) oil concessions be reconstituted to reflect current national aspirations." See also Holden and Johns, *op. cit.*, pp. 187–88: "Warning signals began to appear here and there that the encouragement of Nasser might be dangerous for Saudi Arabia. More and more he talked of revolution as the Arab necessity, and spoke of Arab oil as the sinew of Arab strength. A three-day strike by Aramco members in the spring of 1956 showed where that sort of thing might lead."

9. p. 171 "The State Department was basically pro-Nasser during the early years of his revolution": Elie Kedourie, *Islam in the Modern World* (Mansell Publishing, 1980), p. 237, writes of U.S. "support for Nasser after 1952" and cites American pressure on the British to yield to Nasser and give up their bases on the Suez Canal, as well as U.S. thwarting of the Anglo-French-Israeli expedition against Suez in 1956.

10. p. 171 ". . . Oil and allied business interests effectively turned around both the executive branch and Congress . . . emasculating the pro-Nasser influence in the State Department, weakening congressional support . . . influencing Eisenhower . . . and helping persuade the Kennedy-Johnson administrations . . .": For the oil company lobby's success in squelching the State Department's "soft on Nasser" attitude and stiffening up the White House against the Nasserite wave, see Copeland, *op. cit.*, pp. 193–94, 206–07. For the oil lobby's success in turning Congress sour on aid to Nasser, see *ibid.*, pp. 220–23. For the U.S. role in supporting the anti-Nasser palace coup of Faisal against pro-Nasser King Saud, see Stork, *op. cit.*, pp. 98–99. For a review of the U.S. 1960's anti-Nasserite policy, see *ibid.*, pp. 98–102, 108–12.

11. p. 172 "Nasser ran terrible risks": See Stephens, *op. cit.*, p. 562, on Nasser's risks and why he had to run them. Interestingly, in the early years of his rule, Nasser launched an attempt to bring about a political settlement with Israel that foundered when fighting erupted between Israelis and Palestinians in the Gaza Strip (Bernard Gwertzman, "Quaker Reveals Nasser Sought Israeli Accord," *The New York Times*, November 28, 1982).

12. p. 173 ". . . trying to stand by Syria in its provocations of Israel while heading off the consequences through quiet negotiations with Washington . . .": See Copeland, *op. cit.*, pp. 234–38, on Nasser's brinkmanship.

13. p. 174 ". . . or hang back . . . [revealing] . . . his own leadership of the Arab union an empty imposture": The Jordanians had already charged Nasser with cowardice and with hiding behind the United Nations peace-keeping force instead of attacking Israel for its reprisal raids against Syria; Syria was openly calling for war (Copeland, *op. cit.*, p. 237).

14. p. 174 "He had plunged into the unknown . . . while trusting that the goddess of war would smile on him": Nasser biographer Robert Stephens, *op. cit.*, p. 504, concludes: "He had risked war above all because he believed that sooner or later the Arabs and particularly Egypt must turn and face the power of Israel or live forever at the mercy of Israel's political will and her supposed international backers. He had run the risk and failed."

15. p. 177 ". . . by signs from across the Arab world that his vision of Arab unity might yet be vindicated": Some of those signs came even from Saudi Arabia. When Nasser charged U.S. participation with Israeli forces, "In Dhahran, anti-U. S. riots erupted on 7 June. . . . A mob made up mainly of students . . . stormed the military part of the airport and wrecked the American officers' club. . . . It then moved on to attack the American Consulate. . . . The unruly mob then invaded the perimeter of the Aramco senior staff camp, wrecking a number of houses. . . . On the same day in Riyadh several thousand people stormed down University Street. . . . Oil production stopped . . ." (Holden and Johns, *op. cit.*, pp. 252–53).

16. p 178 ". . . oil . . . 90 percent of all world energy exports . . . Europe's

dependence . . . was approximately 95 percent": Uzi B. Arad, Chapter 8, in Szyliowicz and O'Neill, *op. cit.,* pp. 142–43.

17. p. 179 ". . . the amount of Arab oil production cut back had reached 10 million barrels a day!": Statistic from Weisberg, *op. cit.,* p. 96.

18. p. 179 ". . . furnished 3 million barrels a day to Europe—90 percent of its normal demand": Figures from Shoshana Klebanoff, *Middle East Oil and U.S. Foreign Policy* (Praeger, 1974), p. 141.

19. p. 180 ". . . to quickly increase production for export . . . by 1 million barrels a day . . .": As it turned out, U.S. oil production rose by "an unprecedented one million barrels a day" during the relatively brief cutback/boycott (Rand, *op. cit.,* p. 102).

20. p. 180 ". . . Oil . . . had kept its Arab hosts on limited rations": "Most of them absorbed the revenues as quickly as they received the money . . . the low 'government takes' were in fact determined by the oil industry which maintained firm control over the market . . ." (Arad, *op. cit.,* p. 144).

21. pp. 181–182 For the various cracks in the oil-state front recounted on these pages, see Rand, *op. cit.,* pp. 95–96; and Don Peretz, in Szyliowicz and O'Neill, *op. cit.,* p. 90–93.

22. p. 183 ". . . Yamani pressed the view that the cutback/boycott was hurting the Arabs themselves more than anyone else . . .": Zuhayr Mikdashi, *The Community of Oil Exporting Countries* (Cornell University Press, 1972), p. 85.

23. p. 183 ". . . oil companies . . . could supply oil to the West indefinitely without the Arabs": ". . . the 1967 embargo never resulted in a reduction of crude availability" (Weisberg, *op. cit.,* p. 97).

24. p. 184 "But now he had to bend the knee to Faisal . . .": Faisal, of course, saw the justice of the affair differently. "Faisal had no sympathy for Nasser whom he blamed for the debacle and the devastating blow to Arab pride. The Lion of Araby had got his just deserts as far as he was concerned. Faisal's attitude was well summed up in one terse comment he made to an Arab diplomat: 'If someone throws stones at a neighbor's windows he should not be surprised or complain if the owner comes out and beats him with a stick' " (Holden and Johns, *op. cit.* p. 253).

25. p. 188 ". . . no real challenge was foreseeable to the oil status quo . . .": Henry Kissinger: "Conventional wisdom was that the oil producers could not force the price up by restricting production. The fiasco of the Arab oil embargo of 1967 . . . seemed to confirm this" (Kissinger, *op. cit.,* p. 855). Joe Stork: "Western control of Middle East oil in 1969–70 seemed more secure and stable than could have been predicted in June 1967" (Stork, *op. cit.,* p. 145).

CHAPTER SEVEN

1. p. 193 The growth figures for Libyan oil production—and its pulling abreast of Saudi Arabia and Iran—are shown in Rand, *op. cit.,* Appendix I.

2. p. 193 The rise of North Africa as oil supplier to Western Europe, and the

corresponding decline of the Middle East, is reviewed in Bernard Abrahamson, Chapter 4, in Szyliowicz and O'Neill, *op. cit.,* pp. 74–76.

3. p. 193 ". . . and to 5 million barrels a day in 1973! This leap of 2 million barrels a day . . .": Blair, *op. cit.,* p. 234. Dr. Blair examines how the rise in Libyan oil production was eroding the market for Middle East oil, pp. 211–20.

4. p. 200 " 'Once the regime is stable it will launch a frontal attack on the oil industry . . . the regime will use every possible means of persuasion' ": This Oasis internal memo is cited by Ruth First, *Libya, the Elusive Revolution* (Penguin Books, 1974), p. 200.

5. p. 201 The quote of J. B. Kelly: From his book *Arabia, the Gulf and the West, loc. cit.,* p. 332.

6. p. 201 ". . . elementary in the view of Henry Kissinger": The Kissinger quote is from his memoirs, *Years of Upheaval, loc. cit.,* p. 861.

7. p. 204 For U.S. withdrawal from Wheelus, see Multinationals Subcommittee Report, p. 121: "[The United States] therefore acquiesced to the demands of the Libyan Government for the closing of Wheelus Air Force Base and U. S. military personnel were removed. . . ."

8. p. 205 The excerpts summarizing the Newsom testimony are from the Multinationals Subcommittee Report, p. 121.

9. p. 207 The Kissinger quote: From *Years of Upheaval, loc. cit.,* p. 859.

10. p. 207 The J. B. Kelly quote: From *Arabia, the Gulf and the West, loc. cit.,* p. 332.

11. p. 208 "We cannot really know whether the Libyan government was more corrupt than . . . other fabulous contenders . . .": J. B. Kelly describes Saudi Arabia as "a country where the securing of concessions and contracts by bribery had almost attained the status of a national industry" (*ibid.,* p. 337). Or see Jonathan Spivak, writing of Nigeria in the *Wall Street Journal,* July 12, 1982: " 'This is the most corrupt country I have ever worked in,' declares the German manager of a large construction project. Other foreigners who do business here echo this sentiment."

12. p. 209 The quotations of Fuad Kabazi, along with an explanation of the oil policy pursued by the Idris regime, are contained in the Multinationals Subcommittee Report, pp. 97–100.

13. p. 211 "On their charts the Libyan hole would be plugged by 1973 . . . just in time, they calculated": For an analysis of the oil majors' attempts to balance Middle East production against Libyan, see Blair, *op. cit.,* pp. 211–20.

14. pp. 211–212 "There were desultory discussions . . . a consensus precisely along those lines": Kissinger, *op. cit.,* pp. 859–60.

15. p. 214 The role and rationale of the American Embassy in Tripoli in thwarting the mercenary expedition against Qaddafi are acknowledged in the Multinationals Subcommittee Report. A far more detailed account is given in John Cooley, *Libyan Sandstorm: The Complete Account of Qaddafi's Revolution* (Holt Rinehart & Winston, 1982), pp. 86–93.

16. p. 214 The Kissinger quote is from his memoirs, *Years of Upheaval, loc. cit.,* p. 860.

CHAPTER EIGHT

1. p. 219 " 'What we wish to emphasize . . .' ": Mabruk's address is quoted by Stork, op. cit., p. 156.

2. pp. 221 " 'People who have lived for five thousand years . . . their legitimate right' ": Qaddafi's quote is from Knowles, *op. cit.,* pp. 82–83.

3. p. 221 " 'The just demands we seek here . . .' ": Stork, *op. cit.,* p. 156.

4. p. 222 " 'What the major corporations most wanted . . .' ": Myra Wilkens "The Oil Companies in Perspective," in Vernon, *op. cit.,* p. 168.

5. p. 223 ". . . the true price had dropped to 92 cents a barrel . . .": Adelman, "Is the Oil Shortage Real?" *loc. cit.,* p. 71.

6. p. 223 ". . . long-term oil delivery contracts . . . featured lower prices . . .": *Ibid.*

7. p. 226 " 'Richard Nixon came into office . . .' ": Kissinger, *op. cit.,* p. 855.

8. p. 227 "Saudi Arabia was in the midst of exploration finds . . . by 100 billion barrels": *BPDB,* September 1982, Section II, Table 1.

9. pp. 227–228 ". . . Saudi sheikhs . . . press Aramco to raise production . . . to 20 million barrels a day": Holden and Johns, *op. cit.,* p. 321.

10. p. 228 ". . . the Iranian Consortium was discovering eighteen major fields . . . shah's constant exhortation . . . 10 million": Rand, *op. cit.,* p. 153.

11. p. 228 ". . . spare capacity . . . that amounted to 5 million barrels a day": Weisberg, *op. cit.,* p. 72, Figure 3.

12. p. 228 ". . . an offer . . . from the shah of Iran . . . 1 million barrels a day . . . $1 a barrel . . .": Kissinger, *op. cit.,* p. 857.

13. p. 228 ". . . Saudi Arabia would approach the White House . . .": See p. 311 of *Fiasco.*

14. p. 235 ". . . 9 out of 10 wells drilled will be dry holes, of which there were already 700,000 in the United States alone . . .": Knowles. *op. cit.,* pp. 43–45; also, U.S. Senate statement by Senator Gale McGee (D-Wyo.), *Congressional Record—Senate,* December 1, 1969, p. 36221.

15. p. 241 "Per barrel rates . . . rose from $1.10 . . . to $3 . . .": *Petroleum Economist,* December 1970, p. 475.

16. p. 241 " 'The oil companies will negotiate only if *they* are losing money' ": Knowles, *op. cit.,* p. 83.

17. pp. 241–244 The discussions between Occidental's Williamson and Exxon's Wynne and between Hammer and Jamieson are summarized in the Multinationals Subcommittee Report, pp. 122–23.

18. p. 244 "Qaddafi not only had . . . purchased Soviet tanks, engaged East German security forces, and was trying to get hold of a nuclear bomb from Communist China": For more than a decade my syndicated column, "The Washington Merry-Go-Round," has been keeping regular tabs on such Qaddafi activities as these.

19. p. 247 "... the actual production cost of Libyan oil ... was 7.5 cents a barrel ...": Weisberg, *op. cit.,* p. 16.

20. pp. 247–248 The Adelman quote is from "Is the Oil Crisis Real?" *loc. cit.,* p. 79–80.

21. p. 248 "Occidental had to make a signed admission of this offense ...": "The Occidental settlement was not a compromise, but rather a complete capitulation to the Libyan demands.... The justification for the violation of 50-50 profit sharing was that the companies were to make tax payments retroactive to 1965 in order to compensate for the underposting of Libyan crude since that date, and the government required explicit acknowledgment of that fact" (Weisberg, *op. cit.,* p. 47).

22. pp. 251–252 Sir David Barran's presentation to Douglas-Home: Described by Sampson, *op. cit.,* p. 214.

23. p. 252 "... Akins ... delivered himself of an hour-long lecture ... oil companies should direct their attention to a just settlement of the Palestinian problem ...": *Ibid.;* also Kelly, *op. cit.,* p. 337.

CHAPTER NINE

1. p. 260 "... Barran was determined that this time the oil companies should be organized ...": For Barran's role, see his letter to Senator Frank Church, August 16, 1974, Multinationals Subcommittee Hearings, Part 8, p. 722.

2. p. 261 "The results were equal to ... the hopes [Barran] had conveyed to Hammer ...": For a brief summary of the results of the oil barons' conference, see Weisberg, *op. cit.,* p. 55; for an exhaustive account, see McCloy's letter to the Department of Justice, Multinationals Subcommittee Hearings, Part 6, pp. 231–46.

3. p. 261 "... State Department approved ... Justice Department gave ... waiver of prosecution": Multinationals Subcommittee Hearings, Part 9, pp. 46–49.

4. p. 262 "McCloy requested ... 'high-ranking' representative ... Rogers agreed ...": Kelly, *op. cit.,* p. 345.

5. p. 263 "But the two companies ... refused ...": Weisberg, *op. cit.,* p. 56.

6. pp. 263–264 "Dr. Kissinger describes these arguments ...": Kissinger, *op. cit.,* p. 863.

7. p. 264 " 'Though I leaned to ... a more active government role ... I took the official's time-honored way out ...' ": *Ibid.*

8. p. 265 "But today his mission's purpose was ...": Irwin testimony, Multinationals Subcommittee Hearings, Part 5, pp. 147–49.

9. pp. 265–266 The appraisals of Ambassador MacArthur by McCloy and Schuler are from their Senate testimony. See Kelly, *op. cit.,* p. 349.

10. p. 266 Undersecretary Irwin's presentation to the shah is recounted in his later testimony (Multinationals Subcommittee Hearings, Part 5, pp. 147–49). Kissinger's reaction is from his memoirs, *Years of Upheaval, loc. cit.,* p. 864.

11. pp. 266–267 The dialogue between Irwin and the shah and the contribution of Amouzegar are recounted in detail by Kelly, *op. cit.,* pp. 347–48.

12. p. 268 "Irwin's cable to Secretary Rogers . . .": *Ibid.*

13. p. 268 ". . . MacArthur . . . 'deeply suspicious and afraid that the companies intended to play OPEC members against one another' ": Multinationals Subcommittee Hearings, Part 6, p. 69.

14. p. 268 For a review of U.S. oil policy in the Mossadegh period and thereafter, see pp. 8–14 of *Fiasco.*

15. pp. 269–274 The description in these pages of the several efforts of the shah to elicit pressure from the U.S., British, and French governments to force oil company submission to his demands and the consistent refusal of the U.S. government to undermine the oil companies, and on the contrary to support them strongly, is drawn principally from the Multinationals Subcommittee Report, pp. 103–16.

16. pp. 273–274 The two oil initiatives from the shah to Nixon-Kissinger are based on Kissinger's account, *op. cit.,* pp. 857–59.

17. p. 274 ". . . Rogers endorsed it [the Irwin-MacArthur position]": Weisberg, *op. cit.,* p. 57.

18. pp. 274–275 "Kissinger explains . . .": The long quote is from Kissinger, *op. cit.,* p. 864.

19. p. 276 "Kissinger . . . 'the preordained outcome was that the companies must yield' ": *Ibid.*

20. p. 276 ". . . Amouzegar *did* know of the U.S. defection": Weisberg, *op. cit.,* p. 57; Kelly, *op. cit.,* p. 352.

21. pp. 277–278 ". . . MacArthur undertook no less than to suborn Co-chair Strathalmond into repudiating his instructions . . .": See Multinationals Subcom Hearings, Part 6, p. 79.

22. pp. 278–279 The account of the U.S. participation in the OECD Paris is based on Kelly, *op. cit.,* p. 351, and Adelman, "Is the Oil Crisi *cit.,* pp. 80–81.

23. p. 278 The Schuler quote is from his testimony, Multination Hearings, Part 5, p. 123.

24. p. 282 "Kissinger . . . 'a solemn promise that must h . . .": Kissinger, *op. cit.,* p. 865.

CHAPTER TEN

1. p. 285 ". . . because they feared not being abl *op. cit.,* p. 168.

2. p. 287 The figures on the improved profit forth in the testimony of Dillard Spriggs (Mult 4, pp. 61, 77).

3. p. 289 ". . . almost two million barrels a day . . . would not be there": Estimates of the shipment of Alaskan oil to the lower forty-eight ranged from 350,000 to 600,000 barrels a day in the first year of pipeline operation, reaching about 1 million barrels a day in the second year and rising rapidly thereafter to 2 million. Higher delivery could be achieved in the earlier years if the need was clear. See Charles J. Cicchetti, *Alaskan Oil: Alternative Routes and Markets* (Johns Hopkins University Press, 1972), p. 8.

4. p. 289 ". . . rapid shift . . . from American coal to imported oil . . . would add more than 1 million barrels a day to U. S. oil demand . . .": "Nationwide, had coal provided the same share of electric generating fuel in 1972 as it had, say, in the mid-sixties, 1.1 million barrels of oil per day would have been 'saved' representing some 7 percent of United States oil consumption or 23 percent of oil imports in that year" (Darmstadter and Landsberg, *op. cit.,* p. 28).

5. p. 290 ". . . underavailability of natural gas by 1973 would increase the consumption of oil . . .": Relying on a study done by Paul W. MacAvoy and Robert S. Pindyck, Professor Edward J. Mitchell writes: "The shortfall in natural gas supplies alone—caused by severe Federal Power Commission restrictions—is estimated at the equivalent of 1,800,000 barrels of oil per day in 1972. While not all unmet gas demand was converted into additional oil imports, much of it was, since the Persian Gulf producers were the residual energy suppliers to this nation" (Edward J. Mitchell, *Perspectives on U.S. Energy Policy* [Praeger 1976], p. 16). "From 1.5 to 2 million barrels per day had been added to [oil] imports by 1973 owing to the shortage of gas" (McKie, *op. cit.,* p. 75). This is concurred in by Darmstadter and Landsberg, *op. cit.,* p. 29, who say, "A halt to the expansion of natural gas output not unconnected with the controversies over continued regulation of the natural gas price, may have added 1½ million barrels per day to the 1973 oil demand."

6. p. 290 "A Libya in hostile hands . . . 2 million . . . 3.7 million . . . nearly 5 million . . .": Blair, *op. cit.,* p. 220; Weisberg, *op. cit.,* p. 38; Rand, *op. cit.,* Appendix I. For ›erspective, see Blair, *op. cit.,* pp. 211–20, 234.

p. 291 "You can't explain economics . . .": Evans and Novak, *op. cit.,* p.).

›. 291–292 "If the automobiles . . . almost three-quarters of a million barrels "Continuation of the lower 1960–65 growth rate would have 'freed' another 0 barrels of oil per day in 1972 . . . [more in 1973] (Darmstadter and Landsberg, p. 28).

". . . a 10 percent reduction of oil use . . . could be borne indefinitely ut major dislocation": Report of the President's Cabinet Task Force on Oil ntrol, p. 53.

". . . Congress ignored the energy message . . . quietly laid aside": *cit.,* pp. 410–11.

". . . There was no sense of urgency . . .": Kissinger, *op. cit.,* p. 866.

". . . to allow Middle Eastern oil imports to rise above 5 percent of ption before the late 1970's and above 10 percent *ever.* . .": Report Force, pp. 98–108.

CHAPTER ELEVEN

1. p. 304 "... a confidential State Department report ...": John Maffre, *National Journal,* May 13, 1972, pp. 808–18.

2. p. 304 "... John Irwin began a round of visits ... 'first time that energy received top institutional attention'. ..": Lincoln, *op. cit.,* p. 24.

3. p. 305 "U.S. representatives ... insisted on an allocation formula that gave advantage to the United States ...": Weisberg, *op. cit.,* p. 125.

4. p. 307 "... in both instances the United States refused to involve itself beyond lip-service requests ...": *Ibid.,* p. 84.

5. p. 307 "... disunity of the awkward West ... unity of the sophisticated Arabs": *Ibid.,* p. 85.

6. p. 310 "... King Faisal ... was keeping the White House informed ...": Holden and Johns, *op. cit.,* pp. 294–95.

7. p. 311 "On September 30 Yamani made a public proposal ...": *Ibid.,* pp. 321–23.

8. p. 312 "... a terminal rebuff": *Ibid.,* p. 323.

9. p. 312 " '... Nixon thought Middle East policy was a loser ...' ": Kissinger, *op. cit.,* p. 196.

10. p. 313 "... 'My principal assignment ... was to stall' ": *Ibid.*

11. p. 314 " 'Within the space of forty-eight hours ...' ": *Ibid.,* p. 212.

12. p. 314 "It is a measure of Sadat's despair ...": Kissinger acknowledged that "coherence in our government had by now disintegrated" (*Ibid.,* p. 214).

13. pp. 314–317 The instances of Faisal's progressive involvement in attempting to pressure U.S. policy through the oil companies first came to my attention when I secretly obtained some correspondence of Socal—one of the Aramco partners—in 1974. Over the years I published much of this material in "The Washington Merry-Go-Round."

CHAPTER TWELVE

1. p. 321 "... the price ... at 120 times the cost of production ...": See Introduction, Note 17.

2. p. 321 "... by the restriction of supply to the level necessary to support the price ...": The time would come when OPEC spokesmen would openly acknowledge this. "Yamani said at a news conference here that the kingdom would adjust its output to support a price that all OPEC members had endorsed" (Douglas Martin, *The New York Times,* November 1, 1981).

3. p. 321 "... comparatively benign dominion of the oil companies ...": Speaking of the era of oil company dominance, Yergin and Hillenbrand say, "... it is only with some exaggeration that we say oil became almost a free good" (*op. cit.,* p. 95).

4. p. 322 "But the key factor [of monopoly]—purposefully restricted production that forced the price up and kept it up—was indisputably there": For a nutshell review, see Introduction, Note 16. For further corroboration, six quotations from four experts: "[Iran and other] OPEC members . . . have committed themselves to limiting their output" (Walter Levy, "A Warning to the Oil Importing Nations," *Fortune,* May 21, 1979); "The essential condition for competitive price behavior is that each firm is forced to seek its own salvation and cannot attempt to join in working out a solution that would be best for the group as a whole" (Adelman, *World Petroleum Market, loc. cit.,* p. 3); "By 1971 the cartel of producing nations had strengthened its ability to control production and prices. This is the major cause of oil scarcity . . ." (Mitchell, *op. cit.,* p. 15); "This huge differential between prices and costs must be attributed to the success of an immensely powerful monopoly" (Mancke, *op. cit.,* p. 19); "Monopoly power is being exercised when the members of an industry are successful in reducing their output and thereby keeping their product's price higher than the cost of producing additional units" (*Ibid.,* p. 26); "Monopoly, the power to overcharge, is the power to withhold supply" (Adelman, "Is the Oil Shortage Real?" *loc. cit.,* p. 71).

5. p. 322 ". . . 33 million barrels a day": "OPEC output . . . stood at a record 33 million barrels a day immediately prior to the 1973 Arab-Israeli War . . ." (from a CIA document, stamped "SECRET," entitled "Developments Since 1973 Oil Embargo," dated October 16, 1975).

6. p. 322 ". . . scheduled by the oil companies to rise to well over 50 million barrels a day by 1980 . . .": M. A. Adelman wrote in 1972 that: "Production growth in 1971–85 is estimated first according to a recent BP forecast of 7.7 percent [annually] which takes account of rising U. S. imports, to come mostly from the Persian Gulf. Doubtless also it registers, as does a Shell forecast, the cessation of growth in European imports as North Sea oil and gas takes over" ("Is the Oil Shortage Real?" *loc. cit.,* p. 75).

7. p. 322 Quote by Walter Levy: From his article "A Warning to the Oil Importing Nations," *loc. cit.*

8. pp. 322–323 The Central Intelligence Agency analysis referred to was compiled in 1979 and is entitled "The World Oil Market in the Years Ahead."

9. p. 323 The quote by Lichtblau is from his analysis "World Oil: How We Got Here—And Where We Are Going," *loc. cit.*

10. p. 323 ". . . (Saudi Arabia alone was slated for a 10-million-barrel-a-day jump in the latter half of the 1970's) . . .": Kelly, *op. cit.,* p. 372. Also, Saudi Oil Minister Yamani announced in an address at Georgetown University on September 30, 1972, plans for the kingdom to raise oil production by 12 million barrels so as to be producing 20 million barrels by 1980 (Holden and Johns, *op. cit.,* p. 321).

11. p. 323 ". . . the amount of OPEC oil the consuming world could afford was to drop . . .": Alan Greenspan, *op. cit.,* says, "While some of the decline is the result of soft economies, by far the largest part reflects lower consumption due to higher prices."

12. p. 323 ". . . any degree of free market competition . . . would have called forth many millions of barrels a day in increased output . . .": M. A. Adelman, writing in 1972 ("Is the Oil Shortage Real?" *loc. cit.,* p. 100), when oil was priced at around $2 a barrel, said: "The price of crude oil, set by a world monopoly, is many times what is enough to make it worthwhile to expand output. Therefore, even if the price declines

and especially if it rises, there will always be more crude oil available than can be sold, as there is now and always has been."

13. p. 324 ". . . a basic cause of unemployment for millions in the industrial West and deepening destitution for hundreds of millions in the third world": This question is explored in the Introduction to *Fiasco.* For further reinforcement, four quotes: ". . . OPEC was the chief source of the world's economic difficulties during the past decade. Without OPEC, we probably never would have had much stagflation" (Alan S. Blinder, professor of economics, Princeton University, from an article in the Washington *Post,* February 18, 1983); ". . . rich and poor nations alike . . . have been driven into a depression by a greedy cartel" (Hobart Rowen, economics columnist for the Washington *Post,* January 26, 1983); "Oil analyst Laurence Goldmuntz . . . cites estimates that OPEC's monopoly pricing has caused 25 to 30 percent of the unemployment in the industrialized countries" (Washington *Post,* January 26, 1983); "The oil shocks appear to have ended the era of high growth and full employment. . . . In its place, they have initiated a new and uncertain and uncomfortable era of 'stagflation,' a dual visitation of high inflation and low growth" (Yergin and Hillenbrand, *op. cit.,* p. 3).

14. p. 324 ". . . one out of every nine barrels . . .": Yergin and Hillenbrand, *op. cit.,* p. 9.

15. p. 324 ". . . where 6 percent of the world's population consumed more than 30 percent of the world's energy": From Weekly Report, *Congressional Quarterly,* Vol. XXXVI, No. 51 (December 23, 1978), p. 3467.

16. p. 325 ". . . given the posture of nonconfrontation with OPEC in any form . . .": "Until now, we have been stymied by the fear that OPEC, or some of its members, will not welcome a joint policy of importing countries, and will characterize such an approach as confrontation. . . . We have until now been afraid to cope effectively as a group with oil and related problems" (Levy, *op. cit.*).

17. p. 325 ". . . oil imports would rise from 6 million to almost 9 million barrels a day": Total U.S. petroleum imports from last Nixon to first Carter year (million barrels per day): 1974, 6.1; 1975, 6.1; 1976, 7.3; 1977, 8.8 (*BPDB,* September 1982, Section IX, Table 1).

18. p. 325 OPEC's share of U.S. oil imports rises from one-third to two-thirds: *BPDB,* September 1982, Section IX, Table 8A shows a 36 percent share of U.S. imports coming from the Eastern Hemisphere in the last full Nixon year, 1973, and 69 percent in the first Carter year, 1977.

19. p. 327 ". . . Gerald Ford tried to get Congress to decontrol domestic oil prices but he was rebuffed. Jimmy Carter . . . tried to boost U.S. oil prices to the world level . . . but once more the Nixon price control system prevailed": For a brief review of the Ford-Carter efforts to get remedial energy legislation enacted prior to 1979, see Weekly Report, *Congressional Quarterly, loc. cit.,* pp. 3463–70. For a lengthy review, read Neil de Mardin and James L. Cochrane, chapters 7 and 8, respectively, in Goodwin, *op. cit.*

20. p. 328 " 'We're not yet far enough along in the decision process . . .' ": From Weekly Report, *Congressional Quarterly, loc. cit.,* p. 3467.

21. p. 329 ". . . within a year the shah had learned from unexpected events . . . that the West would pay more . . .": Dr. R. C. Weisberg, *op. cit.,* p. 112, illustrates

how Western bidding indiscipline "resulted in a dramatic rise in spot prices for crude. In early November it was reported that there were 200 bidders for a single lot of Nigerian oil . . . and cargoes of Algerian and Nigerian crude were sold at prices of $16 a barrel. In December the National Iranian Oil Company (NIOC) received the staggering price of $17.40 a barrel."

Much of the blame for this was due to the failure of American leadership. "The highest prices paid for oil . . . were those arising out of the purchases of Japanese trading companies and by independent United States refineries. The charges levied against the Europeans on this score therefore appear to be unfounded" (Romano Prodi and Alberto Clo, Part 3, *Europe,* in Vernon, *op. cit.,* p. 109).

22. p. 331 "The holes in the foreign oil-sharing agreement of Kissinger's day were still gaping wide . . .": Not only did it lack enforcement teeth, but, as Walter Levy pointed out in *Fortune, loc. cit.,* there had to be a supply reduction of 7 percent to trigger sharing. "Yet an actual or perceived shortfall of only 2 or 3 percent could set off a price explosion," which proved to be the case in 1979.

23. p. 331 ". . . down only 2.2 million barrels a day . . .": *Petroleum Intelligence Weekly,* May 14, 1979, p. 1; "higher than . . . 1978": *National Journal,* June 23, 1979, p. 1028; "17 million barrels a day, compared to 18.1": *Ibid.,* p. 1030; "8.4 million barrels a day, as against 8.2": *BPDB,* September 1982, Section IX, Table 1.

24. p. 331 "Professor Dankwart Rustow has described . . .": The quotations, facts, and figures of the succeeding four paragraphs on p. 327 are taken from Dankwart Rustow's *Oil and Turmoil—America Faces OPEC and the Middle East* (W. W. Norton, 1982), pp. 184–85.

25. p. 332 The Walter Levy quote is from his *Fortune* article, *loc. cit.*

26. p. 336 "At one-fourth of what was feasible . . . single digits . . .": If the Saudis, instead of repressing Aramco's oil production to about one-sixth of the U.S. production/reserves ratio, had pumped at just one-fourth of it, the additional output of 4 million barrels a day would have forestalled the artificial "sellers' market" and the rise of oil prices out of the single-digit range.

27. p. 337 ". . . even at a profit of 3,000 percent . . .": If the cost of doing business per barrel were $1, and the price received $31, the profit would be 3,000 pecent. Of course, the cost for most OPEC members was less than $1 and the price was more than $30, but when we peek behind veils in such exotic realms, our profit percentage seems close enough.

28. p. 338 "Overall, oil demand . . . shrank from 52.4 million barrels a day in 1979 to 45.5 million in 1982. . . . Non-OPEC output rose from 15 million barrels a day in 1977 to 22 million in 1982": *Time,* February 7, 1983, p. 43.

29. pp. 338–339 For a thorough treatment of the year-by-year surge in OPEC oil reserves and trade balances, see Kelly, *op. cit.,* Chapter VIII. OPEC's 1980 oil income is calculated at $272 billion by Rustow, *op. cit.,* p. 188. The figures on the recent plunge in the OPEC trade balance are from *Time,* February 7, 1983, p.42.

30. p. 339 ". . . revenues that had multiplied 100 times in ten years . . .": Saudi Arabia's 1970 oil revenues were $1.2 billion (Vernon, *op. cit.,* p. 288, Table A-5); by 1980 Saudi oil revenues had risen to "more than $100 billion," according to oil authority S. Fred Singer, writing in the *Wall Street Journal,* March 18, 1983.

Index